# SNOW AND CLIMATE

The extent and variability of seasonal snow cover are important parameters in the climate system, due to their effects on energy and moisture budgets, and because surface temperature is highly dependent on the presence or absence of snow cover. In turn, snow-cover trends serve as key indicators of climate change.

In the last two decades many new methods and techniques have become available for studying snow–climate relationships. Satellites provided the first capability for monitoring snow-cover extent at continental and hemispheric scales, enabling the investigation of synoptic-scale snow cover–climate linkages. This global view of snow cover has been accompanied by rapid advances in snow modeling physics to represent snow cover and snow processes in Global Climate Models (GCMs). These advances have changed the way we look at snow cover, and the main goal of this book is to provide an up-to-date synthesis of the current state of snow–climate science that reflects this new perspective. This volume will provide an excellent synthesis for researchers and advanced students.

RICHARD ARMSTRONG is a Senior Research Scientist at the National Snow and Ice Data Center, the World Data Center for Glaciology and the Cooperative Institute for Research in Environmental Sciences at the University of Colorado. His current research includes remote sensing and evaluation of fluctuations in snow cover and glaciers as indicators of climate change.

ERIC BRUN is Head of Research at Météo-France and Director of the Centre National de Recherche Météorologiques. He is a specialist in snow and avalanches and developed original methods to assess the impact of climate change on snow cover and alpine rivers.

# SNOW AND CLIMATE

## Physical Processes, Surface Energy Exchange and Modeling

*Editors:*

RICHARD L. ARMSTRONG

*Cooperative Institute for Research in Environmental Sciences,*
*University of Colorado*

ERIC BRUN

*CNRM/GAME*
*(Météo-France, CNRS)*

CAMBRIDGE UNIVERSITY PRESS
Cambridge, New York, Melbourne, Madrid, Cape Town, Singapore,
São Paulo, Delhi, Dubai, Tokyo

Cambridge University Press
The Edinburgh Building, Cambridge CB2 8RU, UK

Published in the United States of America by Cambridge University Press, New York

www.cambridge.org
Information on this title: www.cambridge.org/9780521130653



First published 2008
This digitally printed version 2010

*A catalogue record for this publication is available from the British Library*

ISBN 978-0-521-85454-2 Hardback
ISBN 978-0-521-13065-3 Paperback

Additional resources for this publication at www.cambridge.org/9780521130653

This book is dedicated to the memory of Professor Donald M. Gray, (1929–2005), Chairman of the Division of Hydrology, University of Saskatchewan, and editor (with David Male) of the *Handbook of Snow: Principles, Processes, Management and Use* (1981), which is one of the earliest books to include a complete review of snow physics, accumulation and ablation processes and associated meteorology and climatology. Don pioneered the study of snow in Canada with an insistence that its study include a strong physical and observational basis, the full coupling of the mass and energy equations and consideration of atmospheric as well as surface processes. His contributions to this book are greatly appreciated.

# Contents

# List of contributors

**Mary Remley Albert** U.S. Army Cold Regions Research and Engineering Laboratory, Hanover, New Hampshire, U.S.A.

**Richard L. Armstrong** Cooperative Institute for Research in Environmental Sciences, University of Colorado, Boulder, Colorado, U.S.A.

**Ross Brown** Environment Canada, Climate Research Division, Montréal, Québec, Canada

**Eric Brun** CNRM/GAME (Météo-France, CNRS), Toulouse, France

**Judah Cohen** Atmospheric and Environment Research, Lexington, Massachusetts, U.S.A.

**Richard Essery** School of Geosciences Grant Institute, University of Edinburgh, Edinburgh, UK

**Charles Fierz** WSL Institute for Snow and Avalanche Research SLF, Davos, Switzerland

**Paul Föhn** WSL Institute for Snow and Avalanche Research SLF, Davos, Switzerland

**Don M. Gray** Division of Hydrology, University of Saskatchewan, Saskatoon, Canada, † deceased

**Richard J. Harding** Centre for Ecology and Hydrology, Institute of Hydrology, Wallingford, UK

**Rachel E. Jordan** U.S. Army Cold Regions Research and Engineering Laboratory, Hanover, New Hampshire, U.S.A.

**John C. King** British Antarctic Survey, Natural Environmental Research Council, Cambridge, UK

**Eric Martin** Météo-France/CNRM, Toulouse, France

**Christian Plüss** WSL Institute for Snow and Avalanche Research SLF, Davos, Switzerland, Now at Erdgas Ostschweiz AG

**John W. Pomeroy** Centre for Hydrology, University of Saskatchewan, Saskatoon, Canada

**Zong-Liang Yang** Department of Geological Sciences, University of Texas, Austin, Texas, U.S.A.

# Preface

While the idea of a book with a focus on snow and climate had been discussed for several years previous, it first took formal shape at a meeting of the International Commission on Snow and Ice (ICSI) of the International Association of Hydrological Sciences (IAHS) in Victoria, British Columbia, Canada in 1996. At that time, a group was formed within the ICSI Division of Snow Cover and Avalanches (chaired by Eric Brun) to begin work on an outline of a state-of-the-art book that would focus on the relationships between snow and climate. A basic outline of the book was developed and section leads were assigned to coordinate the development of detailed outlines of each individual chapter. We recognized that the study of snow and climate was a rapidly evolving science and that there were new demands for detailed representations of snow processes to provide an up-to-date understanding of the physical processes in snow and to support modeling for climate change scenarios. The earlier *Handbook of Snow* (Gray and Male, 1981) provided excellent information on the principles, processes, management, and uses of snow, but no similar book had been published in the intervening two decades. Because we envisioned that this book would perhaps be the only example of an overview of snow and climate available, we agreed that the audience should be broad. Our intent is that this new scientific and comprehensive treatment of the subject of snow and climate becomes a fundamental reference text for university undergraduate and graduate students as well as a reference guide for instructors in secondary schools to assist them in teaching about snow and climate. Because of the world wide interest in snow and snow processes among winter recreationists and much of the general public living in cold regions, this book should also offer a source of basic information for those with only a limited scientific background. Regardless of audience, the fundamental purpose of this book is to introduce the basic scientific principles that enable us to understand the relationships between snow and climate.

Researchers have long been aware of the important role of snow in the climate system. For example, studies on snowfields in Scandinavia were undertaken by

Sverdrup (1935) and Wallen (1949). Later, Houghten (1954), Budyko (1956), and Geiger (1959) included the effect of snow in their studies of the surface heat balance and the associated snow–albedo feedback while Lamb (1955) was one of the first to look at the synoptic-scale influence of snow cover. Williams (1975) was one of the first to explore the influence of snow cover on atmospheric circulation and climate change. In the decades since these studies, many new methods and techniques have become available to study snow–climate relationships. Satellites provided the first capability for monitoring snow-cover extent at continental and hemispheric scales, thus enabling the investigation of synoptic-scale snow cover–climate linkages. This global view of snow cover was accompanied by rapid advances in snow modeling physics to represent snow cover and snow processes in Global Climate Models (GCMs). These processes have changed the way we look at snow cover, and the main goal of this book is to provide an up-to-date synthesis of the current state of snow–climate science that reflects this new perspective.

## References

Budyko, M. I. (1956). *Teplovoi Balans Zemnoi Poverkhnosti*. Leningrad: Gidrometeorologicheskoe Izdatel'stvo. (English transl.: Stepanova, N. A. (1958). *The Heat Balance of the Earth's Surface*. Office of Technical Services, U.S. Dept. of Commerce, Washington, DC.)

Geiger, R. (1959). *The Climate near the Ground*. Cambridge, MA: Harvard University Press.

Gray, D. M. and Male, D. H. (1981). *Handbook of Snow*. Toronto: Pergamon Press.

Houghten, H. G. (1954). On the annual heat balance of the Northern Hemisphere. *J. Meteorol.*, **11**, 1–9.

Lamb, H. H. (1955). Two-way relationships between the snow or ice limit and 100–500 mb thickness in the overlying atmosphere. *Q. J. Roy. Meteorol. Soc.* **181**, 172–189.

Sverdrup, H. U. (1935). Scientific results of the Norwegian–Swedish Spitzbergen expedition in 1934. Part IV. The ablation on Isachsen's Plateau and on the 14th of July Glacier. *Geogr. Ann.*, **17**, 53–166.

Wallen, C. C. (1949). Glacial–meteorological investigations on the Karsa Glacier in Swedish Lappland 1942–1948. *Geogr. Ann.*, **30**, 451–672

Williams, J. (1975). The influence of snow cover on the atmospheric circulation and its role in climate change. *J. Appl. Meteorol.*, **14**, 137–152.

# Acknowledgments

The editors wish to express their gratitude to the reviewers of the book sections. Their comments and suggestions were extremely helpful and very much appreciated. We would also like to thank the support and encouragement provided by the International Commission on Snow and Ice (ICSI) of the International Association of Hydrological Sciences (IAHS), as well as by the newly formed (2007) International Association of Cryospheric Sciences (IACS).

The work of Ms. Rachel Jordan was funded primarily by the Director of Research and Development, U.S. Army Corps of Engineers, through projects at ERDC-CRREL, and partly by the National Science Foundation via awards OPP-98-14024 and OPP-00-84190. Dr. Mary Albert's contribution to chapter 2 was supported by the National Science Foundation under grant OPP-01-39988. The work of M. Eric Brun was funded primarily by the Centre National de Recherches Météorologiques at Météo-France.

Rachel Jordan is grateful to Dr. Robert E. Davis and Dr. Edgar L Andreas for their continued support and helpful discussions, as well as to ERDC-CRREL's technical staff for maintaining the CRREL show research site.

Zong-Liang Yang is indebted to Dr. R. C. Bales who suggested that he read the WMO-led snow model survey while designing his questionnaire. He also wishes to thank Dr. R. E. Dickinson for his encouragement and Dr. W. J. Shuttleworth for reading the early draft. All those who took time to complete the snow model questionnaire are gratefully acknowledged.

# Nomenclature

| Symbol | Description | Section | Value | Units (SI) |
|---|---|---|---|---|
| $a$ | Parameter, saturation vapor pressure | 2 | | hPa |
| $a_1$ | Parameter, effective thermal conductivity | 4 | | $W\ m^{-2}\ K^{-1}$ |
| $a_2$ | Parameter, albedo weight function | 4 | | $m^{-1}$ |
| $a_3$ | Parameter, albedo weight function | 4 | | $m^{-1}$ |
| $a_4$ | Parameter, dimensionless sensible heat flux | 3 | | |
| $a_5$ | "Saltation efficiency" coefficient | 3 | 0.68 | $m\ s^{-1}$ |
| $A$ | Specific surface of snow | 2 | | $m^2\ m^{-3}$ |
| $b$ | Parameter, saturation vapor pressure | 2 | | |
| $b_1$ | Parameter, effective thermal conductivity | 4 | | $m^6\ kg^{-1}\ s^{-3}\ K^{-1}$ |
| $B$ | Bowen ratio $H_S/H_L$ | 3 | | |
| $B_*$ | Lower limit for Bowen ratio | 3 | | |
| $c$ | Parameter, saturation vapor pressure | 2 | | °C |
| $c_1$ | Parameter, apparent change in roughness length | 3 | 0.1203 | |
| $c_2$ | New snow drift coefficient | 3 | $0.8 \times 10^{-4}$ | $s^3\ m^{-2}$ |
| $c_{susp}$ | Mass concentration of suspended snow | 3 | | |
| $c_{can}$ | Canopy closure | 3 | | |
| $c_{p,a}$ | Specific heat of air at constant pressure and at 273.15 K | 2, 3 | 1005 | $J\ kg^{-1}\ K^{-1}$ |

| Symbol | Description | Section | Value | Units (SI) |
|---|---|---|---|---|
| $c_{p,i}$ | Specific heat of ice at constant pressure and at 273.15 K | 2, 3 | 2114 | $J\,kg^{-1}\,K^{-1}$ |
| $c_{p,\ell}$ | Specific heat of water at constant pressure and at 273.15 K | 2 | 4217 | $J\,kg^{-1}\,K^{-1}$ |
| $c_{p,s}$ | Specific heat of snow at constant pressure | 2 | | $J\,kg^{-1}\,K^{-1}$ |
| $C$ | Shape factor | 2 | | m |
| $C_{hq}$ | Bulk transfer coefficient | 3 | | |
| $C_D$ | Bulk transfer coefficient for momentum | 3 | | |
| $C_H$ | Bulk transfer coefficient for heat | 3 | | |
| $C_Q$ | Bulk transfer coefficient for water vapor | 3 | | |
| $C_{DN}$, $C_{HN}$, $C_{QN}$ | Parameters, bulk transfer coefficients in terms of the bulk Richardson number | 3 | | |
| $d$ | Geometrical grain size or diameter | 2 | | m |
| $d_{opt}$ | Optical grain size | 2 | | m |
| $D_g$, $D_a$ | Diffusion coefficient of water vapor in air | 2, 3 | | $m^2\,s^{-1}$ |
| $D_s$ | Diffusion coefficient of water vapor in snow | 2 | | $m^2\,s^{-1}$ |
| $D_{bs}$ | Horizontal blowing snow transport | 3 | | $kg\,m^{-1}$ |
| $D_{salt}$ | Saltating transport of blowing snow | 3 | | $kg\,m^{-1}\,s^{-1}$ |
| $D_{susp}$ | Suspension transport of blowing snow | 3 | | $kg\,m^{-1}\,s^{-1}$ |
| $E$ | Sum of sublimation and evaporation rate at surface | 3 | | $kg\,m^{-2}\,s^{-1}$ |
| $E_{subl}$ | Surface sublimation rate | 3 | | $kg\,m^{-2}\,s^{-1}$ |
| $E_{evap}$ | Surface evaporation rate | 3 | | $kg\,m^{-2}\,s^{-1}$ |
| $E_{bs}$ | Sublimation rate of blowing snow | 3 | | $kg\,m^{-2}\,s^{-1}$ |
| $f$ | Parameter, compaction | 2 | | |
| $F$ | Heat flux | 2 | | $W\,m^{-2}$ |
| $f_s$ | Weight function for albedo | 4 | | |
| $g$ | Acceleration due to gravity | 2, 3 | 9.81 | $m\,s^{-2}$ |

| Symbol | Description | Section | Value | Units (SI) |
|---|---|---|---|---|
| $G$ | Ground heat flux | 3 | | $\mathrm{W\,m^{-2}}$ |
| $G_g$ | Enhancement factor for grain growth | 2 | | m |
| $h$ | Capillary rise | 2 | | m |
| $h_{S,b}$ | Portion of bare ground sensible heat advected to a snow patch | 3 | | $\mathrm{W\,m^{-2}}$ |
| $h_*$ | Lower boundary for suspended snow (upper boundary for saltating snow) | 3 | | m |
| $\tilde{h}$ | Mass transfer coefficient | 2 | | $\mathrm{m\,s^{-1}}$ |
| $\mathcal{H}$ | Internal energy of the snowpack | 3 | | $\mathrm{J\,m^{-2}}$ |
| $H_L$ | Latent heat flux | 3 | | $\mathrm{W\,m^{-2}}$ |
| $H_P$ | Energy flux carried by precipitation and blowing snow | 3 | | $\mathrm{W\,m^{-2}}$ |
| $H_S$ | Sensible heat flux | 3 | | $\mathrm{W\,m^{-2}}$ |
| $H_{S,b}$ | Sensible heat flux over bare ground | 3 | | $\mathrm{W\,m^{-2}}$ |
| $H_{S,s}$ | Sensible heat flux over a completely snow-covered fetch | 3 | | $\mathrm{W\,m^{-2}}$ |
| $H'_S$ | Dimensionless sensible heat flux | 3 | | |
| $HN_w$ | New snow depth due to blowing and drifting snow | 3 | | m |
| $HS, H$ | Depth of snowpack, snow depth | 2, 3, 4, 5 | | m |
| $I$ | Snow interception | 3 | | $\mathrm{kg\,m^{-2}}$ |
| $I_1$ | Interception before unloading | 3 | | $\mathrm{kg\,m^{-2}}$ |
| $k$ | Thermal conductivity | 2 | | $\mathrm{W\,m^{-1}\,K^{-1}}$ |
| $k_a$ | Thermal conductivity of air | 3 | | $\mathrm{W\,m^{-1}\,K^{-1}}$ |
| $K_\ell$ | Permeability of the interconnected water pathways in wet snow | 2 | | $\mathrm{m^2}$ |
| $K_F$ | Cryoscopic constant for water | 2 | 1.855 | $\mathrm{K\,kg\,mol^{-1}}$ |
| $k_{cl}$ | Snow clump shape coefficient | 3 | | |
| $k_{eff}$ | Effective thermal conductivity | 2 | | $\mathrm{W\,m^{-1}\,K^{-1}}$ |
| $k_i$ | Thermal conductivity of ice | 4 | | $\mathrm{W\,m^{-1}\,K^{-1}}$ |
| $K$ | Intrinsic or saturated permeability | 2 | | $\mathrm{m^2}$ |
| $K_k$ | Permeability of phase $k$ in unsaturated wet snow | 2 | | $\mathrm{m^2}$ |
| $K_{rk}$ | Relative permeability of phase $k$ in unsaturated wet snow ($=K_k/K$) | 2 | | |
| $L_{i\ell}, L_{\ell i}$ | Latent heat of fusion for ice at 273.15 K | 2, 3 | $3.335 \times 10^5$ | $\mathrm{J\,kg^{-1}}$ |

| Symbol | Description | Section | Value | Units (SI) |
|---|---|---|---|---|
| $L_{iv}, L_{vi}$ | Latent heat of sublimation for ice at 273.15 K | 2, 3 | $2.838 \times 10^6$ | $J\,kg^{-1}$ |
| $L_{\ell v}, L_{v\ell}$ | Latent heat of evaporation for water at 273.15 K | 2, 3 | $2.505 \times 10^6$ | $J\,kg^{-1}$ |
| $\mathrm{L}_{ji}$ | Latent heat of phase change from phase $j$ to phase $i$ | 2 | | $J\,kg^{-1}$ |
| $\mathcal{L}_{L}$ | Snow load | 3 | | $kg\,m^{-2}$ |
| $\mathcal{L}_{L,o}$ | Initial snow load | 3 | | $kg\,m^{-2}$ |
| $\mathcal{L}_{L}^{*}$ | Maximum canopy snow load | 3 | | $kg\,m^{-2}$ |
| $\mathcal{L}_{L,b}$ | Maximum snow load per unit area of branch | 3 | | $kg\,m^{-2}$ |
| $L_{O}$ | Obukhov length | 3 | | m |
| $L_{N}$ | Net flux of longwave radiation, $L\downarrow + L\uparrow$ | 3 | | $W\,m^{-2}$ |
| $LAI$ | Winter leaf and stem area index | 3 | | $m^2\,m^{-2}$ |
| $\mathrm{L}_{L}^{*}$ | Maximum canopy snow load | 3 | | $kg\,m^{-2}$ |
| $L\downarrow$ | Downward component of longwave radiation | 3 | | $W\,m^{-2}$ |
| $L\uparrow$ | Upward component of longwave radiation | 3 | | $W\,m^{-2}$ |
| $m$ | Parameter, water saturation (van Genuchten) | 2 | | |
| $M$ | Molality of species | 2 | | $mol\,kg^{-1}$ |
| $m_{p}$ | Mass of single blowing snow particle | 3 | | kg |
| $\dot{m}$ | Mass growth rate of snow particle | 2 | | $kg\,s^{-1}$ |
| $\mathcal{M}$ | Snow mass per unit surface area | 3 | | $kg\,m^{-2}$ |
| $n$ | Parameter, water saturation (van Genuchten) | 2 | | |
| $Nu$ | Nusselt number | 3 | | |
| $N_{Re}$ | Particle Reynolds number | 3 | | |
| $p$ | Pressure | 2 | | Pa |
| $p_{a}$ | Atmospheric or air pressure | 2 | | hPa or Pa |
| $p_{e}$ | Air-entry pressure | 2 | | Pa |
| $p_{k}$ | Pressure of constituent $k$ | 2 | | Pa |
| $p_{a\ell}$ | Capillary pressure or pressure drop $(p_{a} - p_{\ell})$ across air/water interface | 2 | | Pa |

| Symbol | Description | Section | Value | Units (SI) |
|---|---|---|---|---|
| $p_{ij}$ | Pressure drop $(p_i - p_j)$ across $i/j$ interface | 2 | | Pa |
| $p_{v,sat,k}$ | Saturation vapor pressure over a flat surface with respect to constituent $k$ | 2 | | hPa |
| $p'_{v,sat,k}$ | Saturation vapor pressure over a curved surface with respect to constituent $k$ | 2 | | hPa |
| $P$ | Precipitation rate | 3 | | $kg\ m^{-2}\ s^{-1}$ |
| $\mathcal{P}$ | Accumulated precipitation | 3 | | $kg\ m^{-2}$ |
| $Pe$ | Peclet number | 2 | | |
| $P_{bs}$ | Probability of blowing snow occurrence | 3 | | |
| $q$ | Phase change thermal source term | 2 | | $W\ m^{-3}$ |
| $Q$ | Specific humidity | 3 | | kg/kg |
| $Q_o$ | Specific humidity at the snow surface | 3 | | kg/kg |
| $Ra$ | Rayleigh number | 2 | | |
| $r_{pt}$ | Particle radius | 3 | | m |
| $r_{pt,n}$ | Nominal snow particle radius | 3 | 0.0005 | m |
| $r_p$ | Tube radius or equivalent pore radius | 2 | | m |
| $r_g$ | Snow grain radius | 2, 3 | | m |
| $Ra_{crit}$ | Critical Rayleigh number for onset of convection | 2 | | |
| $Re$ | Reynolds number | 3 | | |
| $Ri_B$ | Bulk Richardson number | 3, 4 | | |
| $R_v$ | Gas constant for water vapor | 2 | 461.50 | $J\ kg^{-1}\ K^{-1}$ |
| $R_F$ | Freezing rate | 3 | | $kg\ m^{-2}\ s^{-1}$ |
| $R_M$ | Melt rate | 3 | | $kg\ m^{-2}\ s^{-1}$ |
| $R_N$ | Net radiative flux at the snow surface, $S_N + L_N$ | 3 | | $W\ m^{-2}$ |
| $R_{runoff}$ | Runoff rate at snow–soil interface | 3 | | $kg\ m^{-2}\ s^{-1}$ |
| $s$ | Liquid water saturation $(\theta_\ell/\phi)$ | 2 | | |
| $s_i$ | Irreducible or immobile liquid water saturation | 2 | | |
| $s^*$ | Effective liquid water saturation $[(s - s_i)/(1 - s_i)]$ | 2 | | |
| $S$ | Phase change mass source term | 2 | | $kg\ m^{-3}\ s^{-1}$ |
| $Sh$ | Sherwood number | 3 | | |

| Symbol | Description | Section | Value | Units (SI) |
|---|---|---|---|---|
| $S_{toa}$ | Solar radiation flux at the top of the atmosphere | 3 | | $W m^{-2}$ |
| $SCA$ | Proportion of snow-covered area | 3 | | |
| $S_N$ | Net flux of shortwave radiation, $S\downarrow + S\uparrow$ | 3 | | $W m^{-2}$ |
| SWE | Snow water equivalent | 2, 3, 4, 5 | | $kg m^{-2}$ (or mm w.e.) |
| $S\downarrow$ | Downward component of solar radiation | 3 | | $W m^{-2}$ |
| $S\uparrow$ | Reflected component of solar radiation | 3 | | $W m^{-2}$ |
| $t$ | Time | 2, 3, 4 | | s |
| $t_h$ | Snow surface age | 3 | | hours |
| $T$ | Temperature | 2, 3 | | K or °C |
| $T_d$ | Melting point depression, $273.15 - T$ | 2 | | K |
| $T_0$ | Melting temperature 273.15 or 0 °C | 2 | | K or °C |
| $T_o$, $T_0$ | Snow surface temperature | 3, 4 | | K or °C |
| $T_{o,sp}$ | Ice sphere surface temperature | 3 | | K |
| $T_a$ | Air temperature | 3, 4 | | K or °C |
| $T_s$ | Snow temperature | 3 | | K or °C |
| $T'$ | Snow temperature gradient | 2 | | $K m^{-1}$ or $°C m^{-1}$ |
| $T_{mean}$ | Mean air temperature in layer of depth $\zeta_{ref}$ | 3 | | K |
| $u$ | Wind speed | 3 | | $m s^{-1}$ |
| $u_{10}$ | Average hourly 10-m wind speed | 3 | | $m s^{-1}$ |
| $\bar{u}$ | Mean wind speed | 3 | | $m s^{-1}$ |
| $\bar{u}_{crest}$ | Mean wind speed measured on a crest | 3 | | $m s^{-1}$ |
| $u_*$ | Friction velocity | 3 | | $m s^{-1}$ |
| $u_{*n}$ | Friction velocity applied to small-scale non-erodible elements | 3 | | $m s^{-1}$ |
| $u_{*t}$ | Threshold friction velocity | 3 | | $m s^{-1}$ |
| $U_l$ | Snow unloading coefficient | 3 | | $s^{-1}$ |
| $v_i$ | Velocity of ice matrix | 2 | | $m s^{-1}$ |
| $v_k$ | Flow (filter) velocity of fluid $k$ (a = air, $\ell$ = water) | 2 | | $m s^{-1}$ |

| Symbol | Description | Section | Value | Units (SI) |
|---|---|---|---|---|
| $V$ | Volume per unit mass or reciprocal density | 2 | | $m^3\ kg^{-1}$ |
| $w$ | Vertical (wind) velocity | 3 | | $m\ s^{-1}$ |
| $w_*$ | Dimensionless vertical velocity | 3 | | |
| $x, y, z$ | Coordinate system with $z$ positive upward relative to the ground | 2, 3 | | m |
| $x$ | Spatial coordinate along the direction of flow | 2 | | m |
| $x_M, x_H$ | Parameters, similarity function | 3 | | |
| $X$ | Dimensionless distance downwind of snow patch | 3 | | |
| $z_{\text{ref}}$ | Reference height above ground | 3 | | m |
| $z_{\text{b}}$ | Upper boundary for suspended snow | 3 | | m |
| $z_{\text{o}}, z_0$ | Momentum (or aerodynamic) roughness length; surface roughness length | 3, 4 | | m |
| $z_H$ | Roughness length for heat | 3 | | m |
| $z_Q$ | Roughness length for water vapor | 3 | | m |
| $z_*$ | Scale of roughness elements | 3 | | m |
| $\alpha$ | Snow (surface) albedo | 2, 3 | | |
| $\alpha_{\text{b}}$ | Albedo of snow-free (bare) surface | 4 | | |
| $\alpha_{\text{max}}$ | Upper limit on snow albedo | 4 | | |
| $\alpha_{\text{min}}$ | Lower limit on snow albedo | 4 | | |
| $\alpha_{\text{s}}$ | Albedo of deep, homogeneous snow | 4 | | |
| $\beta$ | Coefficient of solar absorption | 2 | | $cm^{-1}$ |
| $\tilde{\beta}$ | Coefficient of thermal expansion | 2 | | $K^{-1}$ |
| $\beta_{1H}, \beta_{1M}$ | Parameters, similarity function | 3 | | |
| $\beta_1$ | Parameter, similarity function, $\beta_1 = \beta_{1M} = \beta_{1H}$ | 3 | | |
| $\Delta t$ | Time interval | 4 | | s |
| $\Delta x$ | Characteristic distance | 2 | | m |
| $\gamma_1, \gamma_2$ | Parameters, similarity function | 3 | | |
| $\delta$ | Standard deviation of wind speed | 3 | | $m\ s^{-1}$ |
| $\varepsilon$ | Snow infrared emissivity | 3 | | |
| $\varepsilon_{\text{eff}}$ | Effective emissivity for the atmosphere | 3 | | |
| $\varepsilon$ | Exponent, phase permeability | 2 | | |

| Symbol | Description | Section | Value | Units (SI) |
|---|---|---|---|---|
| $\zeta$ | Height above snow surface | 3 | | m |
| $\zeta_{ref}$ | Reference height above snow surface | 3 | | m |
| $\zeta_{ij}$ | Radius of curvature of *i*/*j* interface | 2 | | m |
| $\eta$ | Newtonian viscosity of snow | 2 | | Pa s |
| $\eta_0$ | Newtonian viscosity of snow at $T = 0$ °C and $\rho_s = 0.0$ kg m$^{-3}$ | 2 | | Pa s |
| $\eta_k$ | Dynamic viscosity of fluid *k* | 2 | | Pa s |
| $\theta$ | Contact angle between solid and liquid | 2 | | rad. |
| $\theta_k$ | Volume fraction of constituent *k* | 2, 3 | | |
| $\kappa$ | Von Kármán constant | 3 | | |
| $\kappa_s$ | Turbulent diffusion coefficient for snow particles | 3 | | m$^2$ s$^{-1}$ |
| $\lambda$ | Wavelength | 2 | | μm |
| $\lambda_p$ | Pore-size distribution index | 2 | | |
| $\Lambda$ | Leaf area index | 4 | | |
| $\Lambda$ | Thermal diffusivity ($k/\rho_a \cdot c_{p,a}$) | 2 | | m$^2$ s$^{-1}$ |
| $\mu_a$ | Kinematic viscosity of air ($\eta_a$ / $\rho_a$) | 3 | | m$^2$ s$^{-1}$ |
| $\rho_a$ | Density of air | 2, 3 | | kg m$^{-3}$ |
| $\rho_{0,a}$ | Reference density of air | 2 | | kg m$^{-3}$ |
| $\rho_i$ | Density of ice | 2, 3 | 917 | kg m$^{-3}$ |
| $\rho_k$ | Density of fluid or solid *k* | 2 | | kg m$^{-3}$ |
| $\rho_\ell$ | Density of water | 2, 3, 4 | 1000 | kg m$^{-3}$ |
| $\rho_{max}$ | Maximum permitted snow density | 4 | | |
| $\rho_s$ | Density of snow | 2, 3, 4, 5 | | kg m$^{-3}$ |
| $\rho_v$ | Density of water vapor | 2 | | kg m$^{-3}$ |
| $\rho_{v,a}$ | Density of water vapor in air | 3 | | kg m$^{-3}$ |
| $\rho_{v,pt}$ | Density of water vapor at particle surface | 3 | | kg m$^{-3}$ |
| $\rho_{v,sat}$ | Saturation density of water vapor | 2, 3 | | kg m$^{-3}$ |
| $\sigma_{SB}$ | Stefan–Boltzmann constant | 3 | $5.6697 \times 10^{-8}$ | W m$^{-2}$ K$^{-4}$ |
| $\sigma_{a\ell}$, $\sigma_{\ell v}$ | Surface tension of air/water interface | 2 | 0.076 | N m$^{-1}$ |
| $\sigma_{i\ell}$ | Surface tension of ice/water interface | 2 | 0.028 | N m$^{-1}$ |
| $\sigma_{iv}$ | Surface tension of ice/vapor interface | 2 | 0.104 | N m$^{-1}$ |
| $\sigma_{ij}$ | Surface tension of *i*/*j* interface | 2 | | N m$^{-1}$ |

| Symbol | Description | Section | Value | Units (SI) |
|---|---|---|---|---|
| $\tau_o$ | Surface stress | 3 | | $N\,m^{-2}$ |
| $\tau_{atm}$ | Atmospheric shear stress | 3 | | $N\,m^{-2}$ |
| $\tau_n$ | Atmospheric shear stress on non-erodible surface | 3 | | $N\,m^{-2}$ |
| $\tau_p$ | Atmospheric shear stress on saltating bed of snow particles | 3 | | $N\,m^{-2}$ |
| $\tau_s$ | Atmospheric shear stress on stationary erodible surface | 3 | | $N\,m^{-2}$ |
| $\tau_\alpha$ | Time constant for albedo | 4 | | s |
| $\tau_\rho$ | Time constant for density | 4 | | s |
| $\phi$ | Snow porosity | 2 | | |
| $\Phi$ | Surface-layer similarity function | 3 | | |
| $\chi_I$ | Exponent for bulk transfer coefficient | 3 | | |
| $\psi$ | Angle between direction of flow and downward vertical | 2 | | rad. |
| $\Psi_H$ | Integrated form of surface-layer similarity function for heat | 3 | | |
| $\Psi_M$ | Integrated form of surface-layer similarity function for momentum | 3 | | |
| $\Psi_Q$ | Integrated form of surface-layer similarity function for water vapor | 3 | | |

## Subscripts

| Subscript | Description |
|---|---|
| a | Air or atmosphere |
| b | Bare ground |
| i | Ice |
| $i$ | Phase |
| $j$ | Phase |
| $k$ | Constituent |
| $\ell$ | Liquid water |
| s | Snow |
| sat | Saturated or equilibrium state |
| v | Vapor |

# 1

# Introduction

Richard L. Armstrong and Ross Brown

Snow cover is a part of the "cryosphere," which traces its origins to the Greek word *kryos* for frost. The cryosphere collectively includes those portions of the earth system where water is in a solid form and includes sea ice, river and lake ice, glaciers, ice caps and ice sheets, frozen ground (including permafrost), and snow cover. The cryosphere is an integral part of the global climate system with important linkages and feedbacks generated through its influence on surface energy and moisture fluxes, precipitation, hydrology, and atmospheric and oceanic circulation (Fig. 1.1). In terms of spatial extent, snow cover is the second largest component of the cryosphere after seasonally frozen ground (~65 million km$^2$) with a mean maximum areal extent of 47 million km$^2$, about 98% of which is located in the Northern Hemisphere where temporal variability is dominated by the seasonal cycle (Fig. 1.2).

## 1.1 Basic properties of snow

The following discussion of basic snow properties is presented to provide the uninitiated reader with the background to understand the discussion of snow physics and modeling in Chapters 2–4.

Snow originates in clouds at temperatures below the freezing point. As moist air rises, expands and cools, water vapor condenses on minute nuclei to form cloud droplets on the order of 10 microns in radius. When cooled below 0 °C such small droplets do not necessarily freeze and may "super cool" down to −20 °C and occasionally down to −40 °C. Once a droplet has frozen it grows quickly at the expense of the remaining water droplets because of the difference in saturation vapor pressure between ice and water. The form of the initial ice crystal, columnar, platelike, dendritic, etc. (see Fig. 1.3) depends on the temperature at formation, but subsequent

*Snow and Climate: Physical Processes, Surface Energy Exchange and Modeling*, ed. Richard L. Armstrong and Eric Brun. Published by Cambridge University Press. 

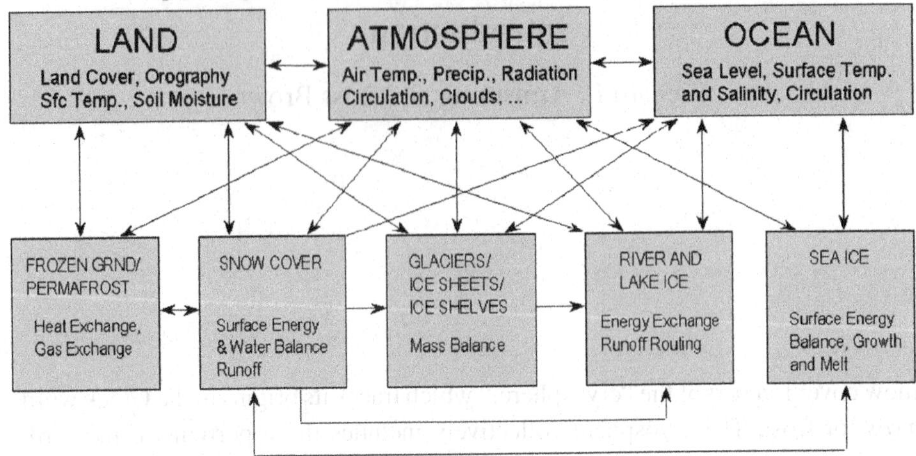


Lists in upper boxes indicate important state variables.
Lists in lower boxes indicate important processes involved in interactions.
Arrows indicate **direct** interactions.

Figure 1.1. Schematic diagram outlining a number of the important interactions between the cryosphere and other major components of the climate system. *Source*: G. Flato, CliC Science and Coordination Plan, 2001.

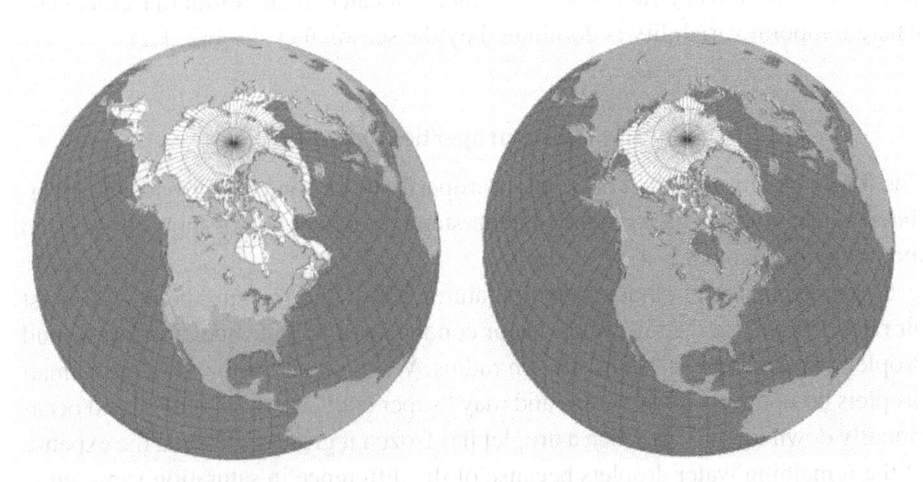

Figure 1.2. Mean seasonal variation in snow (gray) and sea-ice cover (white) between February (left) and August (right) as derived from satellite data. Data from NSIDC "Weekly Snow Cover and Sea Ice Extent," CD-ROM, NSIDC, 1996. (Plate 1.2.)

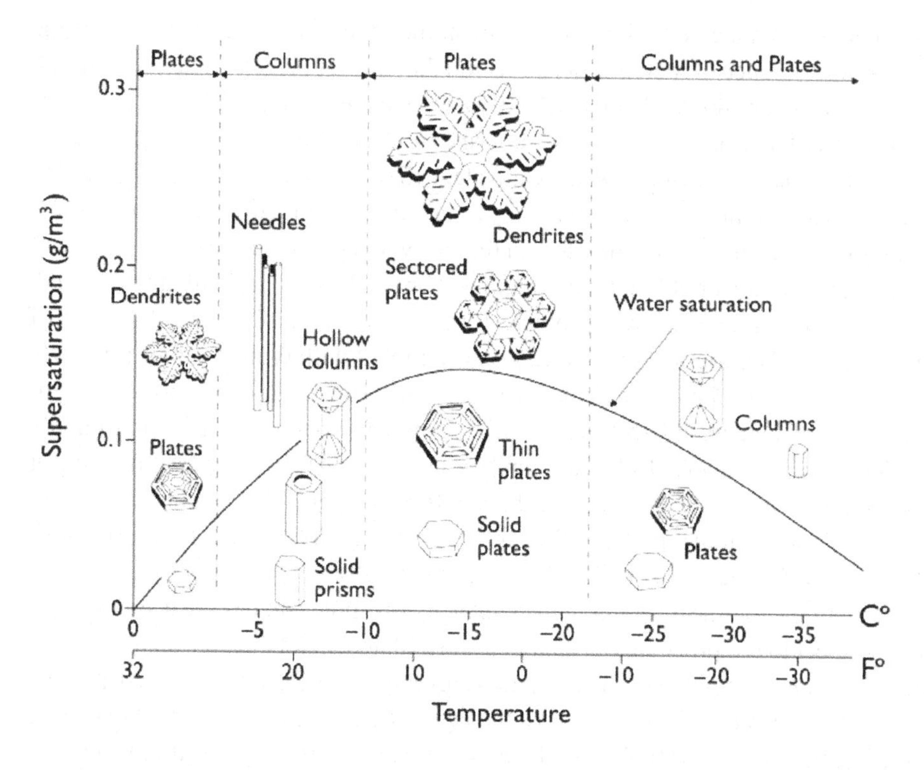


Figure 1.3. Snow crystal morphology diagram showing types of snow crystals that grow at different temperatures and humidity levels, *Source*: Libbrecht, 2003, with permission. Diagram is based on original work by Nakaya (1954).

growth and structural detail also depend on the degree of supersaturation (Hobbs, 1974 and Chapter 2 of this book). During its fall to earth a snow crystal may undergo considerable change due to variations in temperature and humidity with altitude. The character of a surface layer after a snowfall depends on the original form of the crystals and on the weather conditions during deposition. For example, when a snowfall is accompanied by strong winds, crystals are broken into smaller fragments favorable for close packing. After deposition snow may dissipate rapidly by melting or sublimation or it may persist for long periods. If it persists it will undergo metamorphism, changing its grain texture, size, and shape, primarily as a result of the effects of temperature and overburden pressure as it becomes buried by subsequent snowfalls. Snow metamorphism can occur rapidly because the crystals are thermodynamically active due to their large surface area to volume ratio (complex shape) and because their temperature is at, or proportionally close to, the melting temperature. Over the winter the typical snow cover accumulates and develops as a complex layered structure made up of a variety of snow grains, reflecting both the weather and climate conditions prevailing at the time of deposition as well as the persisting influence of metamorphism within the snow cover over time (Armstrong, 1977; Colbeck, 1986; Colbeck *et al.*, 1990 and Chapter 2, Fig. 2.7).

The three basic properties used to describe snow cover are the related parameters of depth, density, and snow water equivalent (SWE). "Snow depth" refers to the thickness or height of snow, typically expressed in centimeters. Maximum snow depths range from a few centimeters in regions with ephemeral snow cover, to several meters in moist cold mountain regions. "Snow density," as with any material, is simply the ratio of mass to volume for a given sample. The standard unit of measurement is kilograms per cubic meter (kg $m^{-3}$) with typical values for newly fallen snow of 30–150 kg $m^{-3}$ increasing to a maximum seasonal snowpack density of approximately 400–500 kg $m^{-3}$. However, wind-deposited snow may rapidly achieve densities of 300–400 kg $m^{-3}$ and crusts that form following the refreezing of melting snow may have densities of about 700–800 kg $m^{-3}$. For reference, the density of pure ice (bubble free) is 917 kg $m^{-3}$ and the density of water is 1000 kg $m^{-3}$. Thus the bulk of the seasonal snow cover is typically composed of 50 percent or more of air by volume throughout the winter. This simple fact has great significance with respect to the processes of metamorphism that depend directly on the amount of water vapor contained in the air spaces surrounding the snow grains (Armstrong *et al.*, 1993). SWE is the thickness of the layer of water resulting from the melting of the initial volume or thickness of snow and is typically expressed in kg $m^{-2}$ or mm. Further details on these basic snow properties and the instrumentation used to measure them are provided in Chapter 5.

## 1.2 Importance of snow in the climate system

Several fundamental physical properties of snow modulate energy exchanges between the surface and the atmosphere. The most important properties are the surface reflectance (albedo), the thermal insulating properties of snow, and the ability to change state (latent heat). Physical properties of a snowpack such as crystal structure, density, and liquid water content are also important for transfers of heat and water. These basic properties also determine the mechanical state of the snow cover, which is important for over-snow transportation and avalanche potential. The following paragraphs adapted from the *EOS Science Plan* (Goodison *et al.*, 1999) outline the importance of these properties for the climate system:

The *surface reflectance* of incoming solar radiation is important for the surface energy balance (Wiscombe and Warren, 1981). Typical albedo values for non-melting snow-covered surfaces are high (~80–90%) except in the case of forests (see Table 1.1). The higher albedo for snow causes rapid shifts in surface reflectivity in autumn and spring in high latitudes. However, the overall climatic

Table 1.1 *Typical ranges for surface albedo.*

| | |
|---|---|
| Fresh, dry snow | 0.80–0.95 |
| Old, dry snow | 0.70–0.80 |
| Wet snow | 0.50–0.70 |
| Melting ice/snow | 0.25–0.80 |
| Snow-covered forest | 0.25–0.40 |
| Snow-free vegetation/soil | 0.10–0.30 |
| Water (high solar elevation) | 0.05–0.10 |

significance is spatially and temporally modulated by cloud cover (planetary albedo is determined principally by cloud cover), and by the small amount of total solar radiation received in high latitudes during winter months. The high reflectivity of snow generates positive feedbacks to surface air temperature through the so-called "snow–albedo feedback," e.g. an initial warming results in a retreat in snow cover, lower albedo, higher absorbed solar energy, and warmer air temperatures. Groisman *et al.* (1994a,b) observed that snow cover exhibited the greatest influence on the earth radiative balance in the spring (April to May) period when incoming solar radiation was greatest over snow-covered areas.

The *thermal properties* of snow also have important climatic consequences. Snow on the ground typically has a density in the range of 100–500 kg $m^{-3}$ and the significant air fraction means snow is very effective at cutting off (or de-coupling) heat and moisture transfers from the ground surface to the overlying atmosphere. The thermal conductivity (a measure of the ability to conduct heat) of fresh snow is $\sim$0.1 W $m^{-1}$ $K^{-1}$ which is 10–20 times lower than values for ice or wet soil. Snow also has important influences on heat flow through ice (e.g. river, lake, or sea ice). The flux of heat through thin ice continues to be substantial until it attains a thickness in excess of 30–40 cm. However, even a small amount of snow on top of the ice will dramatically reduce the heat flux and slow down the rate of ice growth. The insulating effect of snow also has major implications for the hydrological cycle. In non-permafrost regions, the insulating effect of snow is such that only near-surface ground freezes and deep water drainage is uninterrupted (Lynch-Stieglitz, 1994).

The large amount of energy required to melt ice (*the latent heat of fusion*, $3.34 \times 10^5$ J $kg^{-1}$ at 0 °C) means that snow retards warming during the melt period. However, the strong static stability of the atmosphere over areas of extensive snow tends to confine the immediate cooling effect to a relatively shallow layer, so that associated atmospheric anomalies are usually short-lived and local to regional in scale (Cohen and Rind, 1991; Cohen, 1994). In some areas of the world such as Eurasia, the cooling associated with a heavy snowpack and moist spring soils is known to play a role in modulating the summer monsoon circulation (e.g.

Vernekar *et al.*, 1995). Gutzler and Preston (1997) presented evidence for a similar snow–summer circulation feedback over the southwestern United States.

Snow–climate feedbacks such as the snow–albedo feedback operate over a wide range of spatial and temporal scales and the feedback mechanisms involved are often complex and incompletely understood. A major thrust of recent snow modeling work (see Chapter 4) is the correct representation of important snow processes in climate models in order to properly simulate the response of the climate system to external forcing such as increased greenhouse gases. A review of earlier efforts to model snow–climate interactions and feedbacks was provided by Groisman (in Jones *et al.*, 2001).

## 1.3 Importance of snow in natural and human systems

In addition to their impacts on the climate system, snow-cover variability and change have important consequences for a wide range of natural and human systems. In many semi-arid regions of the world such as central Asia and western North America, runoff from mountain snowpacks represents the major source of water for stream flow and groundwater recharge. For example, over 85% of the annual runoff from the Colorado River basin in the southwestern United States originates as snowmelt. In these areas, snow accumulation is a critical resource for drinking water, irrigation, hydro-electrical generation as well as natural river ecosystems. A series of low snow accumulation years in these areas is typically associated with increased risk of forest fires, widespread crop failure, and difficulties meeting local demand for electricity. Worldwide, it is estimated that over one billion people depend on snow accumulation for water resources.

Jones *et al.* (2001) provided a major review of the role of snow in ecological systems. A primary influence of snow is in moderating winter meteorological conditions within and beneath the snow cover. Soil temperature and soil freezing/thawing processes have a great impact on ecosystem diversity and productivity. Beneath even 30 cm of snow the organisms and soil are well protected from the extreme diurnal temperature fluctuations occurring at the snow surface. Exchanges of carbon, methane, and other gases between the land surface and the atmosphere can also continue during the winter period because of the insulating effect of the snow cover (Sommerfeld *et al.*, 1993). Gaseous emissions under the snow may represent a significant part of the annual flux of carbon fixed by photosynthesis. Snow influences on soil temperature are also important for hydrology. When soil moisture freezes, the hydraulic conductivity is reduced leading to either more runoff due to decreased infiltration or higher soil moisture content due to restricted drainage. Knowing whether the soil is frozen is important in predicting surface runoff and spring soil moisture reserves (Zhang and Armstrong, 2001).

Snow cover is also an important recreation and tourism resource in mid-latitudinal mountain regions of the world (e.g. the Rocky Mountains, the Appalachians, and the European Alps). In general, winter tourism in the Rocky Mountains and New England is estimated to provide an economic benefit that exceeds $8 billion (Adams *et al.*, 2004). In contrast, snow can also be a major hazard, causing delays in ground and air transportation, damage to crops and livestock, and accidents, injuries and loss of life during extreme storms. It is estimated that even 5 cm of snowfall in urban areas can increase fuel consumption by 50 percent. In the United States alone the annual cost for snow removal from highways and airport runways exceeds $2 billion with the economic impact of flight delays and airport closures adding another $3.2 billion annually (Adams *et al.*, 2004). McKelvey (1995) provides a history of snow removal and its costs in United States cities. Mergen (1997) and Kirk (1998) provide fascinating coverage of the highly varied and often historically important influences of snow on human activities. In mountainous areas of the world, snow avalanches are an ever-present hazard with the potential for loss of life, property damage, and disruption of transportation. During the period 1985–2005, avalanche fatalities have averaged about 20–30 per year in North America but have averaged approximately 120 per year in the European Alps (McClung and Schaerer, 2006). A National Academy of Sciences/National Research Council report summarized snow avalanche hazards and approaches to mitigation in the United States (Voight *et al.*, 1990).

## 1.4 Climate change implications

As noted above (Section 1.2) snow-cover extent and temperature are negatively correlated through the snow–albedo feedback mechanism, with the strongest feedbacks operating during the spring period. Chapin *et al.* (2005) showed that pronounced summer warming in Alaska was associated with a lengthening of the snow free season caused by earlier snowmelt. Satellite data suggest that Northern Hemisphere snow-cover extent has decreased by about 5% over the 1966–2004 period (IPCC, 2007), and climate model simulations for the next 100 years suggest an extensive northward retreat of NH snow cover, particularly over Eurasia (Fig. 1.4).

Climate warming impacts on snow cover (e.g. earlier spring melt, shorter snow-cover season, lower peak accumulations, and higher potential for rain-on-snow and thaw events) will have far-reaching effects on the natural and human systems described above. Shallow snow cover at lower elevations in temperate regions is the most sensitive to temperature fluctuations and hence the most likely to experience increased melt (Scherrer *et al.*, 2004; IPCC, 2007). Changes to the snow cover and snow melt regime will have major implications for water resources (e.g. reduced

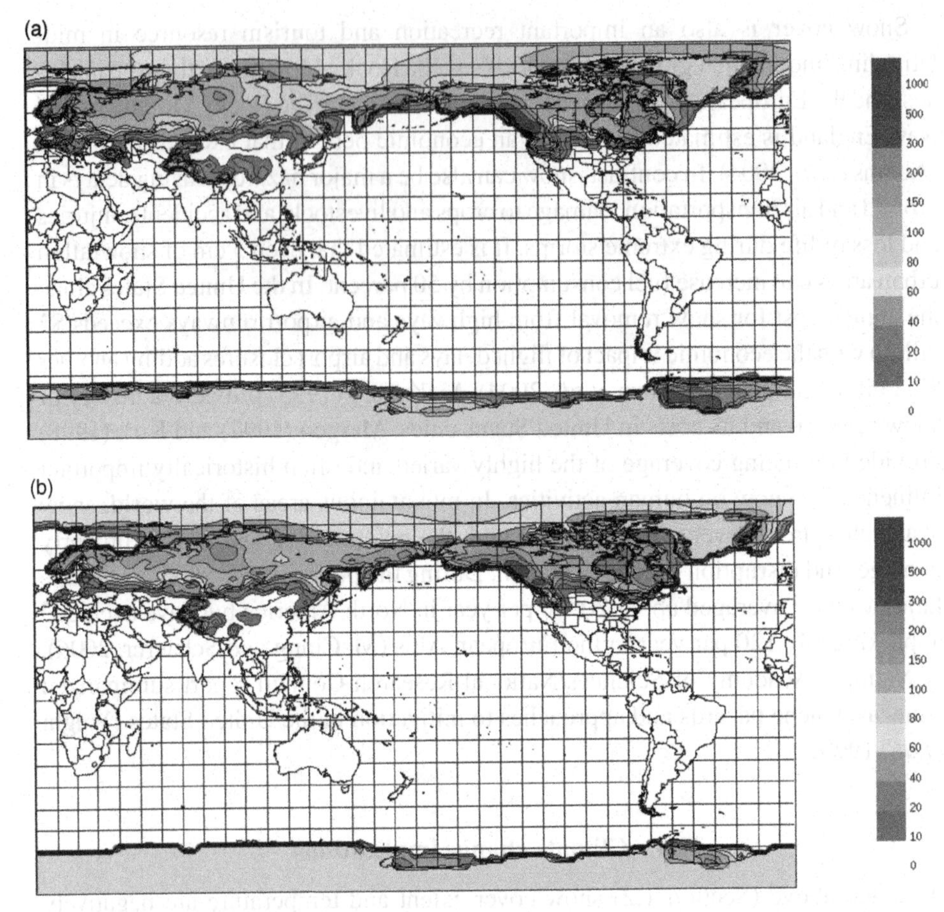


Figure 1.4. Comparison of mean March winter SWE (mm) simulated by the Canadian coupled global climate model (CGCM3) for the 1981–2000 "current climate" period (a) with simulated mean SWE for the 2081–2100 period (b) based on the SRES A2 emission scenario. Data courtesy of the Canadian centre for climate modeling and analysis. (Plate 1.4.)

spring runoff, increased potential for evaporation) and water resource-sensitive industries such as hydro-electric power and agriculture (Barnett *et al.*, 2005). IPCC (1998) provide an assessment of the impacts and vulnerabilities to changes in hydrology and water resources. In mountainous areas, for example, the snow line is likely to rise, but this may be partially compensated by higher precipitation above the freezing level. Ski resorts located in temperate mountain ranges (e.g. western North America, New Zealand, and the European Alps) already experience mean monthly winter temperatures that are only slightly below freezing, and any significant increase in air temperature will adversely impact the length of the ski season. In Austria, for example, it is estimated that an increase in temperature of

1.5 °C will shorten the ski season by about 15 days (Breiling, 1998). Scott *et al.* (2003) showed that improvements in snowmaking capacity reduced the vulnerability of the ski industry in southern Ontario (Canada) to the impact of warmer temperatures.

In summary, snow has substantial impacts, both positive and negative, on the natural environment and human activities. Documenting and understanding these impacts represents an important challenge, and one that is essential for adapting to a changing snow-cover climate.

## 1.5 Layout of book

It is against this backdrop of impending large-scale changes to snow cover on the surface of the earth that scientists are working to provide new and improved information for decision makers. The essence of this process is captured in this book with a review of the latest understandings of *Physical processes within the snow cover and their parameterization* in Chapter 2, a review of *Snow–atmosphere energy and mass balance* in Chapter 3, a history of snow model development in *Snow-cover parameterizations and modeling* in Chapter 4, and a look at *Snow-cover data: measurement, products, and sources* in Chapter 5.

## References


Adams, R. M., Houston, L. L., and Weiher, R. F. (2004). *The Value of Snow and Snow Information Services*. Chanhassen, MN: NOAA's National Operational Hydrological Remote Sensing Center.

Armstrong, R. L. (1977). Continuous monitoring of metamorphic changes of internal snow structure as a tool in avalanche studies. *J. Glaciol.*, **19**(81), 325–334.

Armstrong, R. L., Chang, A., Rango, A., and Josberger, E. (1993). Snow depths and grain size relationships with relevance for passive microwave studies. *Ann. Glaciol.*, **17**, 171–176.

Barnett, T. P., Adam, J. C., and Lettenmaier. D. P. (2005). Potential impacts of a warming climate on water availability in snow-dominated regions. *Nature*, **438**, 303–309.

Breiling, M. (1998). The role of snow cover in the Austrian economy during 1965 and 1995 and possible consequences under a situation of temperature change. In *Conference of Japanese Snow and Ice Society*, October, 1998, Niigata, Japan.

Chapin, F. S., III, Sturm, M., Serreze, M. C., *et al.* (2005). Role of land-surface changes in Arctic summer warming. *Science*, **310**, 657–660.

Cohen, J. (1994). Snow cover and climate. *Weather*, **45**(5), 150–156.

Cohen, J. and Rind, D. (1991). The effect of snow cover on climate. *J. Climate*, **4**, 689–706.

Colbeck, S. C. (1986). Classification of seasonal snow crystals. *Water Resources Res.*, **22**(9), 59S–70S.

Colbeck, S., Akitaya, E., Armstrong, R., *et al.* (1990). *International Classification for Seasonal Snow on the Ground*. Boulder, CO: International Commission on Snow and Ice (IAHS) World Data Center-A for Glaciology, University of Colorado, CB 449.

Flato, G. (2001). *Climate and Cryosphere (CliC) Project, Science and Coordination Plan.* World Climate Research Programme (WCRP) No. 114, WMO/TD No. 1053 (ed. Allison, I., Barry, R. G., and Goodison, B.).
Goodison, B. E., Brown, R. D., and Crane, R. G. (1999). Cryospheric systems. In *EOS Science Plan* (ed. King, M. D.). Greenbelt, MD: NASA Goddard Space Flight Center, pp. 261–306.
Groisman, P. Y., Karl, T. R., and Knight, R. W. (1994a). Observed impact of snow cover on the heat balance and the rise of continental spring temperatures. *Science*, **263**, 198–200.
Groisman, P. Y., Karl, T. R., and Knight, R. W. (1994b). Changes of snow cover, temperature and radiative heat balance over the Northern Hemisphere. *J. Clim.*, **7**, 1633–1656.
Gutzler, D. S. and Preston, J. W. (1997). Evidence for a relationship between spring snow cover in North America and summer rainfall in New Mexico. *Geophys. Res. Lett.*, **24**(17), 2207–2210.
Hobbs, P. V. (1974). *Ice Physics*. Oxford: Clarendon Press.
IPCC (1998). *The Regional Impacts of Climate Change* (ed. Watson, R. T., Zinyowera, M. C., Moss, R. H., and Dokken, D. J.). Cambridge: Cambridge University Press.
IPCC (2001). The Third Assessment Report of the Intergovernmental Panel on Climate Change. In *Climate Change 2001: The Scientific Basis* (ed. Houghton, J. T., Ding, Y., Griggs, D. J., *et al.*). Cambridge: Cambridge University Press.
IPCC (2007). Climate Change 2007: The Physical Science Basis (ed. Solomon, S. *et al.*). Cambridge: Cambridge University Press.
Jones, H. G., Pomeroy, J. W., Walker, D. A., and Hoham, R.W, eds. (2001). *Snow Ecology*. Cambridge: Cambridge University Press.
Kirk, R. (1998). *Snow*. Seattle, WA: University of Washington Press.
Libbrecht, K. and Rasmussen, P. (2003). *The Snowflake: Winter's Secret Beauty*. Stillwater, MN: Voyageur Press.
Lynch-Stieglitz, M. (1994). The development and validation of a simple snow model for the GISS GCM. *J. Climate*, **7**, 1842–1855.
McClung, D. and Schaerer, P. (2006). *The Avalanche Handbook*, 3rd edn. Seattle, WA: The Mountaineers.
McKelvey, B. (1995). *Snow in the Cities: a History of America's Urban Response*. Rochester, NY: University of Rochester Press.
Mergen, B. (1997). *Snow in America*, Washington, DC: Smithsonian Institution.
Nakaya, U. (1954). *Snow Crystals: Natural and Artificial*. Cambridge, MA: Harvard University Press.
Oke, T. R. (1987). *Boundary Layer Climates*. London: Methuen.
Scherrer, S. C., Appenzeller, C., and Laternser, M. (2004). Trends in Swiss alpine snow days – the role of local and large scale climate variability. *Geophys. Res. Lett.*, **31**, L18401, doi:10.1029/2004GL020255.
Scott, D., McBoyle, G., and Mills, B. (2003). Climate change and the skiing industry in southern Ontario (Canada): exploring the importance of snowmaking as a technical adaptation. *Clim. Res.*, **23**, 171–181.
Sommerfeld, R. A., Moisier, A. R., and Musselman, R. C. (1993). $CO_2$, $CH_4$, $N_2O$ flux through a Wyoming snowpack and the implication for global budgets. *Nature*, **361**, 140–142.
Vernekar, A. D., Zhou J., and Shukla, J. (1995). The effect of Eurasian snow cover on the Indian Monsoon. *J. Climate*, **8**(276), 248–266.

Voight, B., Armstrong, B. R., Armstrong, R. L., *et al.* (1990). *Snow Avalanche Hazards and Mitigation in the United States*. Panel on Snow Avalanches, National Academy of Sciences and National Research Council. Washington, DC: National Academy Press.

Wiscombe, W. J. and Warren, G. S. (1981). A model for the spectral albedo of snow. I. Pure snow. *J. Atmos. Sci.* **37**, 2712–2733.

Zhang, T. and Armstrong, R. L. (2001). Soil freeze/thaw cycles over snow-free land detected by passive microwave remote sensing, *Geophys. Res. Lett.*, **28**(5), 763–776.

# 2

# Physical processes within the snow cover and their parameterization

Rachel E. Jordan, Mary R. Albert, and Eric Brun

## 2.1 Introduction

Although the poetic view of snow is one of a uniform white blanket softly covering the ground, snow is, in fact, far from homogeneous. Initially snow crystals precipitate in a wide variety of shapes, depending on the atmospheric conditions where they form and on the temperature and wind speed near the ground where they are deposited. Patterns of snow deposition and topography also contribute to variability in snow accumulation. Once deposited, snow crystals bond together to form a new material – the snow cover. With each snowfall the snow cover is refreshed with a new layer whose properties may be quite different from the older snow beneath it. As the snow cover ages, its physical properties continue to evolve in response to weather conditions and to thermodynamic stresses within the ice–water–vapor system. These changes alter physical and chemical processes within the pack, which then affect the climate through complicated feedback mechanisms. For example, modifications in air temperature and radiation at the surface change temperature gradients within the snow that drive grain growth and metamorphism. Larger grain sizes, in turn, decrease the snow albedo and cause more heat to be retained by the earth's surface, which, on a larger scale, can increase atmospheric moisture and produce heavier snowfalls. Ablation of the polar snow cover in summer exposes bare ice and leads to more absorption of incident sunlight and further melting. Post-depositional metamorphism also changes the nature of the interstitial air space, thereby altering the permeability of snow to flows of air and water. Increased ventilation of the snow cover can lead to increased sublimation and crystal change. Increased water flow accelerates crystal growth, which in turn increases the permeability and accelerates snow-cover runoff. Snow is an extremely effective insulator, trapping heat in the ground and slowing sea-ice growth in winter. But as

*Snow and Climate: Physical Processes, Surface Energy Exchange and Modeling*, ed. Richard L. Armstrong and Eric Brun. Published by Cambridge University Press. 

snow compacts, its thermal conductivity increases, which leads to cooling of the ground and to warming of the atmosphere.

Snow is, therefore, a dynamic, complicated medium, and its microstructure plays a key role in the behavior of snow on many scales. The large-scale behavior or appearance of snow is often due to its small-scale properties. Avalanches, for example, can be launched because of thin weak hoar layers, and remotely sensed signals from satellites orbiting the planet are sensitive to snow crystal type and size.

This chapter focuses on snow physics. Our purpose is to provide a unified introduction to the subject, which addresses both recent issues and the results of past investigations. It is fundamental to the physics of snow that all three phases of water may coexist in relationships that are strictly governed by laws of thermal and mechanical equilibrium. While much of our understanding of snow-cover processes derives from observation, it is also important to examine the microscale and theoretical background for these behaviors. We begin with a description of the origin and characteristics of deposited snow, followed by a discussion of snow classification and metamorphism, which engages the theory of phase equilibria. The next sections on thermal and fluid flow discuss heat and mass transfer within the snowpack and its interaction with the evolving snow characteristics. They include a relatively new presentation of the role of air flow in thermal and vapor transport and of capillary forces and unstable wetting in water flow. We conclude with a section on radiative characteristics, which includes a detailed discussion of snow albedo – a key parameter controlling surface heat exchange.

Snow investigators now have the benefit of detailed computer models, which serve as tools to synthesize and examine current parameterizations of snow properties and processes. Comparisons of computer simulations with the measurement of bulk properties (snow depth and mass), in-snow profiles (snow density, temperature, grain size, and liquid water content), and boundary fluxes (surface exchange and basal outflow) validate the models and highlight areas of snow physics that require further study. A shortcoming of these point models is that they do not treat two- and three-dimensional processes. While we discuss important multi-dimensional processes, such as windpumping, large-scale phenomena are, in general, not addressed by this chapter.

### 2.1.1 Cloud formation and precipitation

Moisture in the atmosphere occurs principally in its gaseous phase, as water vapor, but also condenses to form clouds of water droplets or ice crystals. Vapor condenses when its partial pressure exceeds a saturation value, as determined by equilibrium conditions between the vapor and liquid or vapor and ice phases (Section 2.2.2). Since the saturation pressure decreases nearly exponentially with temperature, warm air at 25 °C can hold about fifty times more vapor than subzero air at −25 °C. Thus, the amount of vapor (measured as the depth of precipitable water vapor in a column extending upwards from the earth's surface) varies widely with season and latitude, from around 1 mm in Arctic continental air in winter to around 60 mm over southern Asia during the monsoon in summer (Barry and Chorley, 2003). The lapse in atmospheric temperature with altitude causes a sharp drop in vapor content and, therefore, about 80% of water vapor is contained within the 1000–700 hPa layer or the lowest 3 km of the atmosphere.

Clouds generally form when warm surface air moves upward, cools adiabatically to saturation, and condenses on condensation or ice nuclei. Precipitation occurs when the cloud particles grow large enough to fall and reach the ground before evaporating or sublimating. Much of the precipitation in high and mid-latitudes begins as snow at higher altitudes and melts to rain as it falls. The occurrence and intensity of precipitation depends on the availability of water vapor and on the concomitant mechanisms for nucleating and growing the particles. Open water in oceans and large lakes forms the major source of moisture in winter. Maximum winter evaporation rates occur when cold continental air blows across warm ocean currents – conditions most often found over the western North Pacific and North Atlantic oceans (Barry and Chorley, 2003). Although snow cover provides an unlimited source of moisture, its evaporation rate is restricted by the low saturation pressures associated with temperatures below 0 °C. Likewise, snowfall amounts in polar regions are relatively limited by the low vapor content of extremely cold air.

Precipitation and clouds are often classified according to the type of vertical air movement leading to their formation. These include (a) frontal or cyclonic lift of air in association with widespread low-pressure systems, (b) orographic lift over topographic barriers, and (c) convective (or buoyant) lift due to heating of the lower atmosphere by land or water surfaces. Table 2.1 (from Schemenauer *et al.*, 1981) lists the typical kinds of snowfall associated with various cloud types. Nimbostratus clouds produce heavier and more persistent snowfalls, while brief but heavy snow showers come from cumulonimbus clouds. When extremely cold conditions prevail at the surface, such as in Antarctica, snow precipitation may occur without significant vertical uplift, just from the condensation of humid air cooled in the vicinity of the surface. Snow crystals then precipitate without any observable clouds. This phenomenon is called "diamond dust."

### 2.1.2 Snow formation and crystal type

Snow consists of particles of ice that form in the clouds, grow initially by vapor deposition, and then reach the ground before evaporating or melting. This definition

Table 2.1 *Types of snowfall associated with various cloud types. (from Schemenauer* et al., *1981, with permission from The Blackburn Press).*

| Front | Cloud type | Possible snowfall |
|---|---|---|
| Warm | Cirrus and derivatives (Ci)<br>Altocumulus (Ac)<br>Altocumulus castellanus (Acc) | Usually virga (snow trails); light snow showers may occur from the Acc. |
| | Altostratus (As) | Light continuous or intermittent snow from As; however, when the snow is heavy, the As has probably graduated to a nimbostratus cloud. |
| | Nimbostratus (Ns) | Continuous snow. Virga occurs from both Ns and As. |
| | Stratocumulus (Sc) | Intermittent light powdery snow (fine flakes). |
| | Stratus (St) | Continuous light powdery snow. (This is the frozen precipitation analogue of warm precipitation drizzle.) |
| Cold | Cumulus (Cu) and towering cumulus (Tcu) | Light snow showers possible; more likely from the Tcu. |
| | Cumulonimbus (Cb) | Moderate to heavy snow showers. |

excludes hail and sleet, which form through the freezing of water, as well as frost, which forms through vapor deposition on the ground. Snow occurs as single crystals or as aggregations of many crystals joined together to form snowflakes. The shape or habit of a single crystal varies from simple hexagonal columns or plates to the complex dendritic or star shapes so favored by artists.

Conditions for snow to form include atmospheric temperatures less than 0 °C and the presence of supercooled water. Although bodies of fresh water (such as lakes) freeze near 0 °C, cloud droplets of pure water can coexist with ice particles down to temperatures as low as −40 °C. Snow begins as ice crystals, which nucleate from the chance aggregation of water molecules into stable ice-like structures called ice embryos. Nucleation can occur either homogeneously or heterogeneously. In the latter process, crystals nucleate onto the surfaces of ice nuclei, which serve to lower the free energy barrier to ice formation. Homogeneous nucleation does not occur in the atmosphere, except perhaps in cirrus clouds, where supercooled drops may freeze spontaneously into cirrus ice (Heymsfield and Miloshevich, 1993; Pruppacher, 1995). Pruppacher and Klett (1997) discuss four mechanisms for heterogeneous nucleation, including the freezing of supercooled water droplets and the direct deposition of water vapor onto ice nuclei.

Major sources of aerosol particles for ice nucleation are dust (commonly silicate minerals of clay) and combustion products from industrial plants, volcanoes, and forest fires. While the concentrations of ice nuclei are generally higher in continental air masses, global measurements show no systematic variation with location

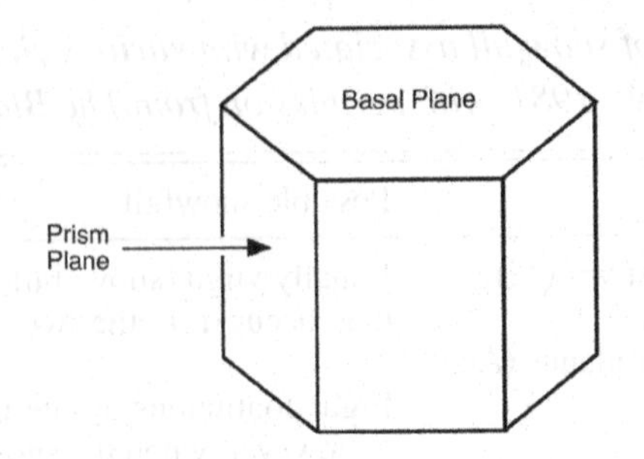


Figure 2.1. Basic hexagonal ice crystal form.

and suggest that nuclei are active far from their origins (Pruppacher and Klett, 1997). Splinters of ice broken from existing snow particles by wind can serve as secondary nuclei, thereby multiplying ice particle concentrations by many orders of magnitude. Unlike condensation nuclei, ice nuclei are generally restricted to substrates similar in structure to ice and are only effective at larger sizes, with diameters typically ranging between 0.1 and 15 μm. Because the ice-forming process is so selective, the fraction of aerosol particles acting as ice nuclei may be as small as one in $10^7$. Since the critical size of the ice embryo decreases with increasing supersaturation (see Table 2.3 in Section 2.2.2), smaller nuclei become activated at lower temperatures. Based on a review of published data, Fletcher (1962) suggested that the number of ice nuclei increases nearly exponentially with decreasing temperature, from typical concentrations of $0.01\,l^{-1}$ at −10 °C to $100\,l^{-1}$ at −30 °C.

Once nucleated, ice crystals grow by mass diffusion of water vapor onto their surfaces in a mechanism called deposition. Ice crystals in a supercooled cloud will grow at the expense of water droplets, because the vapor pressure over ice is less than that over water. Vapor is thus directed from the droplet to the ice surface (see Section 2.2.2). Ice crystals in the early stages of growth are usually less than 75 μm in diameter and are simple in shape (Schemenauer *et al.*, 1981). The basic shape common to all ice crystals is a hexagonal prism with two basal planes and six prism planes (see Fig. 2.1), which originates because of the covalent bonding of oxygen to hydrogen within the water molecule (Hallett, 1984). The relative growth rates of the basal and prism faces vary with temperature and supersaturation, giving rise to a wide variety of crystal shapes. As indicated by the diagram in Fig. 2.2 (Pruppacher and Klett, 1997, Fig. 2.36: b), crystal habit switches back and forth between plates (radial growth) and columns (axial growth) at transition temperatures of around −3 to −4, −8 to −10, and −20 to −25 °C (Mason, 1971; Frank, 1982; Hallett, 1987; Colbeck *et al.*, 1990). The degree of supersaturation determines secondary crystal features and the crystal growth rate. Higher saturations favor increasingly hollow

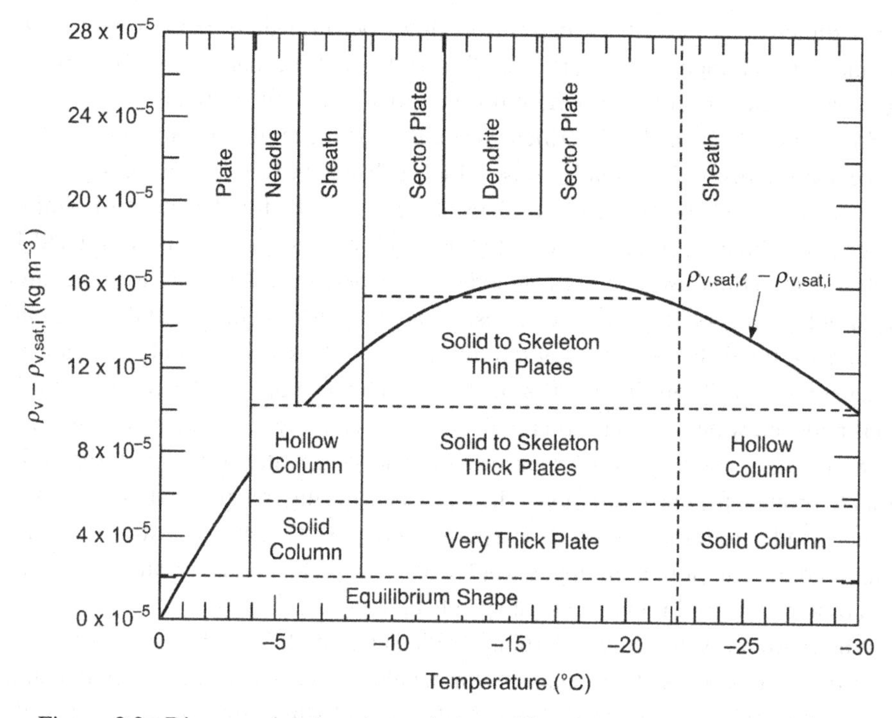


Figure 2.2. Diagram showing the variation of ice crystal shapes with temperature and with excess vapor density (after Pruppacher and Klett, 1997, Fig. 2.36:b, with kind permission from Kluwer Academic Publishers; based on laboratory observations of Kobayashi, 1961, with permission of Taylor & Francis Ltd, http://www.tandf.co.uk/journals; and Rottner and Vali, 1974, copyright 1974 American Meteorological Society).

and skeletal forms, which appear toward the top of the diagram. The difference in vapor pressure between water and ice saturation peaks around −12 °C (see Fig. 2.5 below), contributing to higher growth rates for crystals with dendritic features.

When an ice crystal grows to a size where it has a significant downward velocity and can survive sublimation during its fall to the ground, it becomes a snow crystal. Snow crystals that continue to grow by vapor deposition tend to keep their characteristic shape and proportions. The bulk densities of most crystals are less than that for pure ice, particularly for the more complex or hollow forms. Detailed observations as well as theoretical models have resulted in dimensional and mass–diameter relationship for various crystal types. Pruppacher and Klett (1997) provide a useful synopsis of these relationships. Individual unrimed crystals fall at velocities between about 10 and 80 cm s$^{-1}$ (Kajikawa, 1972), with larger and simpler crystals falling the fastest. Snow crystals reaching the ground typically range by two orders of magnitude in their maximum dimensions from 50 μm to 5 mm. At larger dimensions, growth by deposition slows because there is more surface to expand per increment of particle size. Two additional mechanisms, accretion and aggregation, then become important for the growth of snow particles.

Snow crystals exceeding a critical size can grow through collision and coalescence with supercooled cloud drops, which subsequently refreeze, in a process referred to as riming or accretion. The observed width for the onset of riming varies from about 30 μm for needles to 800 μm for dendritic crystals. Wang and Ji (2000) successfully use a numerical model to compute the collision efficiencies and threshold diameters of hexagonal plates, broad-branched dendrites, and hexagonal columns. Growth by accretion is relatively fast, with single crystals growing to dimensions of 1–2 mm in 10–20 min (Schemenauer *et al.*, 1981). Because collisions between water droplets and raindrops are less likely to result in coalescence, snow crystals grow larger through accretion than their water-drop counterparts. In order to provide for the dissipation of heat released during freezing, riming typically occurs at subzero temperatures, between −5 and −20 °C. Cases of extreme riming result in the formation of graupel particles or snow pellets. Large graupel particles with dimensions of 3 mm can fall at speeds up to 3 m $s^{-1}$.

When snow crystals or snow particles collide and "stick" together, snowflakes are formed by the process of aggregation. Snowflakes can consist of between two and several hundred snow crystals. Because of their radiating arms, dendritic crystals aggregate more readily than other crystal types. The aggregation mechanism is most efficient at temperatures near 0 °C, where liquid-like films promote the formation of bonds (sintering) between the particles. Snowflakes therefore have their largest dimension near 0 °C and aggregation is mostly limited to ambient temperatures above −10 °C. Hobbs (1974) predicts that a 1 mm snowflake falling through a cloud of smaller snow crystals can grow to 10 mm diameter in about 20 min. Because of their high air resistance, large snowflakes of up to 15 mm in diameter have fall rates of around 1–2 m $s^{-1}$.

In summary, snow consists of an intricate variety of snow crystals, as well as rimed and aggregate versions of these forms. The ICSI international classification scheme for snow on the ground (Colbeck *et al.*, 1990) distinguishes eight types of frozen precipitation by shape and growth habitat: columns, needles, plates, stellar dendrites, irregular crystals, graupel, hail, and ice pellets. It is now the standard reference for snow researchers. The older and more elaborate classification scheme of Magono and Lee (1966) is generally supported by recent observations and is recommended for those seeking more detail. Bentley and Humphreys (1931, 1962) and Nakaya (1954) provide large collections of fine photographs of these particles. More in-depth material on snow formation can be found in Mason (1971), Hobbs (1974), Schemenauer *et al.* (1981), Rogers and Yau (1989), and Pruppacher and Klett (1997).

### 2.1.3 Snow deposition

Snowfall amounts are measured both by depth and by snow water equivalent depth or SWE, which is the depth of the snow if it were melted. Typical snowfall rates are 1 cm $h^{-1}$ of depth or 0.8 mm $h^{-1}$ of SWE. The most intense storms occur when a persistent flow of very cold air traveling across open water warms rapidly, picks up moisture, and creates a steep lapse rate. Cumulonimbus clouds form in the unstable air and deposit heavy snowfall on the lee side of the body of water. These storms can be further intensified by orographic uplift.

The density of newly fallen snow varies between about 20 and 300 kg $m^{-3}$, but is typically 60–120 kg $m^{-3}$ for dry snow falling in low to moderate winds. Higher densities occur for warm, wet snow, sleet, and wind-packed snow. In general, smaller crystals with simpler shapes pack the most efficiently and lead to snow layers with higher density. In the case of mixed rain and snow precipitation, the new-snow density can exceed 300 kg $m^{-3}$. Because snow falling in strong winds is typically fragmented and small-grained, density generally increases by about 20 kg $m^{-3}$ for each m $s^{-1}$ of wind speed (Jordan *et al.*, 1999a). Large, dendritic snow crystals falling under calm, cold conditions produce the lightest snow covers.

Blowing and drifting snow and the topography of the underlying ground can lead to considerable spatial variability in snow depth. Pressure-induced variations in wind speed lead to scouring of snow on the upwind side and deposition on the lee side of hills or other convex barriers. Snowfall interception by tree canopies reduces snow accumulation on forest floors. As much as 60% of winter snowfall, for instance, may be intercepted by boreal forests (Pomeroy and Schmidt, 1993; Hedstrom and Pomeroy, 1998 ). Canopy snow coverage is subsequently lost through evaporation/sublimation, melting, or unloading, which forms cone-shaped accumulation rings around the peripheries of coniferous trees (Sturm, 1992; Hardy *et al.*, 1997).

## 2.2 General characteristics

After being deposited on the ground or on a previous snow layer, snow crystals accumulate and give birth to a new snow layer. The initial structure of this layer depends on the shape and size of the crystals and on the stress applied to the bonds that link them together. Ice forms a solid matrix that delimits pores filled with humid air and, in the case of wet snow, with liquid water. In snow, most pores are interconnected. Thus, snow belongs to the great family of porous media whose members generally present complex physical properties. Compared to other porous media, the complexity of snow physical properties is increased by the fact that the solid, liquid, and gaseous phases of the principal component of snow – water – may

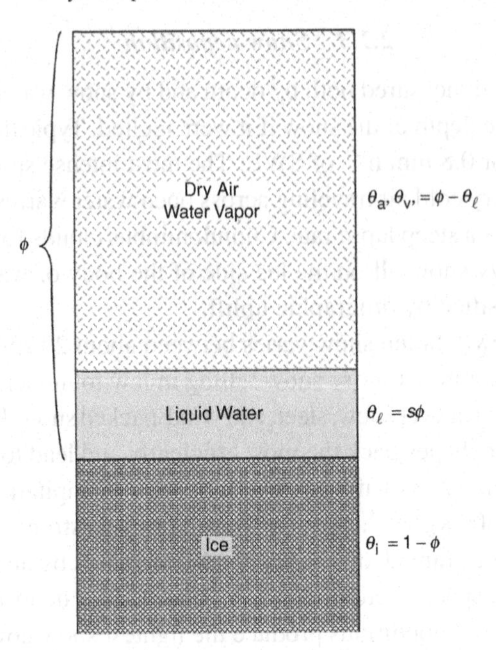


Figure 2.3. Diagram of volume fractions of the four components of snow (from Jordan, 1991).

coexist in the medium. Snow also differs from most other porous media in that the snow grains are bonded together. Because of the high activity of water thermodynamics around the triple point, the solid matrix of snow is continuously and sometimes rapidly evolving, thus making snow a unique, dynamic, and complex component of the earth's surface.

### *2.2.1 General concepts of a porous medium*

*Constitution*

A general feature of porous media is that the solid matrix presents physical and mechanical properties that significantly differ from the properties of the fluid filling the pores. In the case of snow, the fluid may consist of one phase (humid air) or of two immiscible phases (humid air and liquid water). Here we use volume fractions, $\theta_k$, to specify the mixing ratio of the four components in snow, where $k$ becomes i, a, v, and $\ell$, for ice, air, water vapor, and liquid water, respectively (see Fig. 2.3 and also Morris, 1983). The volume ratio between the solid matrix and the fluids is a key determinant of the physical and mechanical properties of the medium (Scheidegger, 1974; Dullien, 1992). For that reason, porosity $\phi$ (or $1 - \theta_i$) is the most basic parameter for describing snow. The density $\rho_s$ of dry snow is expressed in terms of porosity as $\rho_i(1 - \phi)$, where $\rho_i$ is the density of ice ($= 917$ kg m$^{-3}$). Liquid water content is also of primary interest for describing the constitution of a snow sample. Although it can be characterized by mass or volume, here we use the volume fraction $\theta_\ell$ or the liquid saturation $s$, which is the ratio of the liquid volume to the pore volume, or $\theta_\ell/\phi$. Snow porosity (with the exception of ice layers) generally ranges between 0.40 and 0.98 for seasonal snow covers. Snow density therefore ranges well over an order of magnitude, which leads to a range of over two orders of magnitude for some of its physical or mechanical properties.

### *Texture*

Independent of porosity, the texture (i.e. the distribution of shape and size of the grains, pores, and water menisces present in the medium) affects most physical and mechanical properties of snow (Arons and Colbeck, 1995; Shapiro *et al.*, 1997; Golubev and Frolov, 1998). As a result of metamorphism, snow texture evolves rapidly and shows a very large variability. Radiative properties are particularly influenced by grain shape and size, whereas thermal properties depend primarily on the bond structure and fluid properties on the structure of the pore space. The ratio of bond size to grain size and the coordination number (the average number of bonds per grain) are critical parameters for snow mechanical and thermal properties.

### *Snow classification*

Because of the large variability in porosity and texture observed in snow, the scientific community uses an international classification to describe the different snow types encountered in seasonal snowpacks (Colbeck *et al.*, 1990). Since no unique parameter alone describes a snow sample, the classification is based on a qualitative description of the shape and the size of the grains that may be determined with a microscope or a magnifying lens. Snow is classified into nine classes and various subclasses. The six main classes observed in seasonal snowpacks are described in Table 2.2. The classification criteria make it possible to classify a snow sample from field observations but do not provide quantitative information. Stereological analysis of thick or thin sections of snow samples (Good, 1987) and analysis of three-dimensional high-resolution gray-level images obtained from X-ray microtomography (Coléou *et al.*, 2001; Flin *et al.*, 2003) are relevant methods to derive characteristics of snow microstructure such as grain and bond size distribution or the coordination number. Compared to other porous media, snow is very brittle and such analyses require sophisticated preparation processes. It is almost impossible to prepare sections of snow with a porosity larger than 0.90 without disturbing the microstructure. Figure 2.4 presents a thick section showing ice, pores, and frozen liquid menisces of a snow type 6a.

Table 2.2 *Six classification schemes for snow types (according to Colbeck* et al., *1990).*

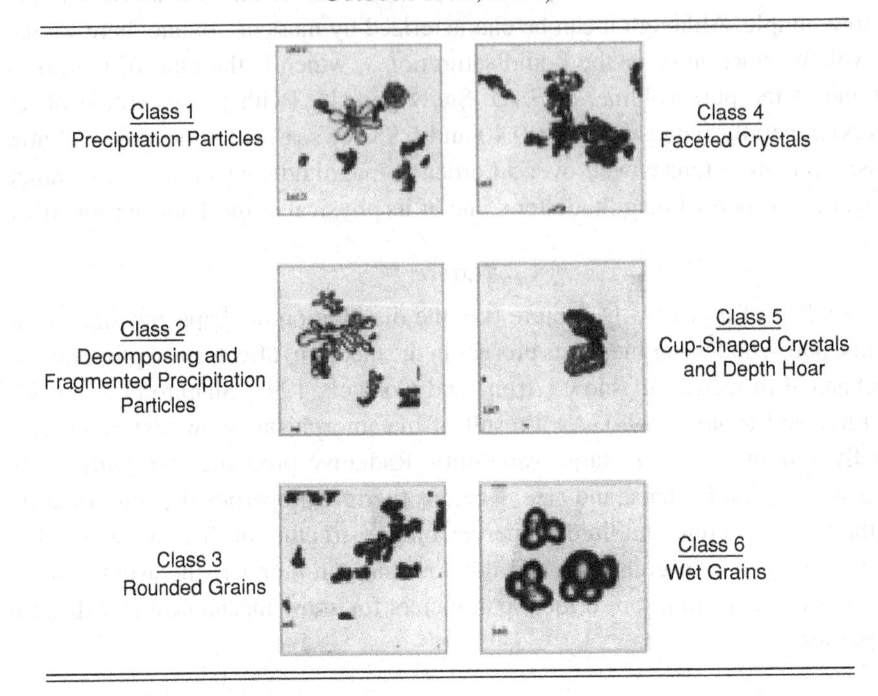


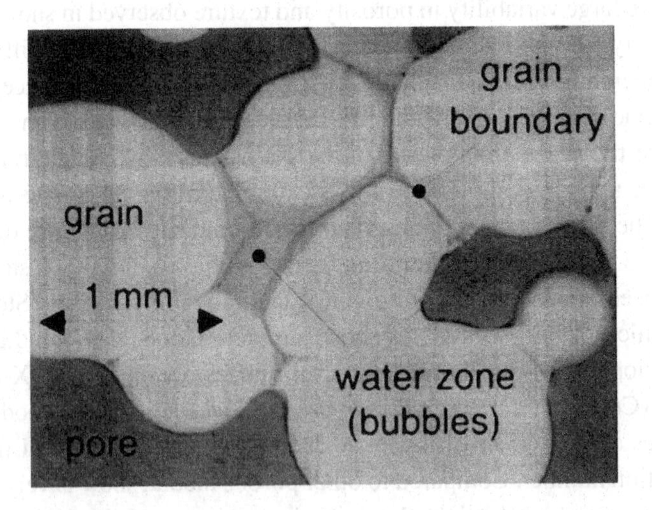


Figure 2.4. Image of a thick section showing ice, pores, and frozen liquid menisces of a snow type 6a (from Brzoska *et al.*, 1998).

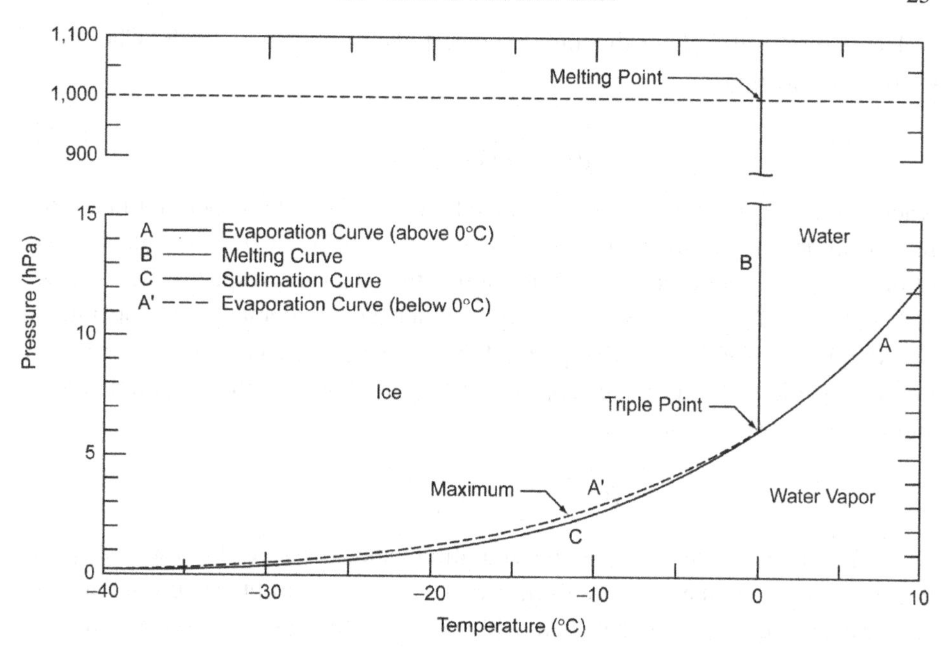


Figure 2.5. $p$—$T$ phase diagram for bulk-water, based on integral forms of Equation (2.1).

### 2.2.2 *Thermodynamics of phase equilibria in snow*

*Bulk equilibrium temperature in a pure water system*

Thermodynamic relationships among the three water phases determine the grain growth and metamorphism of snow. Whether the compound $H_2O$ exists as ice, water, or water vapor depends on its temperature ($T$) and pressure ($p$), as shown in the phase diagram in Fig. 2.5. The curves for evaporation (A), sublimation (B), and melting (C) trace the points where two bulk-water phases coexist in thermodynamic equilibrium. All three phases coexist at the triple point, where the temperature is 0.01 °C and the vapor pressure is 6.1112 hPa. This is not the same as the ordinary melting point of ice, which occurs at a temperature of 0 °C and at 1 atmosphere of pressure (1013.25 hPa). Vapor is in metastable equilibrium with respect to supercooled water along the dotted line A′, which extends the evaporation curve below 0.01 °C.

Air filled with vapor at equilibrium is saturated and the addition of more vapor will cause condensation. Below the triple point, the saturation vapor pressure at equilibrium with respect to water always exceeds that with respect to ice and thus favors the growth of snow crystals, as discussed in Section 2.1.2. Excess saturation vapor pressure reaches a maximum at −11.8 °C (see Fig. 2.5). Supersaturation relative to ice for vapor at equilbrium with respect to water increases from 0% at 0 °C to about 46% at −40 °C.

The curves in the phase diagram are determined by integrating the Clausius–Clapeyron equation

$$\frac{dp}{dT} = \frac{L_{ji}}{T(V_i - V_j)}, \tag{2.1}$$

where $i$ and $j$ are the two phases, $T$ is in kelvin, $V$ is the volume per unit mass (or reciprocal density) of the phase, and $L$ is the latent heat. At 0 °C the latent heats of fusion ($L_{i\ell}$), evaporation ($L_{\ell v}$), and sublimation ($L_{iv}$) are, respectively, $3.335 \times 10^5$, $2.505 \times 10^6$, and $2.838 \times 10^6$ J kg$^{-1}$. $L$ depends moderately on temperature (see Pruppacher and Klett, 1997, p. 97). For an ideal gas and constant $L_{kv}$, integration of Equation (2.1) provides approximate expressions for the evaporation and sublimation curves of the form

$$p_{v,sat,k} = a \exp\left(\frac{bT}{c+T}\right) \tag{2.2}$$

where $T$ is in °C. Here, $p_{v,sat,k}$ is the saturation vapor pressure (hPa) with respect to ice or water, and the coefficients $a$, $b$, and $c$ are empirical fits for given temperature ranges. For the range between −40 and 0 °C, Buck (1981) recommends the expression

$$p_{v,sat,\ell} = [1.0007 + (3.46 \times 10^{-6} p_a)]6.1121 \exp\left(\frac{17.966T}{247.15+T}\right) \tag{2.3}$$

for water saturation and between −50 and 0 °C

$$p_{v,sat,i} = [1.0003 + (4.18 \times 10^{-6} p_a)]6.1115 \exp\left(\frac{22.452T}{272.55+T}\right) \tag{2.4}$$

for ice saturation, where $p_a$ is atmospheric pressure (hPa). The initial coefficient in both expressions is a slight correction factor used when the gas phase is moist air rather than pure water vapor.

### *The effect of curvature on phase equilibrium in snow*

The $T$–$p$ curves derived from the Clausius–Clapeyron equation are for phases separated by plane surfaces. These equilibrium conditions change when the surface is curved, as is the case with ice grains or water menisci in wet snow (see Fig. 2.4). Because work is needed to extend interfacial films, the phases behind convex interfaces experience a higher energy and pressure. The pressure difference ($p_{ij} = p_i - p_j$) across a curved surface is given by Laplace's equation as

$$p_{ij} = \frac{2\sigma_{ij}}{\zeta_{ij}}, \tag{2.5}$$

where $\sigma_{ij}$ is the interfacial surface tension and $\zeta_{ij}$ is the mean radius of curvature (Defay *et al.*, 1966, p. 6 Dullien, 1992, pp. 119–122). The surface tensions $\sigma_{i\ell}$, $\sigma_{iv}$,

Table 2.3 *Ratio of saturation pressure over ice spheres to that over plane surfaces.*

| $r_g$ ($1 \times 10^{-3}$ m) | $10^{-6}$ | $10^{-5}$ | $10^{-4}$ | $10^{-3}$ | $10^{-2}$ | $10^{-1}$ |
|---|---|---|---|---|---|---|
| $p'_{v,sat,i}/p_{v,sat,i}$ 0 °C | 6.05 | 1.20 | 1.018 | 1.002 | 1.0002 | 1.000 02 |
| –20 °C | 6.97 | 1.21 | 1.020 | 1.002 | 1.0002 | 1.000 02 |

and $\sigma_{a\ell}$ are, on average, 0.028, 0.104, and 0.076 N m$^{-1}$ at 0 °C and vary strongly with temperature (Pruppacher, 1995). Higher surface tension and tighter curvature result in larger pressure differentials.

For curved surfaces, the Clausius–Clapeyron equation generalizes to the Gibbs–Duhem equation (Defay *et al.*, 1966, Colbeck, 1980)

$$\frac{V_i dp_i}{dT} - \frac{V_j dp_j}{dT} = \frac{L_{ji}}{T}. \tag{2.6}$$

Relating the phase pressure through Laplace's equation, Gibbs–Duhem's equation integrates to Kelvin's expression for the equilibrium vapor pressure $p'_{v,sat,k}$ over a curved surface

$$p'_{v,sat,k} = p_{v,sat,k} \exp\left(\frac{2\sigma_{kv}}{\zeta_{kv}} \frac{1}{R_v \rho_k T}\right). \tag{2.7}$$

Here $p_{v,sat,k}$ is the equilibrium pressure over a flat surface, $\rho_k$ is the density of water or ice, $R_v$ is the gas constant for water vapor ($= 461.50$ J kg$^{-1}$ K$^{-1}$), and $T$ is in kelvin. Equation (2.7) predicts that the vapor pressure will be higher over smaller ice particles or fine-structure with higher curvature. Table 2.3 shows $p'_{v,sat,k}/p_{v,sat,k}$ for ice spheres of varying radii ($\zeta_{iv} = r_g$) at temperatures of 0 and $-20$ °C.

The Gibbs–Duhem equation also predicts the melting temperature of snow. In the case of a two-phase water–ice system, such as water-saturated snow or pockets of water-saturated snow within the pack (grain A in Fig. 2.6a), pressure in the water computes from capillary pressure, $p_{a\ell} = p_a - p_\ell$, and that in ice from Laplace's equation. We discuss capillary pressure in Section 2.4.2. The melting point depression of water-saturated snow thus decreases directly with capillary pressure and inversely with grain radius as

$$T_d = \frac{273.15}{L_{i\ell}} \left[ p_{a\ell} (V_\ell - V_i) + \frac{2\sigma_{i\ell} V_i}{r_g} \right]. \tag{2.8}$$

Accordingly, larger ice grains have higher melting temperatures and will grow at the expense of smaller grains. In a totally saturated snowpack, the air/water interface is flat and the first term in Equation (2.8) drops out.

At lower water contents, liquid is usually configured as pendular rings about two-grain contacts or as veins in three-grain clusters, as in Figs. 2.6b and c

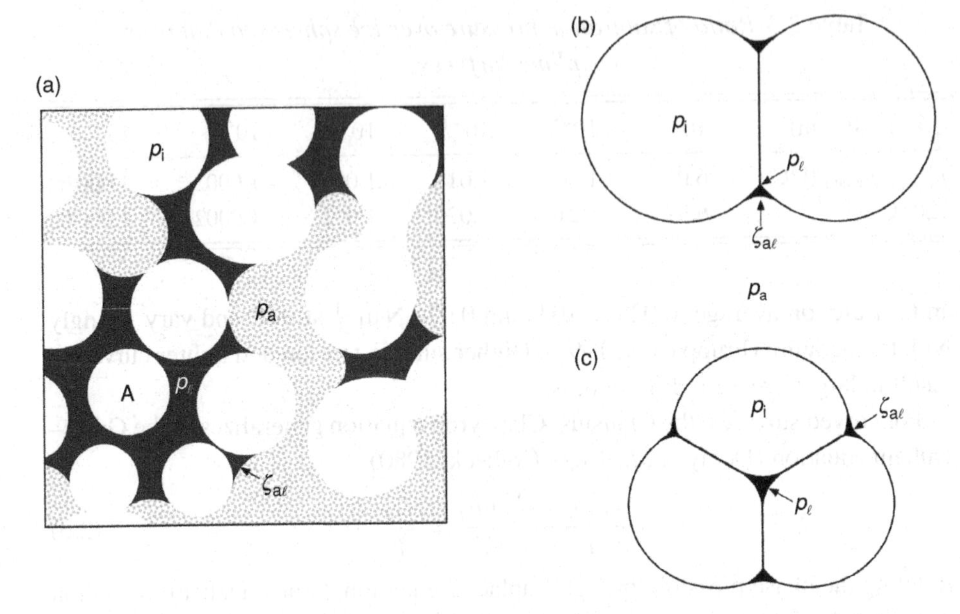


Figure 2.6. Pressure relationships and shape of water inclusions in wet snow, where $p_k$ is the pressure and $\zeta_{a\ell}$ is the radius of curvature for the air/water interface. (a) Saturated zones in snow with high water content (not to scale). (b) Two-grain contact with pendular inclusion in snow with low water content (reprinted with modifications from Colbeck, 1979; copyright 1979, with permission from Elsevier). (c) Three-grain cluster with vein-fillet inclusion in snow with low water content (reprinted from Colbeck, 1979, as in b).

(Colbeck, 1979). With air occupying much of the pore space, pressure in the ice computes from Laplace's equation applied to the ice/air interface and the melting point depression becomes (Defay *et al.*, 1966; Colbeck, 1979)

$$T_d = \frac{273.15}{L_{i\ell}}\left[p_{a\ell}V_\ell + \frac{2\sigma_{iv}V_i}{r_g}\right]. \tag{2.9}$$

Because the second term in (2.9) is small, $T_d$ depends mainly on capillary pressure, which increases with decreasing liquid water.

Melting temperatures predicted by Equations (2.8) and (2.9) are depressed only slightly below the bulk value of 0 °C and thus, for energy balance computations, the effects of surface tension and curvature on temperature are insignificant. They are, however, of prime importance to snow metamorphism and bond strength. At low water contents, the melting temperature of grain surfaces is below that of the ice bonds, directing heat away from the bonds and causing them to grow and strengthen (Colbeck, 1979). When snow grains become totally surrounded by water, there is a thermodynamic reversal and heat flows towards the bonds, causing them to melt. Saturated or "slushy" zones in snow are therefore cohesionless and have little strength. Impurities in the meltwater further depress the melting temperature by $K_F M$, where $K_F$ (=1.855 K kg mol$^{-1}$) is the cryoscopic constant for water and $M$ is the molality of the species (moles of solute per kg of solvent).

### 2.2.3 Snow metamorphism

The great variability in snow microstructure is due in a small part to the initial diversity of precipitating particles but mostly to the various transformations the ice matrix undergoes because of thermodynamic relationships among the water phases. These transformations are called snow metamorphism. Since the temperature of natural snow is generally close to the triple point, mass exchanges between vapor and ice and possibly between ice, vapor, and liquid water are very active. This activity is enhanced by the large specific area of the ice/air interface, which derives from the snow microstructure. Consequently, ice crystals are continuously evolving in snow. The effects and rate of metamorphism differ according to the prevailing thermal and meteorological conditions, which explains why succeeding fresh snow layers evolve differently and why the snowpack becomes stratified. The main cause of differentiation in metamorphism is the absence or presence of the liquid phase. Thus, we distinguish between dry and wet snow metamorphism. From a qualitative point of view, the mechanisms of dry and wet snow metamorphism have been well understood for a long time. The main transformations of snow due to metamorphism are summarized in Fig. 2.7.

#### *Dry snow metamorphism*

Dry snow is defined as snow that contains no liquid water. By this definition, the pores in dry snow are filled only by air that is usually saturated with vapor. This is typically the case when the temperature of a snow layer with few impurities is below −0.01 °C, but pure dry snow may exist at a bulk temperature even closer to 0 °C. Vapor saturation occurs because of the large specific area of the ice/air interface, which facilitates mass exchanges between ice and vapor and thus ensures a macroscopic equilibrium between the two phases when there is no interstitial air movement. At the microscale, equilibrium is not possible because of differences in temperature and in ice crystal geometry.

According to Equations (2.7) and (2.2), microscopic differences in vapor pressure at saturation due to variations in curvature and temperature have the following important consequences for snow metamorphism.

- *Equilibrium growth form.* Because of the crystal's shape, local variations in curvature generate local variations in the equilibrium vapor pressure. If temperature is quasi-uniform, the air around convex crystal surfaces (at the points of crystal branches, for example) tends to be at a higher vapor pressure than the air around concave surfaces (at the bond between two adjacent crystals, for example). Thus, the induced local pressure gradients

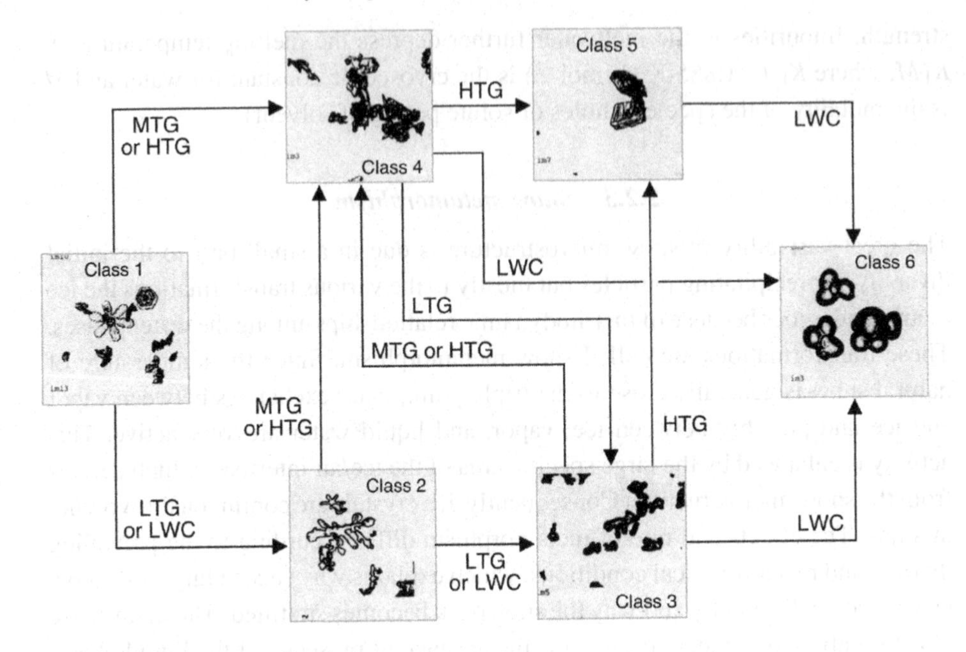


Figure 2.7. Schematic description of the transformation between the main snow classes due to metamorphism.

generate vapor diffusion from convex towards concave regions. To maintain equilibrium (as much as possible), diffusion is partially balanced by sublimation of the convex crystal surfaces and a corresponding deposition over concave or less convex surfaces. If this process continues, highly convex surfaces therefore shrink while concave or less convex surfaces grow. Because the equilibrium form of an individual ice crystal is a sphere (the shape with the smallest specific area or surface-to-volume ratio), grains become rounded over time. If the process continues for several weeks, the average size of snow grains increases, but at a much slower growth rate.

- *Kinetic growth form.* If a temperature gradient is maintained through a snow sample, variations in temperature generate variations in vapor pressure at saturation and thus induce vapor diffusion from the warmest crystal surfaces toward the coldest ones. This diffusion is partially balanced by sublimation of the warmest crystal surfaces and the corresponding deposition over the coldest ones. If the temperature gradient is high enough, the growth of the coldest surfaces is rapid enough to form facets and even strias, which are the respective shapes of faceted crystals (class 4) and depth hoar (class 5). The maintenance of a strong temperature gradient through a snow layer keeps snow far from the equilibrium form and the specific area weakly decreases and even increases if initial snow belongs to classes 3 or 6.

Under natural conditions, curvature and temperature gradient effects work together and compete. Newly deposited precipitation particles (class 1) generally include dendrites or planar crystals, which have numerous sharply convex surfaces. In most situations, these convex features become rapidly rounded and snow metamorphoses into decomposing and fragmented precipitation particles (class 2). At this stage, if the temperature gradient exceeds about 5 °C $m^{-1}$, snow metamorphoses into faceted crystals (class 4) and further into depth hoar (class 5) if it exceeds about 15 °C $m^{-1}$. If the temperature gradient stays below 5 °C $m^{-1}$, rounding continues and snow metamorphoses into rounded grains (class 3).

*Wet snow metamorphism*

When snow holds a significant amount of liquid water (typically more than 0.1% of the volume), mass exchanges between the three water phases must be considered to explain snow metamorphism. In most cases, wet snow is not saturated and mass exchanges are not limited to the solid and liquid phases. If we neglect the effects of impurities (mainly concentrated in the liquid phase) and of air dissolution into water, the Gibbs–Duhem and Laplace equations dictate that the melting point temperature of snow at low liquid water content varies as the sum of negative linear functions of curvature and capillary pressure (Equation 2.9). In snow at higher liquid water content, the melting point temperature varies primarily as a negative linear function of curvature (Equation 2.8). In both cases, the more convex the crystal surface, the lower the melting point temperature around this surface.

Consequently, local temperature gradients within a wet snow sample conduct heat away from the concave or less convex crystal surfaces towards the more convex surfaces. This process generates melting of the most convex surfaces, including the smallest crystals, and refreezing of liquid water (if it is available) onto the concave or less convex surfaces. These mass exchanges lead to rounding and growth of the crystals, in a process that minimizes the specific area and tends towards the equilibrium spherical form. A limiting factor for wet snow metamorphism is the capacity for liquid water to move easily from areas of melting to areas of refreezing. At very low water content, liquid water menisces are rare and disconnected, thus limiting the efficiency of this mechanism. In such cases, exchanges between vapor and solid phases may still dominate. When wet snow freezes, liquid menisces are included in the adjacent ice grains, which explains the rapid growth of grains during melt–freeze cycles.

### 2.2.4 Grain size and growth rate

Since snow texture affects most of its physical properties (among them its albedo), it is of primary importance for snow modeling to have a quantitative knowledge of metamorphism processes. A major difficulty comes from the lack of a unique parameter that may quantitatively describe snow texture. In most cases, the microstructure of a snow sample is described by the grain size and metamorphism is quantified through a growth rate. According to the International Classification, the size of a grain is its greatest extension and the grain size of a snow sample is the average size of its characteristic grains. But for a given physical process, it may be more relevant to consider other textural definitions, such as the geometrical size, the optical grain size, the pore size, or the specific area. In the simplified case where we consider snow as a packing of rounded grains, the effects of curvature on the vapor pressure and on the melting point depression dictate that the smallest grains, which have the most convex surfaces, will always shrink at the profit of larger grains. For that reason, we speak of the growth rate due to metamorphism. In actuality, however, snow shows a large set of possible grain shapes and the use of grain size or growth rate as a texture descriptor is very reductive. Precipitation particles (class 1) illustrate this difficulty: their size may be relatively large (typically one millimeter or more for plates or stellar dendrites), but they are generally very planar and along their perimeter (if we consider them as two-dimensional particles) we find very sharp surfaces that shrink very rapidly. Consequently grain size decreases while the planar crystal transforms into a granular crystal that is larger from the thermodynamic point of view.

Quantification of snow metamorphism has been investigated in three different ways, experimentally, theoretically, and numerically.

- *Experimental.* Different experiments on dry and wet snow metamorphism have been conducted on snow samples submitted to most temperature, temperature gradient, and liquid water content conditions that can be encountered in nature. Wakahama (1968) and Raymond and Tusima (1979) have investigated the growth rate of a population of snow grains immersed in liquid water. Giddings and Lachapelle (1962), Akitaya (1974), and Marbouty (1980) investigated the growth rate of snow submitted to a high temperature gradient. Brun (1989) investigated the growth rate of snow grains as a function of liquid water content. Brun *et al.* (1992) investigated the growth rate and the faceting of fresh snow submitted to a low or medium temperature gradient. These experiments have been conducted in the cold laboratory or in the field and consider the effect of grain size as well as grain shape. Brun *et al.* (1992) summarized these experiments in a complete set of quantitative metamorphism laws that describe the rate of change in growth and shape of snow. These empirical laws use an original formalism that makes it possible to describe snow with a continuous set of parameters {dendricity, sphericity, grain size}. Experiments also have been conducted on sintering and on the growth rate of bonds linking snow grains (Keeler, 1969a). This last process is of little interest for snow atmosphere exchanges but it is of major importance for the evolution of the mechanical properties of snow layers and the stability of the snowpack.

- *Theoretical*. Considering snow as a packing of idealized particles of various sizes and at various temperatures, functions for vapor diffusion between two adjacent grains have been analytically derived. The corresponding growth rate has been deduced, making it possible to compare the relative efficiencies of temperature gradient metamorphism with that due to variations in curvature (Colbeck, 1973, 1980, 1983; Gubler, 1985). Colbeck proposes the following equation to describe the average growth rate of snow particles:

$$\dot{m} = 4\pi C G_g D_g T'(d\rho_{v,sat}/dT), \tag{2.10}$$

where $\dot{m}$ is the mass growth rate of a particle, $C$ a shape factor, $G_g$ an enhancement factor, $D_g$ the diffusion coefficient of water vapor in air, $T$ the temperature, $T'$ the snow temperature gradient, and $\rho_{v,sat}$ the vapor density at equilibrium.

Similarly, the growth rate of grains in water-saturated snow has been calculated. These calculations provide a theoretical background to explain the high growth rates observed at high temperature gradients and at high liquid water contents. The extension of calculated growth rates from dry and wet metamorphism to a population of grains having a distribution of sizes and shapes is difficult but possible. Several snow models use metamorphism laws deduced from these theoretical works to calculate the evolution of grain size of the different snow layers of a snowpack.

- *Numerical*. More recently, high-resolution numerical models have been developed to calculate vapor diffusion caused by temperature gradients inside an idealized snow matrix (Christon *et al.*, 1987). Recent availability of high-resolution three-dimensional images of snow samples (see Fig. 2.8) and the huge increase of computer performance opens fascinating perspectives for the development and validation of snow microstructure models (Flin *et al.*, 2003). These models will allow the establishment of more reliable metamorphism laws in the near future.

### 2.2.5 Snow compaction

#### *Measurement of snow density and liquid water content*

As seen in Section 2.2.1, the mixing ratio between ice, air, and liquid water is a key parameter that explains a significant part of the variability in physical and mechanical snow properties. For dry snow, density is an accurate indicator of this ratio, while for wet snow, an additional measurement of the liquid water content is necessary. Density measurements of snow are relatively easy to perform both in the laboratory and in the field, and are generally obtained by weighing a calibrated snow core. Automatic measurements can be performed in the field by measuring the absorption of gamma rays. In the case of dry snow, measurement of the relative permittivity can also be used to estimate density with reasonable accuracy. Snow density in a seasonal snow cover generally ranges from 50 to 550 kg $m^{-3}$, but densities from 20 to 50 kg $m^{-3}$ and from 550 to 650 kg $m^{-3}$ can be observed under particular conditions.

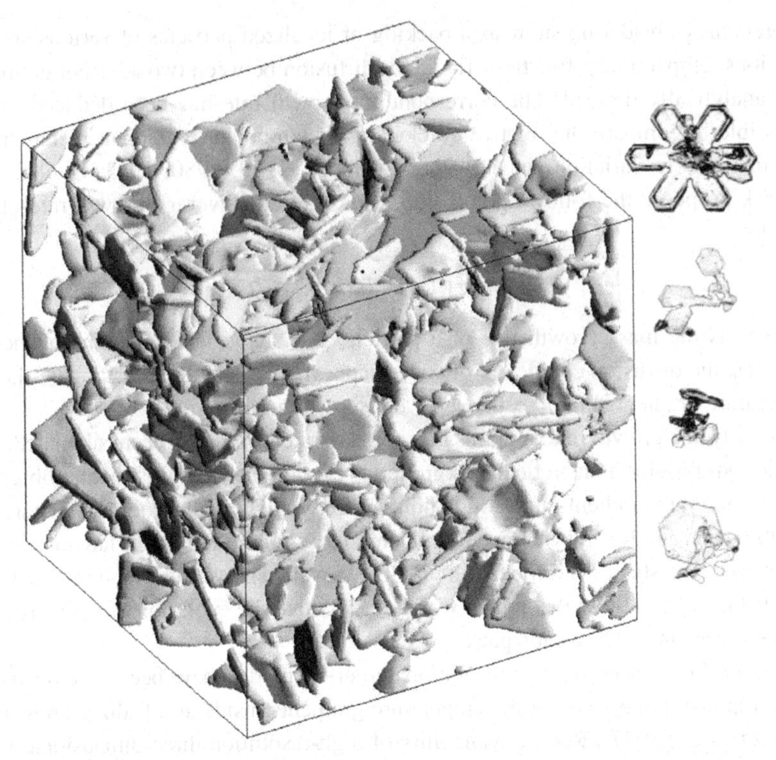

Figure 2.8. High-resolution 3-D image of a snow sample (class 1) obtained by the means of X-ray tomography. The image represents a total volume of 1 $cm^3$ with voxels of 10 μm (from Flin *et al.*, 2003).

Liquid water content can be measured by various procedures. In the cold laboratory, cold calorimetry is frequently used to obtain accurate measurements. In the field, it is difficult to obtain measurements of liquid water content with an accuracy better than 1% per mass. Dielectric measurements, chromatic elution, cold calorimetry, and even hot calorimetry are the techniques generally used. In a well-drained wet snow layer, liquid water content generally ranges from 0 to 15% per mass, but saturated snow layers (almost 100% of pore volume filled with liquid water) can form inside or at the base of a snowpack above low permeability ice or frozen soil layers. In saturated layers, liquid water content can exceed 75% of the snow mass.

### *Compaction rate*

As a result of its importance in the dynamics of snowpacks, snow compaction has been widely investigated in the past. Snow rheological properties are of extreme complexity, showing combined viscous, elastic, and plastic behavior. This behavior is strongly affected by the deformation rate. Here we will consider only the slow deformation of snow within a snowpack under natural conditions.

At the surface or inside a snowpack, compaction of snow occurs under the influence of three main processes that work almost independently with variable efficiency.

- *Snow drift.* As soon as wind velocity exceeds a threshold depending on the snow type of the surface layer, snow drift occurs. During snow drift, sublimation of snow particles and collision between particles transform snow crystals, which rapidly shrink and become more or less rounded (class 2b or class 3). During this process, density rapidly increases because the packing of small and rather rounded particles is denser than that of the precipitation particles, which often include dendrites and plates. In the case of a strong drift event, a fresh snow layer with a density less than 100 kg $m^{-3}$ can transform into a drifted snow layer with a density up to 300 kg $m^{-3}$ within a few tens of hours. This compaction process is of major importance in polar regions where prevailing low temperatures slow down metamorphism (Dang *et al.*, 1997).
- *Metamorphism.* As soon as precipitation particles (class 1) are deposited and form a snow layer, metamorphism starts. In most cases, crystals initially shrink as a result of curvature effects (dendrites disappear and planar crystals become granular) and turn into class 2. This transformation generally requires between a few hours and a week, depending on temperature and liquid water content. Class 2 snow grains are smaller and less dendritic than class 1 particles and consequently constitute a denser packing. Therefore, metamorphism of class 1 snow induces a rapid compaction. In the next stage of metamorphic compaction, the compaction rate depends on the type of metamorphism. If high temperature gradients prevail, snow turns into faceted crystals (class 4) and eventually into depth hoar (class 5), if the process lasts long enough. The grain shape of these snow classes does not facilitate packing, and thus high temperature gradient metamorphism does not involve a significant compaction of snow, except for the initial compaction of precipitation particles. On the contrary, wet snow metamorphism transforms every type of snow into class 6, which consists of rounded and large grains that pack efficiently and densely. Wet snow metamorphism is a very efficient process to compact moist fresh snow layers, even at the surface where the overburden pressure is low. The density of precipitation particles at the surface can increase by up to 250 kg $m^{-3}$ within a few hours during rainfall. Similarly, moistening induces a rapid increase in the density of depth hoar.
- *Deformation strain.* Buried snow layers have to sustain the weight of upper layers. Gravity forces are concentrated in the grain bonds, which break, slide, partially melt, or warp, making the rheological properties of snow very complex (Golubev and Frolov, 1998). These processes are very active in fresh snow, which settles very rapidly during and after snowfall (see Fig. 2.9). Rheological properties of snow have been extensively investigated but, until now, no universal law describing snow rheology is recognized by the scientific community. Deformation strain is very sensitive to the deformation rate. At slow deformation, snow is usually considered as a Newtonian fluid with a viscosity depending on snow density, as well as on snow microstructure and on other parameters, such as

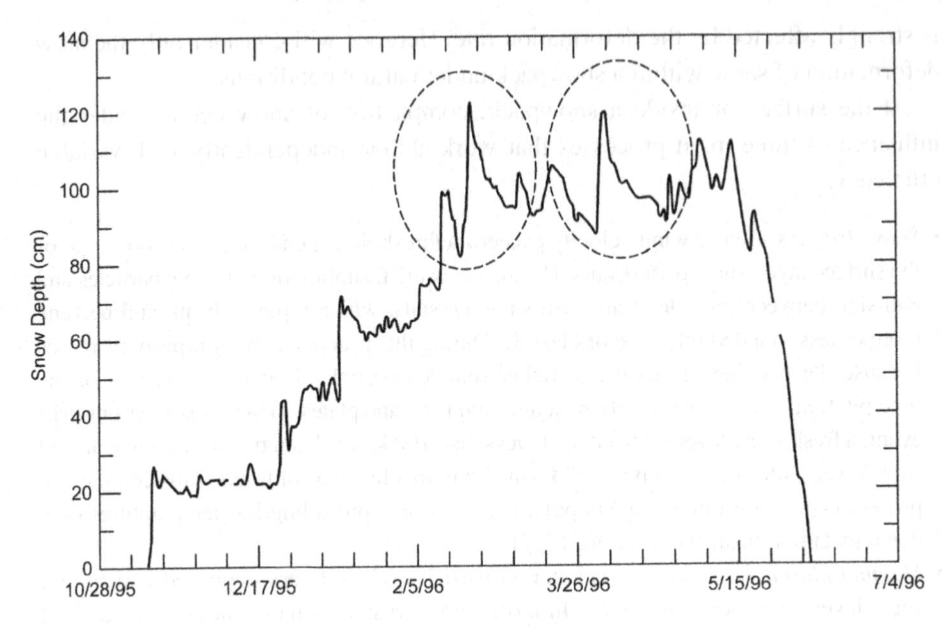


Figure 2.9. Snow depth between November 1995 and June 1996 at Col Agnel (French Alps, 2630 m). Circled areas indicate rapid compaction following new snowfalls.

temperature and liquid water content. Snow belonging to classes 1 and 2 has a low viscosity and compacts rapidly under the pressure of the upper layers. The higher the pressure, the faster the compaction. Lower snow layers, which have sustained a higher overburden over a longer time than upper layers, generally show a higher density. This is not always the case, however: viscosity also depends on snow type, and cold depth hoar layers almost do not compact. They can remain at a density close to 300 kg m$^{-3}$ for several months at the base of a deep snowpack.

Recent papers have reviewed the present knowledge of snow compaction and of the effect of climate on the average density of snowpacks all around the Northern Hemisphere (Sturm and Holmgren, 1998). In most snow models or parameterizations, the effects of metamorphism and deformation strain are combined and simulated by way of a Newtonian viscosity. Snow viscosity is described in more or less sophisticated ways to take into account the effects of temperature, liquid water content, and microstructure.

From laboratory experiments, Navarre (1975) established the following equation to describe the Newtonian viscosity of dry snow as a function of temperature and density:

$$\eta = \frac{\eta_0}{1-f} e^{0.023\rho_s - 0.1(T-T_0)}, \tag{2.11}$$

where $\eta_0$ is a constant equal to $6.0 \times 10^6$ Pa s, $\rho_s$ is snow density, $T$ is snow temperature, $T_0$ is the melting temperature, and $f$ is a constant depending on the snow type, set equal to 0.4 by Navarre. Morris *et al.* (1997) compile viscosity observations for a range of snow types, densities and temperatures. Recently, some snow models have taken into account the effects of snow drift on the compaction of surface layers (e.g. Brun *et al.*, 1997).

## 2.3 Thermal behavior of snow

Because of its relatively low thermal conductivity, the snowpack acts as a thermal blanket, protecting the earth from rapid atmospheric temperature changes. However, the snow is not a passive player in snow–atmosphere interactions. For example, snow grain size is a function of both the accumulation event (e.g. surface hoar growth versus precipitation) and local temperature. Grain type and size have a large impact on the snow albedo, which in turn has a large impact on atmospheric heat and mass transfer processes. A change in snow albedo over a large area can influence atmospheric circulation patterns. Because snow as a material is always relatively close to its melting point under normal environmental conditions, the metamorphic processes have profound effects on its properties. In turn, the properties of the snow affect its thermal response. In order to study snow behavior, then, it is almost always imperative to understand the heat transfer component.

### *2.3.1 Thermal properties*

As in all aggregate materials, the thermal properties of snow depend upon the microstructure; for snow this reflects on crystal type, crystal organization, and connectivity. Because snow undergoes metamorphism with time and ambient conditions, its thermal properties also change. Although the measurements of Izumi and Huzioka (1975) confirm that metamorphism induces anisotropy in physical properties, there are no quantitative analyses that can predict thermal properties based on microstructure. Arons and Colbeck (1995) give a historical perspective and a useful summary of the attempts to quantitatively predict material properties in snow based on the microstructure, and conclude that until improvements are made in the ability to characterize the snow geometry, little progress is likely. Thus, there is a poor physical understanding of the geometric effects of microstructure on the aggregate properties. The most useful estimates to date of the thermal properties of snow are based on macroscopic measurements of the aggregate that are then correlated to the snow type and physical characteristics such as density and grain size.

### *Thermal conductivity*

The thermal conductivity of a medium controls the speed at which heat will be transferred within the medium. In one dimension, the flux of heat at a point is given by the Fourier equation as

$$F = -k_{\mathrm{eff}} \frac{\partial T}{\partial x}, \tag{2.12}$$

where $F$ is the flux, $k_{\mathrm{eff}}$ is the thermal conductivity, $T$ is temperature, and $x$ is the spatial coordinate along the direction of flow. The measurement of thermal conductivity in snow includes the effects of heat conduction through connected grains and through the air space, along with the "hand-to-hand" transfer of latent heat by water vapor. In this process, vapor sublimes from warmer ice grains, diffuses through the pore space, and condenses on colder grains. With this understanding, $k_{\mathrm{eff}}$ as defined above is actually an "effective" thermal conductivity that includes both the diffusion of heat and vapor transport processes. Singh (1999) demonstrates that for temperatures above $-10\,°\mathrm{C}$ and approaching the melting point, the effective thermal conductivity of depth hoar can be one-half to two-thirds due to vapor transport. For snow of higher density and larger bonds, it is likely that most of the effective conductivity is due to conduction through the ice. Singh also notes a marked increase in thermal conductivity with increasing liquid water content. Sturm and Johnson (1992) made measurements of the thermal conductivity of depth hoar and added these measurements to the compilation of Mellor (1977). Figure 2.10 is the compiled figure by Sturm and Johnson (1992), which depicts measurements of the thermal conductivity of snow as a function of snow density. Here, the earlier measurements of Sturm and Johnson are replaced with a quadratic curve from Sturm *et al.* (1997), based on about 500 new measurements of thermal conductivity. Although a general trend of increasing thermal conductivity with increasing density is clear, there is wide scatter in the graph as a result of variations in microstructure and temperature. For a given density, Sturm *et al.*'s (1997) compilation of measurements by others has a standard deviation of about 0.1 $\mathrm{W\,m^{-1}\,K^{-1}}$, and their own measurements have a standard deviation somewhat less than this.

### *Specific heat*

The specific heat of snow reflects on the amount of energy that must be put into a unit amount of snow in order to change its temperature. Because snow is a conglomerate of ice, air, and liquid water, the specific heat is calculated as the weighted average of the parts:

$$c_{\mathrm{p,s}} = (\rho_{\mathrm{a}}\theta_{\mathrm{a}}c_{\mathrm{p,a}} + \rho_{\mathrm{i}}\theta_{\mathrm{i}}c_{\mathrm{p,i}} + \rho_{\ell}\theta_{\ell}c_{\mathrm{p},\ell})/\rho_{\mathrm{s}}, \tag{2.13}$$

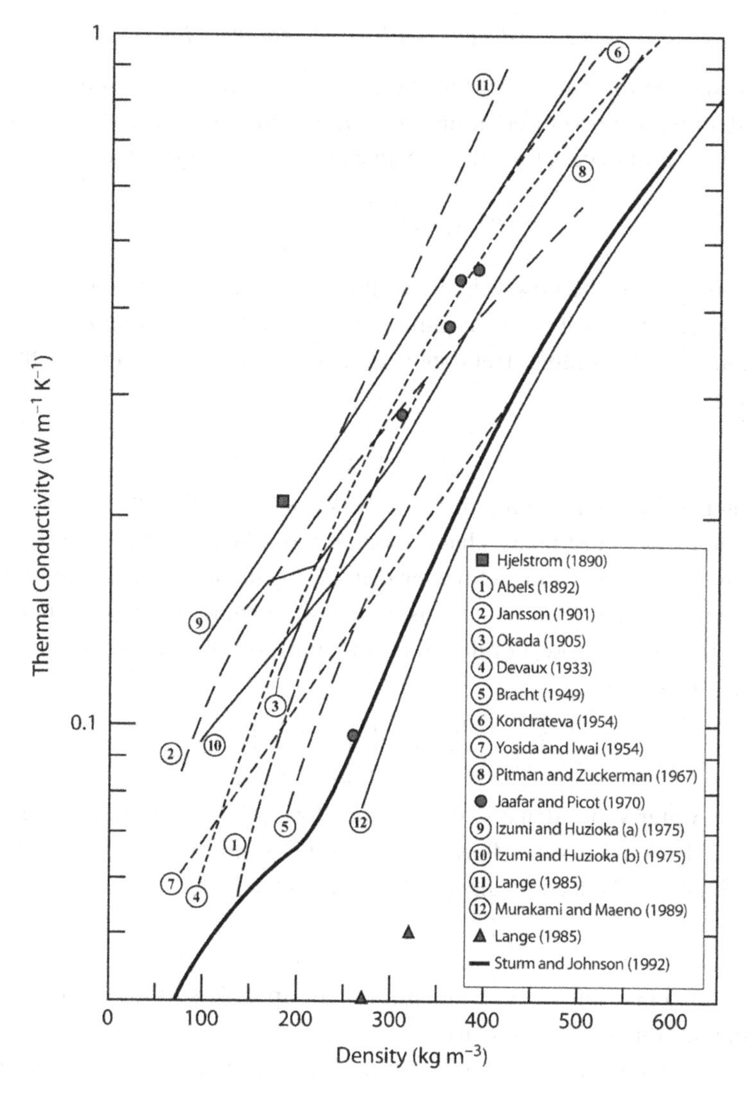


Figure 2.10. Effective thermal conductivity of snow versus snow density (from Sturm and Johnson, 1992, with modifications. Published 1992 American Geophysical Union. Reproduced/modified by permission of American Geophysical Union).

where $c_{p,s}$ is the specific heat of snow, $c_{p,a}$ is the specific heat of air (= 1005 J kg$^{-1}$ K$^{-1}$), $c_{p,i}$ is the specific heat of ice (= 2114 J kg$^{-1}$ K$^{-1}$), $c_{p,\ell}$ is the specific heat of water (= 4217 J kg$^{-1}$ K$^{-1}$), and $\theta_k$ are the constituent volume fractions. Values given for specific heats are at 0 °C and at 1 atmosphere of pressure. Because the constituent specific heats are temperature dependent, the specific heat of snow has a mild temperature dependence.

### 2.3.2 Advective–diffusive heat transfer

For the prediction of heat transfer in snow as a medium, the ice–vapor–air–liquid water system is assumed to be in thermal equilibrium on scales of grain size and less. Heat transport occurs though advective and diffusive processes:

$$\rho_s c_{p,s}\frac{\partial T}{\partial t} + \phi\rho_k c_{p,k} v_k \frac{\partial T}{\partial x} = \frac{\partial}{\partial x}\left[k_{eff}\frac{\partial T}{\partial x}\right] + q, \tag{2.14}$$

where $t$ is time, $v_k$ is the flow velocity of fluid $k$ (a = air, $\ell$ = water), and $q$ is the thermal source term. If there is no fluid (air or water) flow through the snow, the energy transfer is by simple heat conduction. The source term includes the latent heat effects:

$$q = -L_{ik}S. \tag{2.15}$$

When water is the fluid flowing through the snow, $L_{i\ell}$ is the latent heat of fusion and $S$ is the melt–freeze phase change source term. When the fluid flow through the snow is air, $S$ is the vapor source term and $L_{iv}$ is the latent heat of sublimation.

For heat transfer, the thermal effects of vapor transport are more pronounced in dry snow than in wet snow. For dry snow, the transport of water vapor through the snow is described by

$$\frac{\partial \rho_v}{\partial t} + v_a\frac{\partial \rho_v}{\partial x} = \frac{\partial}{\partial x}\left[D_s\frac{\partial \rho_v}{\partial x}\right] + S, \tag{2.16}$$

where $\rho_v$ is water vapor density, $v_a$ is the air flow (ventilation) velocity, $D_s$ is the diffusion coefficient for vapor flow in snow, and $S$ is the source of water vapor due to phase change, enhanced by air flow through the snow as

$$S = \tilde{h}A(\rho_{v,sat} - \rho_v), \tag{2.17}$$

where $\tilde{h}$ is the mass transfer coefficient, $A$ is the specific surface of the snow, and $\rho_{v,sat}$ is the saturation vapor density.

Albert and McGilvary (1992) demonstrated that the heat transfer associated with vapor transport is significant in the determination of the overall temperature profile of a ventilated snow sample, but that the major temperature effects are controlled by the balance between the heat carried by the dry air flow through the snow and heat conduction due to the temperatures imposed at the boundaries. The relative thermal response of the snow to air flow (advection) and imposed temperature gradients (conduction) is characterized by the Peclet number:

$$Pe \equiv \phi v_a \rho_a c_{p,a}\frac{\Delta x}{k_{eff}}, \tag{2.18}$$

where $\Delta x$ is the characteristic distance. Low $Pe$ numbers correspond to heat conduction-controlled temperature profiles, while $Pe$ numbers greater than one

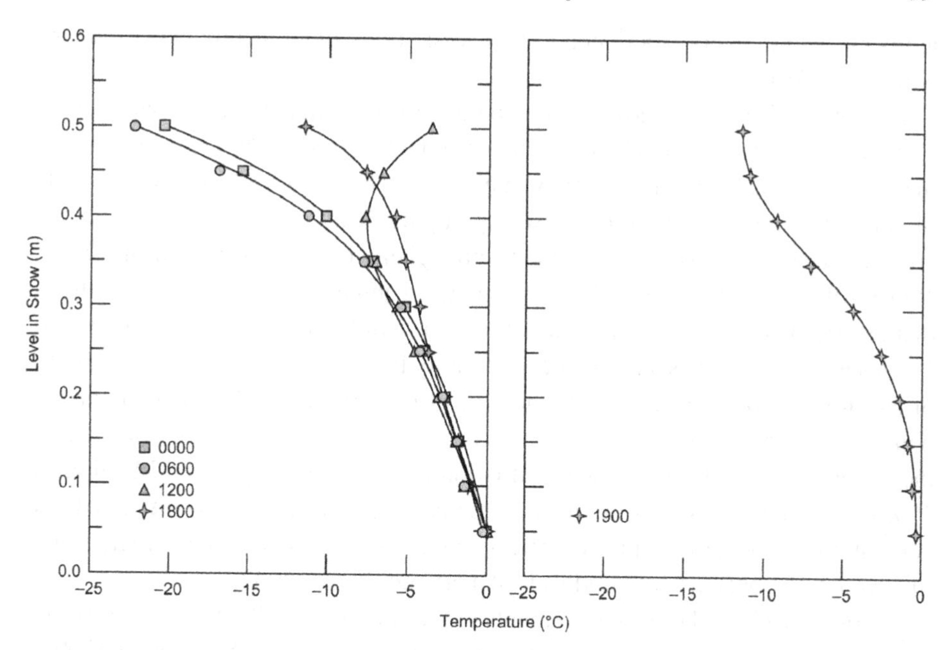


Figure 2.11. (a) In-snow temperature profiles at 6 h intervals for 6 February 1987 in Hanover, NH, during a mostly clear and calm day. (b) The same as for (a), except at 1900 hr on 9 February when the wind speed was 7 m $s^{-1}$. A comparison of this curve with that at 1800 hr on 6 February demonstrates the effects of windpumping.

correspond to advective profiles. For high porosity snows such as fresh snow or hoar, the thermal conductivity is sufficiently low that air flow through snow can cause advection-controlled temperature profiles with $Pe > 1$. This was demonstrated in a field experiment (Albert and Hardy, 1995) where the immediate temperature effects from ventilation appeared when flow was induced in natural seasonal snow. For lower porosity, higher conductivity snow, such as wind-packed snow or firn, the thermal conductivity is sufficiently high that the *Pe* number is almost always low under natural conditions, making it likely that the temperature profile will be dominated by the heat conduction profile, even though there may be significant air flow through the snow (Albert and McGilvary, 1992).

As an example of observed thermal effects in a seasonal snow cover, we compare diurnal temperature profiles for calm and windy days within a 50 cm deep snowpack in Hanover, NH (Jordan *et al.*, 1989) during the winter of 1987. Figure 2.11a shows profiles for 6 February, when wind speeds rarely exceeded 1 m $s^{-1}$ and clear skies prevailed for much of the day. Night-time radiational losses and daytime solar gains dominated the surface energy budget and caused large swings in temperature within the upper 20 cm of the snowpack. Such diurnal temperature patterns are typical for temperate snow covers experiencing sunny conditions and little turbulent exchange with the atmosphere.

In contrast, Fig. 2.11b shows a temperature profile from the same snowpack on 9 February 1987 at 1900 hr, when the wind speed was 7 m $s^{-1}$. The concave shape of the profile, compared with the convex shape for 6 February at 1800 hr, is an example of convective heat transfer through windpumping. Within this two-week data set, concave profiles consistently coincided with higher wind speeds (Jordan and Davis, 1990). Albert and Hardy (1995) present modeled and measured snow temperatures that clearly demonstrate the thermal effects of ventilation in light seasonal snow. Fresh seasonal snow has a thermal conductivity that is low compared to the ability of the snow to advect heat through the flow of interstitial air. This is in contrast to wind-packed snow and polar firn, where the temperature is usually dominated by the conduction profile even when ventilation is occurring (Albert and McGilvary, 1992). Inert tracer gas measurements in polar conditions have shown that natural ventilation can occur to depths of tens of centimeters to meters in polar snow and firn even in the presence of windpack (Albert and Shultz, 2002).

Ventilation affects both measurements of the chemistry (Albert *et al.*, 2002) and measurements of the thermal properties in snow. For example, Sturm *et al.* (2002) show that thermal conductivity values inferred from measured temperature profiles in snow-covered sea ice during the SHEBA experiment (surface heat budget of the Arctic Ocean) are higher than measurements with a needle probe at the same site. Jordan *et al.* (2003) propose that advective heat transport through windpumping may account for this discrepancy and show that high winds ($>10$ m $s^{-1}$) can penetrate dense, wind-packed snow on sea ice down to 10 or 20 cm depth. Further investigation is needed to parameterize the effects of windpumping, and possibly of other multi-dimensional heat transfer mechanisms, for one-dimensional modeling.

## 2.4 Fluid flow behavior in snow

While high porosity and low thermal conductivity make snow a protective blanket against extreme cold, this same openness makes it quite permeable to flows of air and water. Snow acts as both a source and a transmitter of water, quickly routing it through to the ground, which then channels much of the water to streams or rivers. Knowledge of the snow melting rate and snow hydrological processes (Marsh, 1990; Bales and Harrington, 1995) allows a more accurate prediction of flooding and of the distribution of water for agriculture and forest growth. Possible increased rainfall and snowmelt as a result of global warming increase the urgency for accurate runoff predictions (Jones, 1996). As well as its hydrological significance, snow acts as a buffer to chemical species transported by wind, rain, and meltwater (Bales *et al.*, 1989; Davis, 1991; Harrington and Bales, 1998).

Water flow through snow is similar to that through other granular materials but is complicated by freeze–thaw effects, metamorphism of the ice matrix, and textural layering. Because snow is highly permeable, water moves through it rapidly with typical speeds of 1–20 cm $h^{-1}$. Within this coarse material, gravitational forces dominate capillary forces. Low viscosity wetting fluids (such as water) develop instabilties which can concentrate flow in preferential channels, or flow fingers, ahead of the background wetting front. Stratigraphic inhomogenieties in permeability and capillary tension can impede and laterally divert the flow, as well as trigger the formation of flow fingers.

Unlike soils, snow is sufficiently open to allow the movement of air in the interstitial pore space. The interstitial air flow, most often caused by wind-induced pressure variations across the rough surface of snow, is known as ventilation. Only in very high permeability snow, such as hoar, does buoyancy-induced natural convection occur. Both natural and forced convection accelerate transport of water vapor and chemical species through snow and firn. An understanding of the mechanics of chemical transport in snow and firn is essential both to understanding air–snow transfer processes of chemical exchange (Albert *et al.*, 2002) and to interpreting the species concentrations in polar cores, which are vital records of paleoclimate history (e.g. Waddington *et al.*, 1996).

This section describes the theory of saturated (one-phase) and unsaturated (two-phase) flow through snow and discusses the macroscopic properties of permeability and capillarity that affect the flow. Just as thermal conductivity relates to the structure of the ice matrix, flow properties relate to the pore structure. The section concludes with a discussion of the basic features of uniform water flow and of unstable flow through flow fingers.

### 2.4.1 Saturated or one-phase flow

When one immiscible fluid, such as humid air or water, fills the pore space, flow is saturated relative to that phase. Except through very coarse snow, such as large-grained depth hoar, the flow is slow enough to be described by Darcy's law. In this case, the flow velocity of the fluid $v_k$ is proportional to the combined pressure and gravitational forces. Thus,

$$v_k = -\frac{K}{\eta_k}\left(\underbrace{\frac{\partial p_k}{\partial x}}_{\text{pressure gradient}} - \underbrace{\rho_k g \cos\Psi}_{\text{gravitational force}}\right). \tag{2.19}$$

Here, $K$ is the saturated or intrinsic permeability, $x$ is a spatial coordinate along the direction of flow, $\Psi$ is the angle between the flow direction and a downward vertical, g is the acceleration due to gravity, and $p_k$, $\eta_k$, and $\rho_k$ are the pressure, dynamic viscosity, and density of fluid $k$ (a = air, $\ell$ = water) respectively. Since snow is rarely water-saturated (except when water ponds above ice layers or frozen soil), Equation (2.19) is primarily used to model air flow.

*Saturated or air permeability*

Saturated or intrinsic permeability is a property of the pore structure and determines the ease and rate at which snow transmits fluids. $K$ is typically measured by forcing air through a snow sample while measuring the flow rate and pressure drop (Chacho and Johnson, 1987; Albert *et al.*, 2000). It is therefore often referred to as "air" permeability. Air permeability varies widely with snow type and changes over time as a result of metamorphism. In cold sites experiencing little or no melt conditions or wind drifting, the permeability of snow usually increases over time as its texture coarsens. Below the top several meters in firn, however, permeability can decrease as the pressure of overlying layers causes compaction and crystal sintering (Albert *et al.*, 2000). In seasonal snow, the permeability of the surface layers can experience dramatic changes within the span of a day as a result of insolation and surface heating (Albert and Perron, 2000).

Figure 2.12 (after Jordan *et al.*, 1999b) compiles measurements of $K$ from several investigators and compares them with the ice fraction $(1 - \phi)$. Measurements range over two orders of magnitude from $3 \times 10^{-10}$ m$^{-2}$ for fine-grained, wind-packed snow (low porosity and small pore size) to $600 \times 10^{-10}$ m$^{-2}$ for large-grained depth hoar (medium porosity and very large pore size). The dashed lines indicate general ranges of permeability for different snow categories observed by Bader *et al.* (1939). Most measurements fall within the outer boundaries of their classification scheme. Those reported by Sturm (1991) and Jordan *et al.* (1999b) for depth hoar in Alaska and Northern Canada are notably higher and most likely reflect extremely large pores and possibly the presence of vertical mesopores (Arons and Colbeck, 1995). Albert and Perron (2000) also showed that ice layers in seasonal snowpacks are permeable, although their permeability is substantially lower than the surrounding snow.

The scatter in Fig. 2.12 demonstrates that porosity alone is not a useful indicator of $K$. If the pore structure is idealized as bundles of tubes, $K$ becomes proportional to the square of the tube or pore diameter as well as to porosity (Dullien, 1992). Other factors, such as pore shape, pore interconnectivity, size distribution, and tortuosity also influence $K$. While the tubular model works well for soils and denser snow, it is unrealistic for light snow, when particle shape and specific area are dominant factors. Because pore size is difficult to measure directly, it is usually estimated from a combination of porosity (or snow density) and grain diameter. By binning his permeability measurements into equi-density groups, Shimizu (1970)

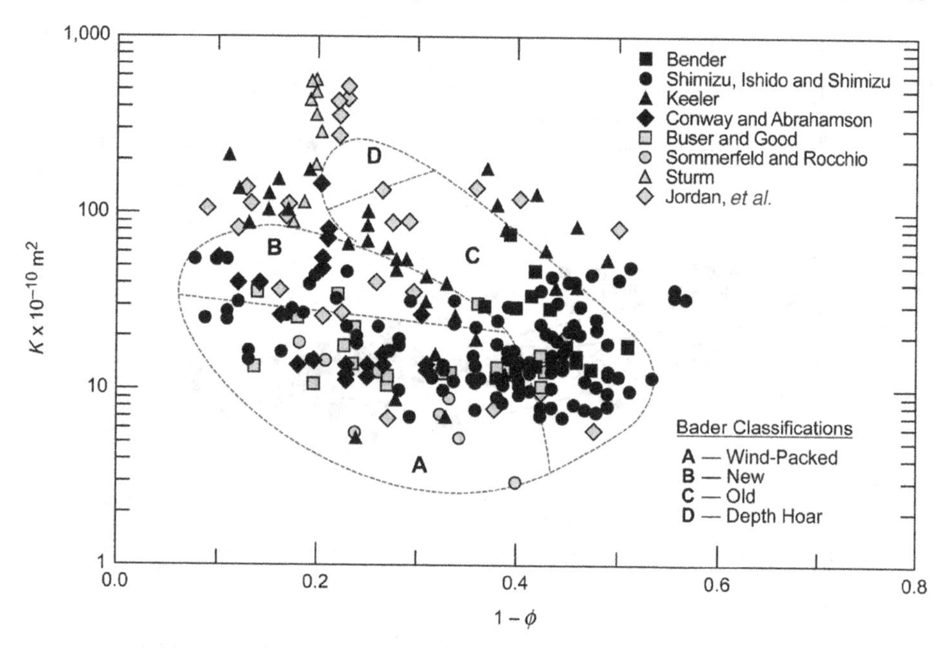


Figure 2.12. Measured permeability $K$ versus ice fraction $(1 - \phi)$ based on the laboratory observations of Bender (1957), Ishida and Shimizu (1958), Shimizu (1960), Keeler (1969b), Shimizu (1970), Conway and Abrahamson (1984), Buser and Good (1986), Sommerfeld and Rocchio (1989), Sturm (1991), and Jordan *et al.* (1999b). Dotted lines indicate Bader *et al.* (1939) classification scheme (after Jordan *et al.*, 1999b, copyright 1999; copyright John Wiley & Sons Limited. Reproduced with permission).

decoupled the relationships with density and grain size and derived the widely used formula:

$$K = 0.077e^{-0.0078\rho_s}d^2 = 0.077e^{-7.153(1-\phi)}d^2. \tag{2.20}$$

It is interesting to note that his empirically derived function has the correct theoretically derived units of length squared. If we normalize $K$ to $d^2$, as shown in Fig. 2.13 (after Jordan *et al.*, 1999b), the resulting relationship shows a clear decrease in $K$ with ice fraction. While Shimizu's data set was limited to fine-grained, wind-packed snow, the trends in Fig. 2.13 suggest that his function is also applicable to other snow types. More recent measurements of Luciano and Albert (2002), however, found that his formula yielded permeability estimates that vary from observations by more than an order of magnitude.

Figure 2.13 also shows theoretical curves for assemblages of thin discs and spheres. For a given porosity and grain radius, beds consisting of thin discs with an aspect ratio of 25 have approximately eight times the specific area of those consisting

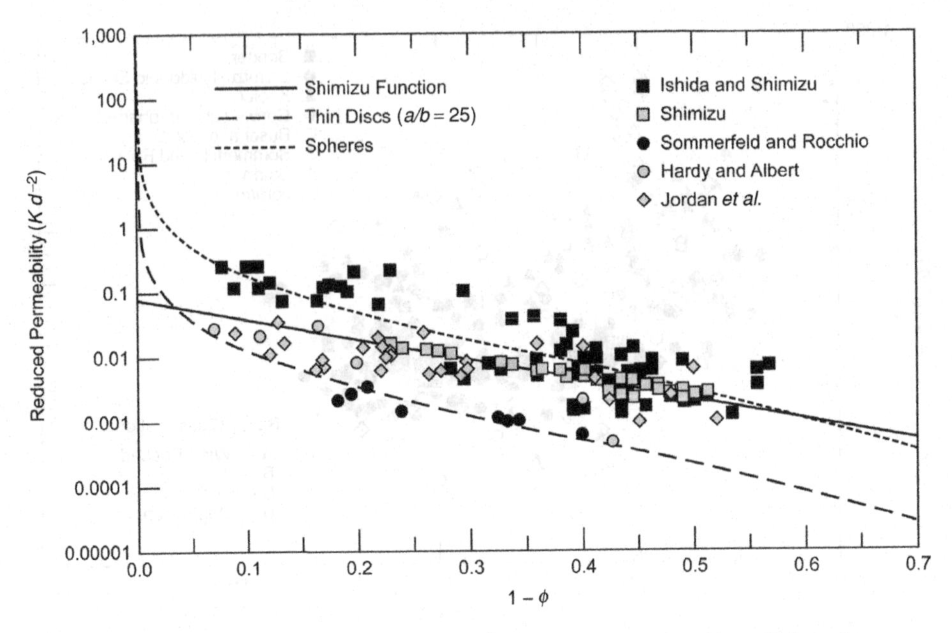


Figure 2.13. Reduced permeability $K\ d^{-2}$ versus ice fraction $(1 - \phi)$ based on laboratory observations of Ishida and Shimizu (1958), Shimizu (1960), Shimizu (1970), Sommerfeld and Rocchio (1989), Hardy and Albert (1993), and Jordan *et al.* (1999b). Lines show the reduced Shimizu function (Eq. 2.20) and theoretical solutions (Jordan *et al.*, 1999b) for beds of thin discs ($a$ = diameter, $b$ = thickness, $a/b = 25$) and spheres. Discs and spheres have equal radii (after Jordan *et al.*, 1999b, copyright 1999; copyright John Wiley & Sons Limited. Reproduced with permission).

of spheres and therefore exert a higher drag on fluids. This suggests that, in addition to grain radius and porosity, the surface-to-volume ratio – typically higher for new snow – is an important parameter in characterizing snow permeability.

The natural stratification of the snow cover usually causes it to be more resistant to flow in the vertical than in the horizontal direction (Ishida and Shimizu, 1958; Dullien, 1992). Luciano and Albert (2002) reported measurements of anisotropy or directionality of permeability in snow and firn field measurements, and concluded that differences in permeability between layers accounts for greater variation than directional differences within a single layer. Because vertical resistances in layered media are computed in series, permeability is controlled by the most resistant (or least permeable) strata.

### *Forced and natural convection of air*

Air movement in porous media can result from two mechanisms: (1) pressure differences that force the flow (windpumping or forced convection) or (2) a temperature gradient that induces buoyancy-driven thermal convection (also termed natural

Table 2.4 *Temperature difference per meter for the onset of natural convection.*

| Snow type | $K$ $\times 10^{-10}$ m$^2$ | $\rho_s$ kg m$^{-3}$ | $k_{eff}$ Wm$^{-1}$K$^{-1}$ | $Ra_{crit} = 17.7$ constant flux $\Delta T_{crit}$ | $Ra_{crit} = 27.1$ constant $T$ $\Delta T_{crit}$ |
|---|---|---|---|---|---|
| Depth hoar | 200 | 200 | 0.05 | 10 | 15 |
| New snow | 80 | 200 | 0.10 | 50 | 77 |
| Old snow | 100 | 300 | 0.20 | 80 | 123 |
| Firn | 40 | 300 | 0.20 | 200 | 308 |
| Wind-pack | 10 | 300 | 0.20 | 800 | 1234 |

convection). There are occasions when forced and natural convection of air can occur within natural snow.

When a large temperature gradient exists between a warm substrate and a cold surface, air density changes and causes buoyancy-driven air circulation within porous media; this is known as *natural convection.* Powers *et al.* (1985) detail the theory of natural convection in snow. The non-dimensional Rayleigh number governs both the onset of convection and its intensity, and is given by

$$Ra = \frac{\rho_{0,a} g \tilde{\beta} \Delta T \, HS \, K}{\eta_a \Lambda}. \tag{2.21}$$

In Equation (2.21), $\rho_{0,a}$ is the reference density of air, $\tilde{\beta}$ is the coefficient of thermal expansion, $\Delta T$ is the driving temperature difference, $HS$ is the depth of the snow, and $\Lambda = k_{eff}/(\rho_a c_{p,a})$.

The critical Rayleigh number $Ra_{crit}$ for the onset of natural convection is approximately 27 when the surface is permeable with constant temperature and the bottom boundary is impermeable with constant temperature. When the bottom boundary is instead impermeable with constant heat flux, the critical Rayleigh number is approximately 18. For common values of permeability and thermal conductivity of various snow types, Table 2.4 lists the temperature difference that would be required for the onset of natural convection in a 1 m deep snowpack under the two bottom boundary conditions. Because natural temperature gradients of 50 °C per meter are very rare, it is highly unlikely that natural convection will occur in natural snow except in the case of packs composed of depth hoar with no intervening layers. Sturm and Johnson (1992) did measure temperatures in highly porous, thin subarctic snowpacks (composed primarily of large depth hoar crystals) that support the existence of natural convection in that case.

*Forced convection* in snow is caused by natural pressure changes. It has been shown that turbulent winds over a flat surface, barometric surface pressure changes, and winds over surface relief can all cause pressure perturbations that propagate vertically into the snow (Colbeck, 1989a; Clarke and Waddington, 1991). Barometric pressure changes may cause slow, low-velocity air movement in near-surface snow. Turbulence in the winds causes high-frequency pressure fluctuations that propagate millimeters or centimeters into the snow. The "form drag" pressure differences, caused by air flow over surface roughness (such as across sastrugi), can cause stronger and more sustained air flow deeper within the snow.

Snow layering will affect ventilation through differences in the permeability of the strata. Albert (1996) showed that, even under the assumption that snow within a given layer is isotropic, differences between the layers affect subsurface air flow fields. For example, buried high-permeability (e.g. hoar) layers can serve as channels for increased lateral flow through the snowpack.

The forced convection of air in snow can affect other processes in snow, such as sublimation (Albert, 2002), chemical changes (Waddington *et al.*, 1996; McConnell *et al.*, 1998; Albert *et al.*, 2002), or heat transfer (Albert and Hardy, 1995; Jordan, *et al.*, 2003; Andreas, *et al.*, 2004).

### 2.4.2 *Unsaturated or two-phase flow*

In unsaturated wet snow, both air and water occupy the pore space and, strictly speaking, flow of both phases should be considered. However, since the volume of water in freely draining snow is usually less than 10% (Colbeck, 1978), air is not confined by the snow and a modified one-phase treatment may be used (Scheidegger, 1974). Replacing water pressure in Equation (2.19) with capillary pressure ($p_{\mathrm{a}\ell} = p_{\mathrm{a}} - p_\ell$) and the intrinsic permeability $K$ with the liquid permeability $K_\ell$, the downward water velocity $v_\ell$ becomes

$$v_\ell = \frac{K_\ell}{\eta_\ell}\left(\frac{\partial p_{\mathrm{a}\ell}}{\partial x} + \rho_\ell g\right). \tag{2.22}$$

The equation for water flow through snow then derives from liquid continuity as

$$\underbrace{\rho_\ell \phi \frac{\partial s}{\partial t}}_{\text{rate of change in water saturation}} = \underbrace{-\frac{\partial}{\partial x}\left[\frac{\rho_\ell K_\ell}{\eta_\ell}\left(\frac{\partial p_{\mathrm{a}\ell}}{\partial x} + \rho_\ell g\right)\right]}_{\text{net water flux}} \underbrace{- \rho_\ell \phi s \frac{\partial v_{\mathrm{i}}}{\partial x}}_{\text{compaction}} + \underbrace{S}_{\text{melt or condensation}}, \tag{2.23}$$

where $s$ is the liquid saturation, $v_{\mathrm{i}}$ is the velocity of the compacting ice matrix, and $S$ is a phase change source term. The compaction rate is determined from the snow viscosity in Equation (2.11). Otherwise, Equation (2.23) contains three unknowns $-s$, $p_{\mathrm{a}\ell}$, and $K_\ell$ – which are related through the $s$–$p_{\mathrm{a}\ell}$ and $s$–$K_{\mathrm{r}\ell}$ constitutive functions. In the $s$–$K_{\mathrm{r}\ell}$ function, $K_{\mathrm{r}\ell}$ is the relative permeability, expressed in terms of the saturated permeability as $K_{\mathrm{r}\ell} = K_\ell / K$.

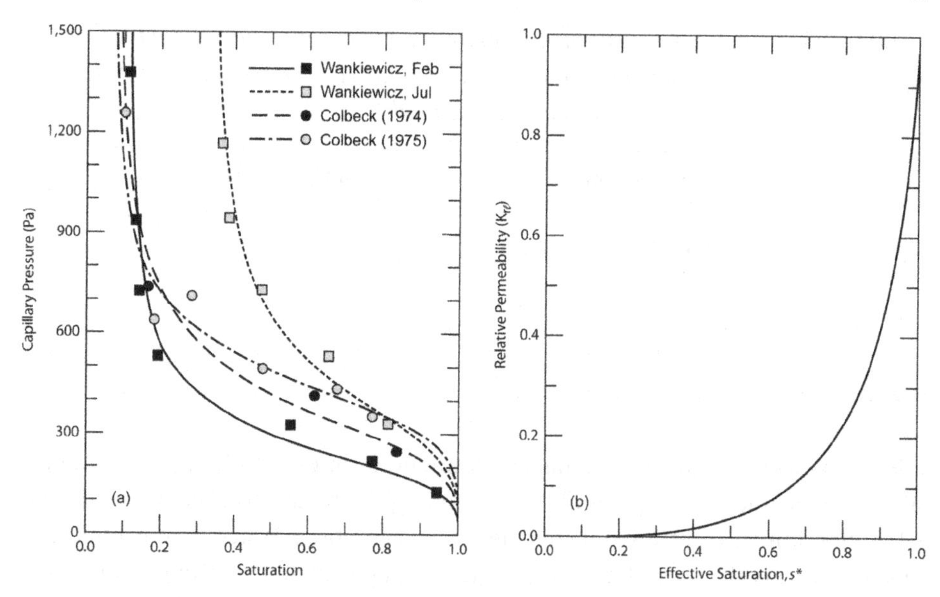


Figure 2.14. (a) Measured saturation–pressure curves from Wankiewicz (1979), Colbeck (1974), and Colbeck (1975). Lines show curves fit with the van Genuchten (1980) function. (b) Saturation-relative permeability curve computed from the Brooks–Corey function with $\varepsilon = 3$.

### *Capillary forces and the pressure–saturation (or $s$–$p_{a\ell}$) curve*

Capillary forces result from pressure drops across the concave water menisci in wet snow (see Figs. 2.4 and 2.6). Such forces suck water from wetter to drier regions of the snow cover and oppose or augment the gravitational forces. Recalling that $\zeta_{a\ell}$ is the radius of curvature of the meniscus, the pore-scale pressure drop or capillary pressure, $p_{a\ell}$, computes from the Laplace equation (2.5) as $2\sigma_{a\ell}/\zeta_{a\ell}$. Although known as capillary "pressure," $p_{a\ell}$ is actually a capillary suction or tension. Because $\zeta_{a\ell}$ relates directly to the pore diameter, capillary tension is highest for small pores and water entering dry snow will fill the smallest pores first. A plot of water saturation versus water tension thus approximates a cumulative size distribution of the pore space. Such plots are referred to as water retention or $s$–$p_{a\ell}$ curves and are standardly used to characterize soils.

The fragility of snow and its loss of cohesion when wet make it very difficult to measure water retention curves. Figure 2.14a shows the few published $s$–$p_{a\ell}$ drying curves for snow (Colbeck, 1974, 1975; Wankiewicz, 1979), which resemble those for coarse sand. Water suction in snow typically ranges from 0.1 to 1 kPa – several orders of magnitude less than in fine-grained soils. Water retention curves exhibit a directional hysteresis, with suctions in draining snow (drying phase) being about one-half that in infiltrated snow (wetting phase). Water pathways disconnect

Table 2.5 *Constitutive relationships between water saturation and capillary pressure, and between water saturation and relative permeability in partially wetted porous media.*

| | Brooks and Corey (1964) | van Genuchten (1980) |
|---|---|---|
| $s$–$p_{a\ell}$ | $s^* = \left(\frac{p_e}{p_{a\ell}}\right)^{\lambda_p} \quad p_{a\ell} \leq p_b$ | $s^* = \left[1 + \left(\frac{p_{a\ell}}{p_e}\right)^n\right]^{-m}$ |
| $s$–$K_\ell$ | $K_{r\ell} = s^{*\varepsilon}$ | $K_{r\ell} = \sqrt{s^*}\left[1 - \left(1 - s^{*(1/m)}\right)^m\right]$ |
| | $\varepsilon = 2.5 + 2/\lambda_p$ | $m = 1 - 1/n; \quad nm \cong \lambda_p$ |

at lower moisture contents and, thus, all four curves in Fig. 2.14a reach an immobile or irreducible saturation limit, $s_i$. Isolated liquid inclusions cannot be reduced below this limit except through freezing or evaporation. Water flow parameters thus scale to an effective saturation, $s^* = (s - s_i)/(1 - s_i)$, so that capillary pressure tends to infinity and relative permeability to 0 as $s$ approaches $s_i$. While the retained immobile saturation after drainage is typically around 0.07, the minimum saturation requirement during infiltration can be much lower (see Denoth *et al.*, 1979; Kattelmann, 1986; Dullien, 1992). Jordan *et al.* (1999a) suggests an "average" value of 0.04 for general modeling purposes, but values as low as 0.01 may be appropriate when modeling infiltration.

Experimental $s$–$p_{a\ell}$ data is typically represented by inverse power functions between $s$ and $p_{a\ell}$. The simplest of these is the Brooks–Corey (1964) function, while the somewhat more complicated van Genuchten (1980) function provides a closer representation of the "knee" that occurs near full saturation. These functions are summarized in Table 2.5. Both functions have two fitting parameters – an air entry or bubbling pressure $p_e$ and an exponent $\lambda_p$, termed the pore-size distribution index. Figure 2.14a includes curve fits of the van Genuchten function to the four sets of experimental data. Air-entry pressures ranged between 240 and 440 Pa and pore-size distribution indices ranged between 2.6 and 3.8. Jordan (1983) obtained a value near 3.0 for $\lambda_p$ using *in situ* tensionmeter and lysimeter measurements. These values for $\lambda_p$ are within the range suggested by Brooks and Corey (1964) for sand, while the values for $p_e$ are somewhat lower.

The wicking height of dye water into a snow sample provides a rough estimate of its entry pressure. Water in snow is under tension and, thus, rises spontaneously to a height where its weight balances the pressure drop across the liquid meniscus. Conceptualizing the pore space as bundles of capillary tubes, the capillary rise, $h$, relates inversely to the pore size as

$$\rho_\ell g h = \frac{2\sigma_{a\ell}\cos\theta}{r_p}. \tag{2.24}$$

Here, $\theta$ is the contact angle between the ice grain and water meniscus, which is usually taken as 0. Rise in natural snow ranges from about 0.5 cm for old, coarse snow to 5.0 cm for ice crusts (see Wakahama, 1968; Wankiewicz, 1979; Coléou *et al.*, 1999; and Jordan *et al.*, 1999b), which corresponds to a range of entry pressures ($= \rho_\ell gh$) between 49 and 490 Pa.

*Liquid and relative permeability*

In wet snow, air and water share the pore space, leaving less cross-sectional area available for flow of either phase. As with saturated permeability, liquid permeability $K_\ell$ relates to the square of the pore size – but the size distribution is now limited to water-filled pores. Since water flow is usually restricted to smaller pores in the snowpack, liquid permeability is typically several orders less than saturated permeability. $K_\ell$ depends as well on tortuosity and interconnectivity among the water-filled pores. Because of the implicit pore-size relationship in the s–$p_{a\ell}$ curve, liquid permeability is parameterized as a power function of $s$. Table 2.5 presents the s–$K_{r\ell}$ functions of Brooks and Corey (1964) and van Genuchten (1980). The simple Brooks and Corey formula is more commonly used for snow modeling and has only the exponent $\varepsilon$ as a fitting parameter.

Figure 2.14b depicts the Brooks and Corey permeability function for $\varepsilon = 3$. While no laboratory measurements have been made for the s–$K_{r\ell}$ curve in snow, Denoth *et al.* (1979) determined values of $\varepsilon$ between about 2 for new snow and 5 for old, clustered snow by observing the discharge rate from cylindrical tubes filled with sieved snow. Colbeck and Anderson (1982) found that a value of 3.3 predicts flow rates for well-metamorphosed snow that are in close agreement with field data.

*Flow with a uniform wetting front*

When a steady rainfall infiltrates a homogeneous snowpack, the solution to the flow equation (2.23) is a traveling wave of fixed shape as shown in Fig. 2.15a (Illangasekare *et al.*, 1990; Gray, 1996; Albert and Krajeski, 1998). Because capillary forces in snow are small, Colbeck (1972, 1978) neglected the pressure term in Equation (2.22) and derived a gravitational solution. For gravitational flow, the front has the shape of a shock wave or step function and the transition between wetted and dry snow regions is abrupt. The downward wavefront velocity through homogeneous snow is easily approximated by balancing the interior flow rate with the surface influx (rain or meltwater) and then computing the time it takes to warm the snow to 0 °C and to satisfy the immobile and equilibrium saturation deficits (Colbeck, 1976; Jordan, 1991; Albert and Krajeski, 1998). Because capillary suction accelerates water movement into the snowpack, the gravitational approximation will somewhat underpredict the wavefront velocity, but the effect is minor. Capillary forces do, however, play a vital role in the formation of capillary barriers and flow fingers and in the wicking of ponded water.

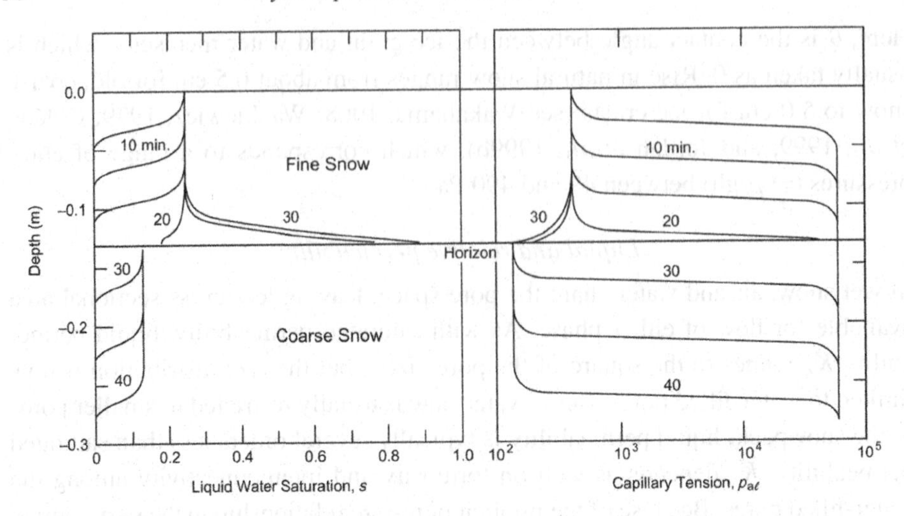


Figure 2.15. Simulated profiles of (a) liquid water saturation and (b) capillary tension at 5 min intervals for an irrigation rate of 5 cm $h^{-1}$. The snowpack consists of fine and coarse layers with respective entry pressures of 262 and 118 Pa. Depth is with respect to the snow surface (after Jordan 1996, with modifications).

Figure 2.16 compares outflow from lysimeters at two snow research sites (Davis *et al.*, 2001) with simulations from the SNTHERM snow model (Jordan, 1991). SNTHERM's assumptions of gravitational flow and a uniform wetting front closely represent outflow in the full melt season, once the snowpack is ripe or totally wetted. Heterogeneity in the snowpack in the early melt season may explain why SNTHERM predicts outflow not captured by the lysimeters. In heterogeneously stratified snow covers, flow is proportionally slower in fine-textured, dense snow or ice crusts, and water saturation is therefore higher to maintain flux continuity. Water levels must also adjust to maintain pressure continuity across textural discontinuities. So-called capillary barriers arise between fine and coarse layers because suction is higher in the finer layer (Jordan, 1996). Infiltrating water thus accumulates above the fine–coarse horizon until it fills pores of the same size (and hence the same tension or pressure) as the underlying layer (see Fig. 2.15). In cold snow, backed-up water refreezes to form ice crusts or lenses, which can further impede the downward flow of water. In sloped terrain, capillary barriers and ice layers can direct rain or meltwater to the bottom of a hill before it infiltrates to the bottom of the snowpack. For very deep snowpacks, this can accelerate the arrival of hydraulic pulses by hours or even by days.

*Unstable flow*

In addition to the effects of layering, the flow field in snow is complicated by the development of vertical channels or fingers, which concentrate flow ahead of the

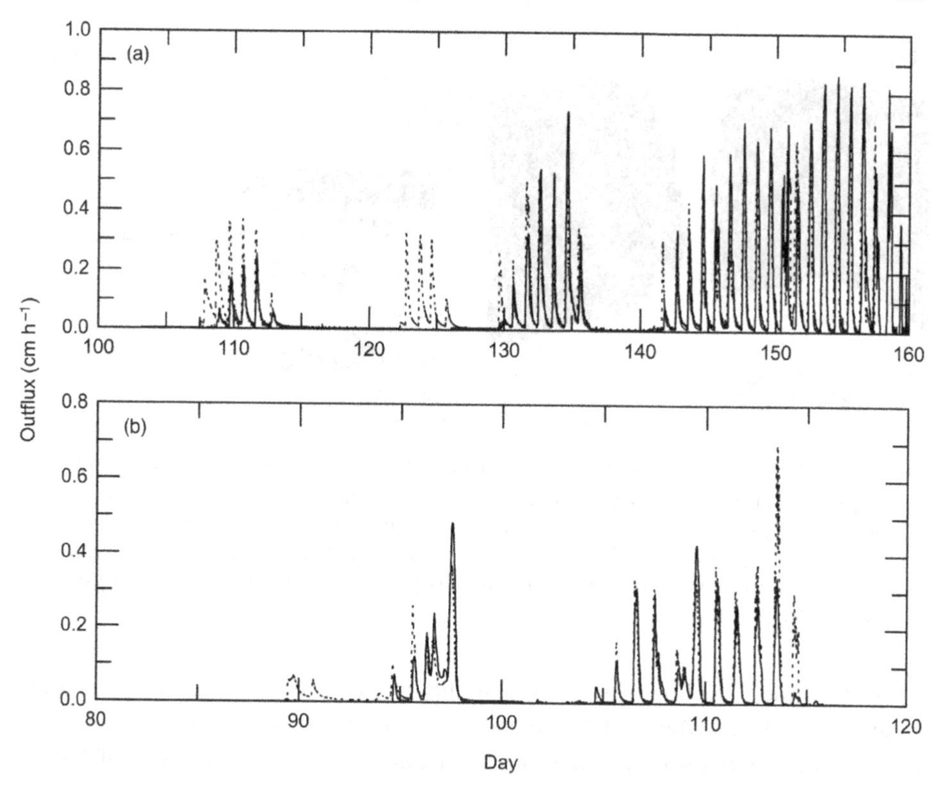


Figure 2.16. Observed outflux (cm $h^{-1}$) from the snow cover and predictions from the SNTHERM snow model: (a) between days 100 and 160, 1994, at Mammoth Mountain, CA and (b) between days 80 and 120, 1997, at Sleepers River Research Watershed, VT. The solid line shows the measurements from lysimeters, while the dotted line shows the simulation results (after Davis *et al.*, 2001, copyright 1999; copyright John Wiley & Sons Limited. Reproduced with permission).

background wetting front. Raats (1973) and Philip (1975) established that flow instabilities develop in soil when the suction gradient is in opposition to the flow, which causes the flow velocity to accelerate with depth. Selker *et al.* (1992) later confirmed their criterion with measurements of the capillary tension within growing instabilities in unsaturated soil. Such conditions occur for flow through fine-over-coarse layers and in homogeneous media when the surface flux decelerates.

Marsh and Woo (1984a) observed flow fingers in snow similar to those reported for soils, which are characteristically narrow and uniform in width and randomly spaced at frequent intervals. In snow at subzero temperatures, fingers often freeze to form ice columns (see Fig. 2.17a). Marsh and Woo (1984a) and Marsh (1988, 1991) reported finger widths in cold arctic snow of from 3.5 to 5 cm thick, with mean spacings between the fingers of 13 cm and areal coverages of 22–27%. About

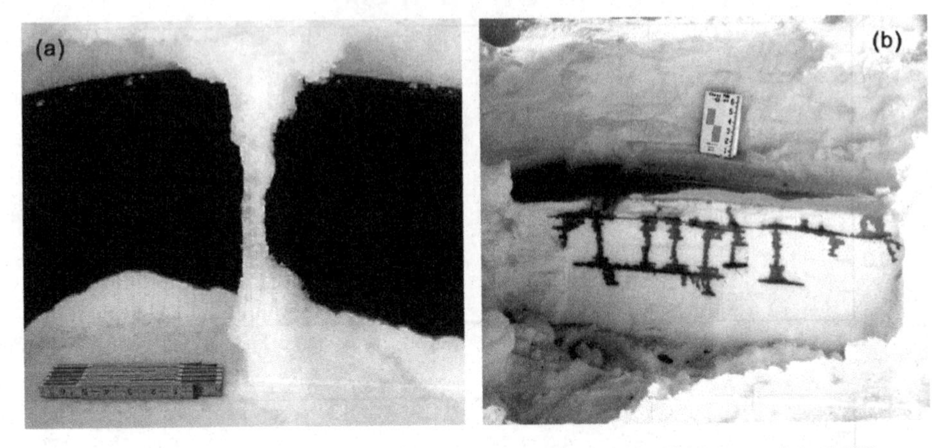


Figure 2.17. (a) Refrozen flow finger (from Albert *et al.*, 1999, copyright 1999; copyright John Wiley & Sons Limited. Reproduced with permission). (b) Dye studies showing flow fingers and pre-melt horizons in layered snow (after Marsh and Woo, 1984a, photo by Philip Marsh, with permission. Published 1984 American Geophysical Union. Reproduced/modified by permission of American Geophysical Union).

one-half of the total flow was carried by the fingers. Using a novel thick-section cutter, McGurk and Marsh (1995) measured mean finger diameters of 1.8–3.0 cm in warm snow, with mean spacings of 1.9–4.6 cm, and a mean wetted area of 4–6%. Using a high-frequency FMCW radar to detect flow fingers in a seasonal snowpack, Albert *et al.* (1999) estimate an areal density of about 3 fingers $m^{-2}$ from rain falling on temperate snow in the early melt season and a diameter range from 2 to 5 cm. Clearly, the surface flux intensity as well as snow temperature and grain size will impact the size and spacing of the fingers.

Marsh and Woo (1984a) and Marsh (1988, 1991) observed that fingers often originate at horizon interfaces (see Fig. 2.17b), regardless of the snow properties above and below the interface. In layered studies of sifted and poured snow, Jordan (1996) found that finger formation was limited to fine–coarse transitions. This inconsistency suggests that natural snow layers develop surface characteristics distinct from their bulk properties (e.g. surface melt crusts or surface hoar) that can affect the flow pattern. Using a multiple dye tracer application, Schneebeli (1995) demonstrated that the position of the finger paths usually changed between subsequent melt–freeze events.

Marsh and Woo (1984b) developed a multiple path simulation model that separates water flow into background and finger front components. Their model also accounts for ice-layer growth at strata horizons and demonstrates that ice-layer growth in subfreezing snow can be sufficiently large to interrupt finger-flow advance for short periods of time. Because much of the finger flow is laterally diverted to growing ice horizons, their simulations predict that the finger front in cold snow is never more than 10–15 cm below the background front. Predicted finger flow is more rapid through warm snow, however, and reaches the base of the pack well ahead of the background wetting front.

Flow fingers become more permeable than the surrounding dry snow because snow coarsens faster in the presence of water. Grain bond weakening within fingers can also cause snow to collapse, leaving "bowl-shaped melt depressions" or dimples in the snow surface (McGurk and Kattelmann, 1988; Marsh, 1991).

Despite the omission of fingering, even-wetting front theory does a reasonable job predicting the timing of basal outflow once the snowpack is ripe (Tuteja and Cunnane, 1997; Davis *et al.*, 2001). There is a certain logic to this, given the trend over time towards homogeneity in well-wetted snowpacks. Thus, while finger flow plays an important role at the beginning of the melt season or for mid-winter rain-on-snow events, the even wetting front approach may be adequate once late season melting is well underway.

## 2.5 Radiative properties of snow

In this section, we limit ourselves to the radiative properties that affect the energy fluxes at the snow/atmosphere interface. Radiant energy incident on the earth's surface occurs principally in two broad bands: solar or shortwave radiation (0.3–2.8 μm) and thermal or longwave radiation (5–40 μm). For climate studies, snow albedo is of major importance and therefore this section principally deals with snow reflectance. However, we also discuss the coefficient of absorption, which effects the penetration of radiation into the snowpack and thus influences the surface temperature. Subsurface penetration is quite important for the snow ecosystem.

### *2.5.1 Reflectance and bi-directional reflectance*

The main feature of snow in the shortwave spectral range is its very high reflectance, especially over the visible spectrum. This explains its white color. Its high reflectance comes from the combination of its microstructure with the optical properties of ice. Over the visible spectrum, pure ice is quasi-transparent (absorption between 0.02 and 0.05 $cm^{-1}$), while in the near-infrared spectrum absorption increases until ice becomes quasi-opaque above 1.5 μm (Hobbs, 1974). The refractive index of ice is close to 1.30. A photon incident on the snow surface is either refracted or reflected at each encounter with an ice/air interface until it is either absorbed or ejected from the layer. At visible wavelengths, and if snow has a very low impurities content, the photon may travel large distances through ice without being absorbed, and therefore may be refracted and reflected many times within the snow cover. Snow therefore behaves as a diffuser. Multiple refraction and reflection drastically change the propagation direction of an individual photon, so that the probability of its ejection from the snow layer is very high. This explains the high reflectance of clean snow at visible wavelengths (Wiscombe and Warren, 1980). If snow contains many impurities, the probability that photons are absorbed before being ejected is much higher and the reflectance is therefore smaller. At visible wavelengths, snow reflectance depends slightly on the grain size and shape, but is mainly affected by carbon soot particles present inside the snow layer and by the surface deposition of dust (Warren and Wiscombe, 1980). Naturally, high reflectance requires a deep enough snow layer to ensure that photons be diffused in depth before being ejected. Many radiative models consider snow as a semi-infinite medium. In practice, this assumption is valid for snow layers deeper than about 10 cm.

Ice absorption increases at longer wavelengths, which decreases the probability for photon ejection and consequently decreases snow reflectance. This probability primarily depends on the ice distance traversed by the photon during multiple refractions and reflections. If snow grains are large and granular rather than small and planar, this distance is longer. This explains why snow reflectance in the near-infrared spectrum generally decreases with increasing grain size, except when facets are growing during temperature gradient metamorphism. Snow reflectance has been investigated in the field as well as in the laboratory and by means of specific models. In most models, the reflectance of pure snow is described only as a function of grain size and impurities content (see Figs. 2.18 and 2.19, Wiscombe and Warren, 1980), while assuming that snow consists of spherical grains. Sergent *et al.* (1998) emphasize that grain shape must also be considered.

According to the relative isotropy of the orientation of ice/air interfaces in snow, shortwave radiation reflected by snow is diffuse and relatively isotropic. Nevertheless, bi-directional reflectance must be considered. Bi-directional effects are sensitive only at high angles of incidence and are more pronounced in the near-infrared spectrum than in the visible spectrum. Bi-directional effects on reflectance are more sensitive for hexagonal particles than for spherical ones (Dozier and Warren, 1982; Leroux, 1996).

Specular reflection can also be observed at a very high angle of incidence when the snow surface presents special features such as a sun crust or firnspiegel (class 9c).

### 2.5.2 *Snow albedo*

Climate scientists are not directly interested in snow spectral reflectance but rather in the reflection of the global incoming solar radiation. Spectrally integrated reflectance is described by the albedo, which is the ratio of the reflected to the incoming global shortwave radiation. Albedo depends not only on snow type, but

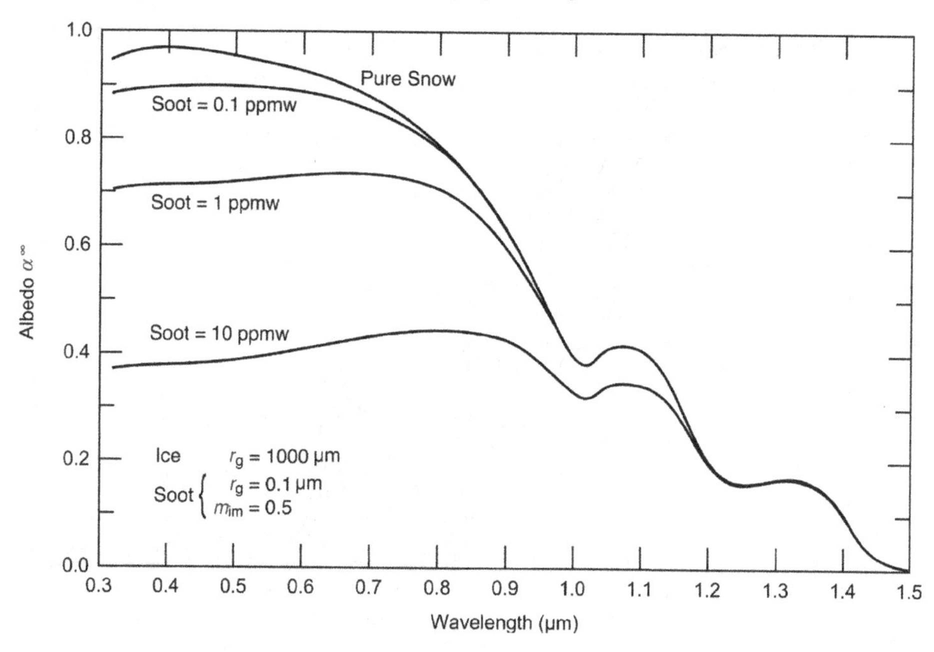


Figure 2.18. Effect of soot concentration on albedo (from Warren and Wiscombe, 1980, copyright 1980 American Meteorological Society).

also on the spectral distribution of the incoming radiation. Over large snow-covered areas, albedo also depends on vegetation and surface roughness.

### *Effects of snow type on the albedo*

For a given spectral distribution of the incoming solar radiation, the albedo of a homogeneous deep enough snow layer depends mainly on snow quality, i.e. on its type and its impurity content. Since grain size and impurity content generally increase with age (through mechanisms of metamorphism and dry deposition), the albedo of a snowpack often decreases with time until it is refreshed by a new snowfall (O'Neill and Gray, 1973; Nolin, 1993). This is particularly the case during the melting period because of the rapid growth and rounding of snow grains throughout the snowpack and because of the progressive emergence of older snow layers, which may have high surface concentrations of scavenged impurities. In a few days, snow albedo can drop from 0.90 to 0.50, causing a fivefold increase in the absorption of solar radiation and a drastic increase in the melting rate. This process is a major factor in the very rapid melting of polar snowpacks in the Northern Hemisphere during late spring. However, albedo does not always decrease with time. When a snowpack is subjected to high temperature gradients for several weeks, its albedo

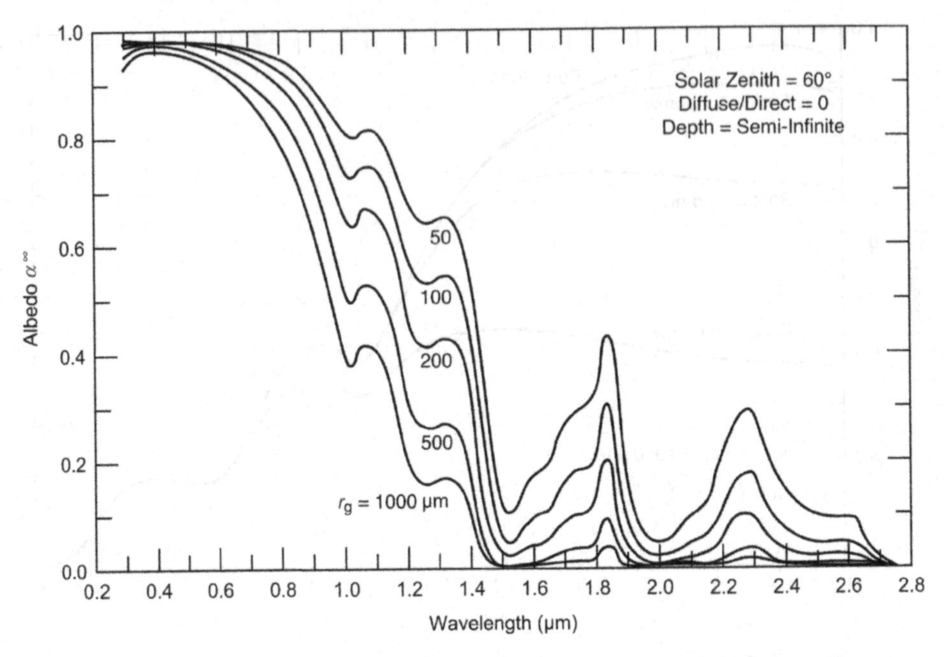


Figure 2.19. Semi-infinite direct albedo as a function of wavelength for various grain radii (from Wiscombe and Warren, 1980, copyright 1980 American Meteorological Society).

can stay constant and sometimes even increase because of the transformation of surface snow into faceted crystals or because of the formation of surface hoar.

Most parameterizations used in snow models take into account the type of snow for calculating its albedo. Often, only ageing of the snow surface is considered, although a few models consider grain size and even grain type. The variation of reflectance with grain size has a functional shape that cannot be expressed analytically over the whole solar spectrum. Therefore, it is more convenient for a parameterization of the albedo and of the coefficient of absorption to consider several spectral bands, such as those used in the model CROCUS (Brun *et al.*, 1992), which have been derived from Sergent *et al.* (1998):

$$\begin{aligned}
&\alpha = \min\left(0.94, 0.96 - 1.58\sqrt{d_{\text{opt}}}\right),\\
&\beta = \max\left(0.04, 0.0192\rho_s/\sqrt{d_{\text{opt}}}\right) \text{ for } 0.3 < \lambda \leq 0.8\ \mu\text{m},\\
&\alpha = 0.95 - 15.4\sqrt{d_{\text{opt}}},\ \beta = \max(1.0, 0.1098\rho_s/\sqrt{d_{\text{opt}}}) \text{ for } 0.8 < \lambda \leq 1.5\ \mu\text{m},\\
&\alpha = 0.88 + 346.3 d_{\text{opt}} - 32.31\sqrt{d_{\text{opt}}},\ \beta = +\infty \text{ for } 1.5 < \lambda \leq 2.8\ \mu\text{m},
\end{aligned} \tag{2.25}$$

where $\lambda$ is the wavelength, $\alpha$ is the albedo in the considered band, $\beta$ is the coefficient of absorption expressed in $\text{cm}^{-1}$, and $d_{\text{opt}}$ is the optical grain size expressed in m, which depends on the grain size and on its shape. The effect of impurities on the albedo is calculated from the age of the snow surface. Table 2.6 gives the values of the albedo and the coefficient of absorption for the main snow types neglecting impurities.

Table 2.6 *Reflectance and absorption coefficients of typical snow samples from the main classes integrated over three spectral bands.*

| | Wave bands | | | | | |
|---|---|---|---|---|---|---|
| | (0.3–0.8 μm) | | (0.8–1.5 μm) | | (1.5–2.8 μm) | |
| Type of snow | $\alpha$ | $\beta$ | $\alpha$ | $\beta$ | $\alpha$ | $\beta$ |
| Class 1 | 0.94 | 0.40 | 0.80 | 1.10 | 0.59 | $+\infty$ |
| Class 2 | 0.94 | 0.40 | 0.73 | 1.36 | 0.49 | $+\infty$ |
| Class 3 | 0.93 | 0.40 | 0.68 | 1.90 | 0.42 | $+\infty$ |
| Class 4 | 0.93 | 0.40 | 0.64 | 1.10 | 0.37 | $+\infty$ |
| Class 5 | 0.92 | 0.40 | 0.57 | 1.12 | 0.30 | $+\infty$ |
| Class 6 | 0.91 | 0.40 | 0.42 | 1.27 | 0.18 | $+\infty$ |

### *Effect of spectral distribution on the albedo*

Since snow reflectance drops from almost 1 to 0 between 0.3 μm and 2.8 μm, snow albedo of a given snow layer is strongly affected by the spectral distribution of incoming solar radiation. This distribution varies a lot according to cloudiness and to the relative contribution of direct radiation and clear-sky diffuse radiation. The spectrum of clear-sky diffuse radiation is focused in the visible part where clean snow always shows a high reflectance. Only a few models take into account this distribution to calculate snow albedo.

### *Large-scale effects on snow albedo*

Snow cover is often heterogeneous and the albedo of a snow-covered surface differs from that of a point snowpack. Except over large ice sheets (Greenland and Antarctica), the albedo of large surfaces is generally much lower than the albedo of the snow covering these surfaces. The main causes for this decrease are given below.

- A patchy snow cover. Because of snow drift and local variations in snowmelt, snow cover is frequently patchy, especially during the melting period. Parameterizations of albedo in climate models take this effect into account by considering that only a part of a considered surface is effectively covered by snow. The ratio of snow-covered surface to the total surface is generally deduced from the average snow depth.
- Vegetation. Interception of radiation by vegetation strongly alters the albedo of snow-covered regions such as the boreal forest. This alteration depends on the type and density of vegetation, on snow deposition on the canopy, and on the incidence of incoming radiation. Most parameterizations of snow albedo in climate models take this effect into account.
- Surface roughness has similar effects as vegetation but at different scales. This is the case for high mountains as well as for smaller relief such as sastruggi.

Summer melting on polar sea ice can also lead to significant variability in albedo, with values running from between 0.1 for deep melt ponds to 0.65 for bare, white ice, when all the snow has melted (Perovich *et al.*, 2002).

### 2.5.3 Snow emissivity

In the thermal spectrum (5–40 μm), snow behaves almost as a perfect black body (Dozier and Warren, 1982; Warren, 1982). This means that it absorbs all the longwave radiation emitted by the atmosphere or by the surrounding obstacles and emits the maximum thermal radiation allowed by its surface temperature. Longwave radiation is completely absorbed in the first millimeter of snow. This property comes from the high emissivity of ice (approximately 0.97) (Hobbs, 1974). The hemispherically averaged snow emissivity is around 0.98. Directional emissivity decreases slightly with lower viewing angle.

The high emissivity of snow, combined with its high reflectance in the shortwave spectrum, plays a major role in the earth's climate and in the large part accounts for the rapid cooling of continental regions in winter.

### 2.5.4 Subsurface heating

The partial transparency of snow in the shortwave spectrum and its high emissivity in the longwave spectrum induces an unusual phenomenon: subsurface heating (Brun *et al.*, 1989; Colbeck, 1989b; Koh and Jordan, 1995). This phenomenon typically occurs under clear-sky conditions and at relatively cold temperatures, when a surface fresh snow layer is submitted to solar radiation. In such conditions, net longwave radiation losses at the surface are only partially balanced by surface absorption of shortwave radiation.

A few centimeters below the surface, snow continues to absorb solar radiation transmitted by the upper layers. This absorption warms a subsurface snow layer until it is balanced by conductive losses to the upper and lower layers. This phenomenon is similar to a greenhouse effect. In some cases, absorption is sufficiently large or conduction sufficiently weak for snow a few centimeters below the surface to reach the melting point, despite surface temperatures that remain a few degrees below 0 °C. If this phenomenon is followed by a nocturnal cooling, it forms a melt–freeze crust (class 9e) below the surface while the surface remains powder-like.

## 2.6 Summary and future directions

Snow consists of particles of ice that form in the clouds, grow initially by vapor deposition, and then reach the ground without evaporating or melting. Conditions for snow to form include atmospheric temperatures less than 0 °C and the presence of supercooled water. Snow begins as ice crystals, which nucleate either homogeneously or heterogeneously onto the surfaces of ice nuclei. The basic shape common to all ice crystals is a hexagonal prism with two basal planes and six prism planes. The relative growth rates of the faces vary with temperature and supersaturation, giving rise to a wide variety of crystal shapes. When an ice crystal grows to a size where it has a significant downward velocity, it becomes a snow crystal. Larger snow crystals continue growing by accretion (riming) or by aggregation into snowflakes. Snow thus consists of an intricate variety of snow crystals, as well as rimed and aggregate versions of these forms.

After being deposited on the ground or on a previous snow layer, snow crystals accumulate and give birth to a new snow layer. Snowfall amounts are measured both by depth and by snow water equivalent depth or SWE, which is the depth of the snow if it were melted. Typical snowfall rates are 1 cm $h^{-1}$ of depth or 0.8 mm $h^{-1}$ of SWE. The density of newly fallen snow is typically between 60 and 120 kg $m^{-3}$ for dry snow falling in low to moderate winds, but wet or wind-blown snow can reach densities of 400 kg $m^{-3}$. Current functions that predict new-snow density from temperature and wind speed alone give uncertain results and can be improved by taking crystal type and size into account. Such functions, for instance, frequently underpredict the density of polar snow, which is usually assumed to be around 300 kg $m^{-3}$. Snow drift, metamorphic settlement, and slow deformation from overburden stress compact the snow cover over time.

Once on the ground, deposited snow particles rapidly bond together to form an ice matrix that delimits pores filled with humid air and, in the case of wet snow, with liquid water. Snow thus belongs to the large family of porous media, which also includes soils. For seasonal snow covers, snow porosity generally ranges between 40% and 98%. Because of the high activity of water thermodynamics around the triple point, the solid matrix of snow is continuously and sometimes rapidly evolving, which makes snow a unique and complex component of the earth's surface. Independent of porosity, the snow texture (or the size, shape, and distribution of the grains) affects most physical and mechanical properties of snow.

The ice matrix undergoes metamorphism in response to thermodynamic stresses among the water phases and continuously evolves towards mechanical equilibrium (Laplace equation) and phase equilibrium (Clausius–Clapeyron and Gibbs–Duhem equations). Activity in snow is enhanced by the large specific area of the ice/air interface, which also explains why snow is usually saturated with vapor. Microscopic variations in curvature and temperature induce local pressure gradients that cause vapor to sublimate from highly convex or warmer surfaces and condense on less convex or colder surfaces. Deposition of vapor causes rounding of snow grains over time and the growth of larger grains and bonds at the expense of smaller grains and fine structure. Recent research suggests that other mechanisms, such as bulk diffusion from the grain boundaries to the ice surfaces, play a role in bond growth. When the temperature gradient exceeds about $5\,^{\circ}\mathrm{C\,m^{-1}}$, this equilibrium growth form is superseded by the kinetic growth of colder ice surfaces to form facets and even strias. In highly wetted snow or slush, the melting point temperature is lower for less convex surfaces. Under these conditions, refreezing of meltwater from smaller grains onto larger grains results in rapid coarsening of the snow.

Both diffusive and advective processes transport heat through the snowpack. In the absence of fluid flow (air or water), heat flow is linearly related to the temperature gradient, where the coefficient of proportionality is thermal conductivity. Measurements of thermal conductivity include the effects of heat conduction through connected grains and through the air space, along with the "hand-to-hand" transport of latent heat by water vapor. Snow thermal conductivity and specific heat depend primarily upon the geometry of the ice matrix and thus vary an order of magnitude over time with snow textural changes. The most useful estimates to date of the thermal properties of snow are based on macroscopic measurements of the aggregate that are then correlated to the snow type and physical characteristics such as density and temperature. As difficulties in characterizing snow microstructure and geometry are overcome, inclusion of these more fundamental properties should improve the parameterizations. Growing evidence suggests that high winds advect air through the upper snowpack, which may explain why high effective thermal conductivities are required to match observed heat transport in windy, polar regions.

While high porosity and low thermal conductivity make snow a protective blanket against extreme cold, this same openness makes snow quite permeable to flows of air and water. Fluid flow through snow is similar to that through other granular materials, but is complicated by freeze/thaw effects, metamorphism of the ice matrix, and the layered nature of snow covers. Unlike soils, snow is sufficiently open to allow the movement of air in the interstitial pore space. Flow of air or water through snow is sufficiently slow that Darcy's law applies. In this case, the fluid velocity relates linearly to combined pressure and gravitational forces, where the coefficient of proportionality is the saturated or intrinsic permeability divided by the fluid viscosity. Intrinsic permeability varies widely with snow type, ranging about two orders of magnitude between fine-grained, wind-packed snow and large-grained depth hoar. It depends on the square of grain size and, to a lesser extent, on porosity. For a given porosity and grain dimension, particle shapes with a larger surface-to-volume ratio exert a higher drag on fluids.

Forced interstitial air flow is known as ventilation or windpumping. Wind-induced turbulence in the surface boundary layer causes high-frequency pressure fluctuations that propagate millimeters or centimeters into the snow. The "form drag" pressure differences, caused by air flow over rough surface features, can cause stronger and more sustained air flow deeper within the snow. Only in very high permeability snow, such as depth hoar with no intervening layers, does buoyancy-induced natural convection occur. Both natural and forced convection accelerate the transport of water vapor and chemical species through snow and firn.

In unsaturated wet snow, both air and water share the pore space and interfacial tensions between the two phases arise. Because of the open pore structure and coarse texture, however, gravitational forces dominate capillary forces during water flow. A gravitational solution is therefore often used to describe water flow through snow, which produces a front with the shape of a shock wave or step function. Because capillary suction accelerates water movement into the snowpack, the gravitational approximation will somewhat underpredict the wave velocity in homogeneous snow, but the effect is minor. Capillary forces do, however, play an important role in the formation of capillary barriers and flow fingers and in the upwards wicking of ponded water. Low-viscosity wetting fluids (such as water) develop flow instabilities and thus infiltration often occurs in preferential flow channels, or flow fingers, rather than as an even wetting front. The effect is particularly noticeable in mid-winter rain-on-snow events or during early season melting, when preferential flow reaches the snow base ahead of the background wetting front. Once heavy melting has ripened the snowpack, however, outflow predictions with the even wetting approach are probably adequate. Stratigraphic inhomogenieties in permeability and capillary tension can impede and laterally divert the water flow, as well as trigger the development of flow fingers.

Radiant energy incident on the earth's surface occurs principally in two broad bands: solar or shortwave radiation (0.3–2.8 μm) and thermal or longwave radiation (5–40 μm). The main feature of snow in the shortwave spectral range is its very high reflectance, especially over the visible spectrum, which explains its white color. The spectrally averaged reflectivity of solar radiation incident on snow, or snow albedo, is of major importance to climate change. Snow albedo ranges between about 0.50 and 0.90, and decreases with grain size, the angle of solar incidence, impurities, and the ratio of direct to diffuse sunlight. Snow albedo is thus higher for new snow and under cloudy skies or when the sun is at a low angle in the sky. According to the relative isotropy of the orientation of ice/air interfaces in snow, shortwave radiation reflected by snow is diffuse and relatively isotropic. Bi-directional effects are sensitive only at high angles of incidence and are more pronounced in the near-infrared spectrum than in the visible spectrum. Nevertheless, bi-directional reflectance is important and warrants further attention in snow models.

In the thermal spectrum (5–40 μm), snow behaves almost as a perfect black body and absorbs and re-emits over 97% of incident longwave radiation within the first millimeters of snow. Emitted radiation is proportional to the fourth power of the surface temperature. Snow emissivity decreases slightly for shallow angles of incidence. Because of the partial transparency of snow in the shortwave spectrum, significant solar energy penetrates up to 10 cm depth in the snowpack. Subsurface solar penetration is quite important for the snow ecosystem.

Snow models to date are mostly one dimensional and their parameterizations depend heavily on aggregate properties, such as snow porosity. Current trends in research towards a better understanding of both micro- and large-scale behavior should lead to more physically based parameterizations for these models. Continued efforts to replace empirically based, and therefore regional, parameterizations with physically based parameterizations will greatly benefit global climate models.

## References

Akitaya, E. (1974). Studies on depth hoar. *Contrib. Inst. Low Temp. Sci., A*, **26**, 1–67.

Albert, M. R. (1996). Modelling heat, mass, and species transport in polar firn. *Ann. Glaciol.*, **23**, 138–143.

Albert, M. R. (2002). Effects of snow and firn ventilation on sublimation rates. *Ann. Glaciol.*, **35**, 510–514.

Albert, M. R., Grannas, A. M., Bottenheim, J., and Shepson, P. (2002). Processes and properties of snow–air transfer in the high Arctic with application to interstitial ozone at Alert, Canada. *Atmos. Environ.*, **36**, 2779–2787.

Albert, M. R. and Hardy, J. P. (1995). Ventilation experiments in a seasonal snow cover. In *International Association of Hydrological Sciences, Publication* No. 288 and *Proc., International Symposium on Biogeochemistry of Seasonally Snow-Covered Catchments*, Boulder, CO (ed. Tonnessen, K. A., *et al.*). Boulder, CO: International Association of Hydrological Sciences, pp. 41–49.

Albert, M. R., Koh, G., and Perron, F. (1999). Radar investigations of melt pathways in a natural snowpack. *Hydrol. Process.*, **13**, 2991–3000.

Albert, M. R. and Krajeski, G. (1998). A fast, physically based point snowmelt model for use in distributed applications. *Hydrol. Process.*, **12**, 1809–1824.

Albert, M. R. and McGilvary, W. R. (1992). Thermal effects due to air flow and vapor transport in dry snow. *J. Glaciol.*, **38**, 273–281.

Albert, M. R. and Perron, F. E., Jr. (2000). Ice layer and surface crust permeability in a seasonal snow pack. *Hydrol. Process.*, **14**(18), 3207–3214.

Albert, M. R. and Shultz, E. (2002). Snow and firn properties and air–snow transport processes at Summit, Greenland. *Atmosph. Environ.*, **36**, 2789–2797.

Albert, M. R., Shultz, E., and Perron, F. (2000). Snow and firn permeability measurements at Siple Dome, Antarctica. *Ann. Glaciology*, **31**, 353–356.

Andreas, E. L, Jordan, R. E., and Makshtas, A. P. (2004). Simulations of snow, ice and near-surface atmospheric processes on Ice Station Weddell. *J. Hydrometeorol.*, **5**(4), 611–624.

Arons, E. M. and Colbeck, S. C. (1995). Geometry of heat and mass transfer in dry snow: a review of theory and experiment. *Rev. Geophys.*, **33**(4), 463–493.

Bader, H., Haefeli, R., Bucher, E., *et al.* (1939). Snow and its metamorphism. *SIPRE Translation* (1954), **14**.

Bales, R. C., Davis, R. E., and Stanley, D. A. (1989). Ionic elution through shallow, homogeneous snow. *Water Resources Res.*, **25**, 1869–1877.

Bales, R. C. and Harrington, R. F. (1995). Recent progress in snow hydrology. *Rev. Geophys.*, **33**(4), 1011–1020.
Barry, R. C. and R. J. Chorley (2003). *Atmosphere, Weather and Climate*, 8th edn. Oxford: Routledge.
Bender, J. A. (1957). *Air Permeability of Snow*. SIPRE Research Report 37. U.S. Army Corps of Engineers, Snow Ice and Permafrost Research Establishment.
Bentley, W. A. and Humphreys, W. J. (1931). *Snow Crystals*. New York: McGraw-Hill.
Bentley, W. A. and Humphreys, W. J. (1962). *Snow Crystals* (reprint). Mineola, NY: Dover Publications.
Brooks, R. H. and Corey, A. T. (1964). *Hydraulic Properties of Porous Media*. Colorado State University Civil Engineering Department, Hydrology Paper no. 3. Fort Collins, CO: Colorado State University.
Brun, E. (1989). Investigation on wet-snow metamorphism in respect of liquid water content. *Ann. Glaciol.*, **13**, 22–26.
Brun, E., David, P., Sudul, M., and Brunot, G. (1992). A numerical model to simulate snow cover stratigraphy for operational avalanche forecasting. *J. Glaciol.*, **38**(128), 13–22.
Brun, E., Martin, E., Simon, V., Gendre, C., and Coléou, C. (1989). An energy and mass model of snow cover suitable for avalanche forecasting. *J. Glaciol.*, **35**(121), 333–342.
Brun, E., Martin, E., and Spiridonov, V. (1997). Coupling a multi-layered snow model with a GCM. *Ann. Glaciol.*, **25**, 66–72.
Brzoska, J-B., Coléou, C., and Lesaffre, B. (1998). Thin-sectioning of wet snow after flash-freezing. *J. Glaciol.*, **44**(146), 54–62.
Buck, A. L. (1981). New equations for computing vapor pressure and enhancement factor. *J. Appl. Meteorol.*, **20**, 1527–1532.
Buser, O. and Good, W. (1986). A rigid frame model of porous media for the acoustic impedance of snow. *J. Sound Vibration*, **111**, 71–92.
Chacho, E. F., Jr. and Johnson, J. B. (1987). Air permeability of snow. *EOS Trans. AGU*, **68**, 1271. (abstract only).
Christon, M., Burns, P., Thompson, E., and Sommerfeld, R. (1987). Water vapor transport in snow. A 2-D simulation of temperature gradient metamorphism. In *Seasonal Snowcovers: Physics, Chemistry, Hydrology* (ed. Jones, H. G. and Orville-Thomas, W. J.). NATO ASI Series C, vol. 211, pp. 37–62.
Clarke, G. K. C. and Waddington, E. D. (1991). A three-dimensional theory of wind pumping. *J. Glaciol.*, **37**(125), 89–96.
Colbeck, S. C. (1972). Theory of water percolation in snow. *J. Glaciol.*, **11**(63), 369–385.
Colbeck, S. C. (1973). *Theory of Metamorphism of Wet Snow.* CRREL Research Report 313. (Hanover, NH: U.S. Army Cold Regions Research and Engineering Laboratory.
Colbeck, S. C. (1974). The capillary effects on water percolation in homogeneous snow. *J. Glaciol.*, **13**(67), 85–97.
Colbeck, S. C. (1975). Grain and bond growth in wet snow. *International Association of Hydrological Sciences Publication 114* (*Snow Mechanics Symposium*, Grindelwald 1974), pp. 51–56.
Colbeck, S. C. (1976). Analysis of water flow in dry snow. *Water Resources Res.*, **12**(3), 523–527.
Colbeck, S. C. (1978). The physical aspects of water flow through snow. In *Advances in Hydroscience,* vol. 11 (ed. Chow, V. T.). New York: Academic Press, pp. 165–206.
Colbeck, S. C. (1979). Grain clusters in wet snow. *J. Colloid Interface Sci.*, **72**, 371–384.
Colbeck, S. C. (1980). Thermodynamics of snow metamorphism due to variations in curvature. *J. Glaciol.*, **26**(94), 291–301.

Colbeck, S. C. (1983). Theory of metamorphism of dry snow. *J. Geophys. Res.*, **88**(C9), 5475–5482.
Colbeck, S. C. (1989a). Air movement in snow due to windpumping. *J. Glaciol.*, **35**(120), 209–213.
Colbeck, S. C. (1989b) Snow-crystal growth with varying surface temperatures and radiation penetration. *J. Glaciol.*, **35**(119), 23–29.
Colbeck, S. C., Akitaya, E., Armstrong, R., *et al.* (1990). *The International Classification for Seasonal Snow on the Ground*. International Commission on Snow and Ice and World Data Center A for Glaciology, Boulder, CO.
Colbeck, S. C. and Anderson, E. A. (1982). Permeability of a melting snow cover. *Water Resources Res.*, **18**(4), 904–908.
Coléou, C., Lesaffre, B., Brzoska, J. B., Ludwig, W., and Boller, E. (2001). Three-dimensional snow images by X-ray microtomography. *Ann. Glaciology*, **32**, 75–81.
Coléou, C., Xu, K., Lesaffre, B., and Brzoska, J. B. (1999). Capillary rise in snow. *Hydrol. Process.*, **13**, 1721–1732.
Conway, H. and Abrahamson, J. (1984). Air permeability as a textural indicator of snow. *J. Glaciol.*, **30**, 328–333.
Dang, H., Genthon, C., and Martin, E. (1997). Numerical modelling of snow cover over polar ice sheets. *Ann. Glaciol.*, **25**, 170–176.
Davis, R. E. (1991). Links between snowpack physics and snowpack chemistry. In *NATO ASI Series G: Ecological Science,* vol. 28, *Seasonal Snowpacks, Processes of Compositional Change* (ed. Davies, T. D., Tranter, M., and Jones, H. G.). Berlin: Springer-Verlag, pp. 115–138.
Davis, R. E., Jordan, R. E., Daly, S., and Koenig, G. G. (2001). Validation of snow models. In *Model Validation; Perspectives in Hydrological Science* (ed. Anderson, M. G. and Bates, P. D.). Chichester: John Wiley & Sons, pp. 261–292.
Defay, R. and Prigogine, I., with the collaboration of Bellemans, A. (translated by D. H. Everett) (1966). *Surface Tension and Adsorption*. New York: John Wiley & Sons.
Denoth, A., Seidenbusch, W., Blumthaler, M., and Kirchlechner, P. (1979). *Study of Water Drainage from Columns of Snow*. Special Report 79–1. U.S. Army Cold Regions Research and Engineering Laboratory, Hanover, New Hampshire.
Dozier, J. and Warren, S. G. (1982). Effect of viewing angle on the infrared brightness temperature of snow. *Water Resources Res.*, **18**(5), 1424–1434.
Dullien, F. A. L. (1992). *Porous Media. Fluid Transport and Pore Structure*, 2nd edn. San Diego, CA: Academic Press.
Fletcher, N. H. (1962). *The Physics of Rainclouds*. Cambridge: Cambridge University Press.
Flin, F., Brzoska, J. B., Lesaffre, B., Coléou, C., and Pieritz, R. A. (2003). Full three-dimensional modelling of curvature-dependent snow metamorphism: first results and comparison with experimental tomographic data. *J. Phys. D*, **36**, A49–A54.
Frank, F. C. (1982). Snow crystals. *Contemp. Phys.*, **23**, 3–22.
Giddings, J. C. and Lachapelle, E. (1962). The formation rate of depth hoar. *J. Geophys. Res.*, **67**(6), 2377–2383.
Golubev, V. N. and Frolov, A. D. (1998). Modelling the change in structure and mechanical properties in dry-snow densification to ice. *Ann. Glaciol.*, **26**, 45–50.
Good, W. (1987). Thin sections, serial cuts and 3-D analysis of snow. In *International Association of Hydrological Sciences Publication 162* (*Symposium*: *on Avalanche Formation, Movements and Effects*, Davos, 1986), pp. 35–48.

Gray, J. M. N. T. (1996). Travelling waves in wet snow. In *Proc., SNOWSYMP-94, International Symposium on Snow and Related Manifestations, Manali, India, 1994.* (ed. K. C. Agrawal), pp. 171–175.

Gubler, H. (1985). Model for dry snow metamorphism by interparticle vapor flux. *J. Geophys. Res.*, **90**(D5), 8081–8092.

Hallett, J. (1984). How crystals grow. *Am. Sci.*, **72**, 582–589.

Hallett, J. (1987). Faceted snow crystals. *J. Opt. Soc. Am. A*, **4**(3), 581–588.

Hardy, J. P. and Albert, D. G. 1993. The permeability of temperate snow: preliminary links to microstructure. In *Proc. 50th Eastern Snow Conference and 61st Western Snow Conference*, Quebec City, 1993, pp. 149–156.

Hardy, J. P., Davis, R. E., Jordan, R., *et al.* (1997). Snow ablation modelling at the stand scale in a boreal jack pine forest. *J. Geophys. Res.*, **102**(D24), 29 397–29 405.

Harrington, R. F. and Bales, R. C. (1998). Interannual, seasonal, and spatial patterns of meltwater and solute fluxes in a seasonal snowpack. *Water Resources Res.*, **34**(4), 823–831.

Hedstrom, N. R. and Pomeroy, J. W. (1998). Measurements and modelling of snow interception in the boreal forest. *Hydrol. Process.*, **12**, 1611–1625.

Heymsfield, A. J. and Miloshevich, L. M. (1993). Homogeneous ice nucleation and supercooled liquid water in orographic wave clouds. *J. Atmos. Sci.*, **50**, 2335–2353.

Hobbs, P. V. (1974). *Ice Physics.* Oxford: Clarendon Press.

Illangasekare, T. H., Walter, R. J., Jr., Meier, M. F., and Pfeffer, W. T. (1990). Modelling of meltwater infiltration in subfreezing snow. *Water Resources Res.*, **26**(5), 1001–1012.

Ishida, T. and Shimizu, H. (1958). *Resistance to Air Flow through Snow Layers, Snow and its Metamorphism.* U.S. Army Snow Ice and Permafrost Research Establishment, Translation 60.

Izumi, K. and Huzioka, T. (1975). Studies of metamorphism and thermal conductivity of snow. *Low Temperature Science* A, **33**, 91–102 (in Japanese with English summary).

Jones, J. A. A. (1996) Predicting the hydrological effects of climate change. In *Regional Hydrological Response to Climate Change* (ed. Jones, J. A. A., Liu, C., Woo, M.-K., and Kung, H.-T.). Dordrecht: Kluwer Academic Publishers.

Jordan, P. (1983). Meltwater movement in a deep snowpack. 1. Field observations. *Water Resources Res.*, **19**(4), 971–978.

Jordan, R. (1991). *A One-dimensional Temperature Model for a Snow Cover.* Technical documentation for SNTHERM.89. U.S. Army Cold Regions Research and Engineering Laboratory, Hanover, NH, Special Report 91–16.

Jordan, R. (1996). Effects of capillary discontinuities on water flow and water retention in layered snow covers. In *Proc., SNOWSYMP-94, International Symposium on Snow and Related Manifestations, Manali, India, 1994.* (ed. K. C. Agrawal), pp. 157–170.

Jordan, R. and Davis, R. (1990). Thermal effects of wind ventilation of snow. American Geophysical Union Fall Meeting, 1990. Abstract. *EOS*, **71** (43), 1328.

Jordan, R., O'Brien, H., and Albert, M. R. (1989). Snow as a thermal background: preliminary results from the 1987 field test. In *Proc., Snow Symposium VII.* U.S. Army Cold Regions Research and Engineering Laboratory, Hanover, NH, Special Report 89–7, pp. 5–24.

Jordan, R. E., Andreas, E. L., and Makshtas, A. P. (1999a). Heat budget of snow-covered sea ice at North Pole 4. *J. Geophys. Res.*, **104**, 7785–7806.

Jordan, R. E., Hardy, J. P., Perron, F. E., Jr., and Fisk, D. J. (1999b). Air permeability and capillary rise as measures of the pore structure of snow: an experimental and theoretical study. *Hydrol. Process.*, **13**, 1733–1753.

Jordan, R. E., Andreas, E. L, Fairall, C. W., *et al.* (2003). Modelling surface exchange and heat transfer for the shallow snow cover at SHEBA. In *7th Conf. on Polar Meteorology and Oceanography*. Hyannis, MA: American Meteorological Society.

Kajikawa, M. (1972). Measurement of falling velocity of individual snow crystals. *J. Meteorol. Soc. Japan*, **50**, 577–584.

Kattelmann, R. C. (1986). Measurements of snow layer water retention. In *Proc., Symposium on Cold Regions Hydrology, Fairbanks, Alaska* (ed. Kane, D. L.). American Water Resources Association, pp. 377–386.

Keeler, C. M. (1969a). The growth of bonds and the increase of mechanical strength in dry seasonal snowpack. *J. Glaciol.*, **8**(54), 441–450.

Keeler, C. M. (1969b). *Some Physical Properties of Alpine Snow*. U.S. Army Cold Regions Research and Engineering Laboratory, Hanover, NH, Research Report 271.

Kobayashi, T. (1961). The growth of snow crystals at low supersaturations, *Philos. Mag.*, Ser. 8, **6**, 1363–1370.

Koh, G. and Jordan, R. (1995). Sub-surface heating in a seasonal snow cover. *J. Glaciol.*, **41**(139), 474–482.

Leroux, C. (1996). Etude théorique et expérimentale de la réflectance de la neige dans le spectre solaire. Application à la télédétection. *Thèse de l'Université des Sciences et Technologies de Lille*.

Luciano, G. L. and Albert, M. R. (2002). Bi-directional permeability measurements of polar firn. *Ann. Glaciol.*, **35**, 63–66.

Magono, C. and Lee, C. W. (1966). Meteorological classification of natural snow covers. *J. Faculty Sci.*, Series VII, *Geophysics*, Hokkaido University, 321–335, reprinted by the U.S. Forest Service, Wasatch National Forest, Alta Avalanche Study Center, 1968.

Marbouty, D. (1980). An experimental study of temperature-gradient metamorphism. *J. Glaciol.*, **26**(94), 303–312.

Marsh, P. (1988). Flow fingers and ice columns in a cold snowcover. In *Proc. 56th Western Snow Conference*, Kalispell, Montana, pp. 105–112.

Marsh, P. (1990). Snow hydrology. In *Northern Hydrology: Canadian Perspectives* (ed. Prowse, T. D. and Ommanney, C. S. L.). NHRI Science Report No. 1, Minister of Supply and Services, Ottawa, Canada, pp. 37–62.

Marsh, P. (1991). Water flux in melting snow covers. In *Advances in Porous Media*, vol. 1, Ch. 9. Amsterdam: Elsevier, pp. 61–122.

Marsh, P. and Woo, M. K. (1984a). Wetting front advance and freezing of meltwater within a snow cover. 1: Observations in the Canadian Arctic. *Water Resources Res.*, **20**(12), 1853–1864.

Marsh, P. and Woo, M. K. (1984b). Wetting front advance and freezing of meltwater within a snow cover. 2: A simulation model. *Water Resources Res.*, **20**(12), 1865–1874.

Mason, B. J. (1971). *The Physics of Clouds*, 2nd edn. Oxford: Clarendon Press.

McConnell, J. R., Bales, R. C., Stewart, R. W., *et al.* (1998). Physical based modelling of atmosphere-to-snow-to-firn transfer of $H_2O_2$ at South Pole. *J. Geophys. Res.*, **103**(D9), 10 561–10 570.

McGurk, B. J. and Kattelmann, R. C. (1988). Transport of liquid water through Sierran snowpacks: flow finger evidence from thick section photography. *American Geophysical Union, EOS*, **69**, 1204.

McGurk, B. J. and Marsh, P. (1995). Finger-flow continuity in serial thick-sections in a melting Sierran snowpack. In *Proc., International Symposium on Biogeochemistry of*

*Seasonally Snow-Covered Catchments*, Boulder, Colorado, July 1–14 (ed. Tonnessen, K. A., *et al.*) Great Yarmouth, IAHS Publication No. 228, pp. 81–88.

Mellor, M. (1977). Engineering properties of snow. *J. Glaciol.*, **19**, 15–66.

Morris, E. M. (1983). Modelling the flow of mass and energy within a snowpack for hydrological forecasting. *Ann. Glaciol.*, **4**, 198–203.

Morris, E. M., Bader, H.-P., and Weilenmann, P. (1997) Modelling temperature variations in polar snow using DAISY, *J. Glaciol.*, **43**, 180–191.

Nakaya, U. (1954). *Snow Crystals: Natural and Artificial.* Cambridge, MA: Harvard University Press.

Navarre, J. P. (1975). Modèle unidimensionnel d'évolution de la neige déposée: modèle perce-neige. *La Météorologie*, **4**(3), 103–120.

Nolin, A. W. (1993). Radiative heating in alpine snow. Ph.D. Thesis, University of California, Santa Barbara, CA.

O'Neill, A. D. J. and Gray, D. M. (1973). Spatial and temporal variations of the albedo of a prairie snowpack. In *The Role of Snow and Ice in Hydrology: Proc. Banff Symposium*, vol. 1, pp. 176–186.

Perovich, D. K., Grenfell, T. C., Light, B., and Hobbs, P. V. (2002). Seasonal evolution of the albedo on multiyear Arctic sea ice. *J. Geophys. Res.*, **107**(C10), DOI 10.1029/2000JC000438.

Philip, J. R. (1975). The growth of disturbances in unstable infiltration flows. *Soil Sci. Soc. Am. Proc.*, **39**, 1049–1053.

Pomeroy, J. W. and Schmidt, R. A. (1993). The use of fractal geometry in modelling intercepted snow accumulation and sublimation. In *Proc. Eastern Snow Conference*, vol. 50, pp. 1–10.

Powers, D. J., Colbeck, S. C., and O'Neill, K. (1985). *Thermal Convection in Snow.* U.S. Army Cold Regions Research and Engineering Laboratory, Hanover, New Hampshire, CRREL Report 85–9.

Pruppacher, H. R. (1995). A new look at homogeneous ice nucleation in supercooled water drops. *J. Atmos. Sci.*, **52**, 1924–1933.

Pruppacher, H. R. and Klett, J. D. (1997). *Microphysics of Clouds and Precipitation*, 2nd edn. Dordrecht: Kluwer Academic Publishers.

Raats, P. A. C. (1973). Unstable wetting fronts in uniform and nonuniform soils. *Soil Sci. Soc. Am. Proc.*, **37**, 681–685.

Raymond, C. F. and Tusima, K. (1979). Grain coarsening of water-saturated snow. *J. Glaciol.*, **22**(86), 83–105.

Rogers, R. R. and Yau, M. K. (1989) *A Short Course in Cloud Physics*, 3rd edn. Oxford: Pergamon Press.

Rottner, D. and G. Vali (1974). Snow crystal habit at small excesses of vapor density over ice saturation. *J. Atmos. Sci.*, **31**, 560–569.

Scheidegger, A. E. (1974). *The Physics of Flow Through Porous Media.* Toronto: University of Toronto Press.

Schemenauer, R. S., Berry, M. O., and Maxwell, J. B. (1981). Snowfall formation. In *Handbook of Snow: Principles, Processes, Management and Use* (ed. Gray, D. M. and Male, D. H.). Toronto: Pergamon Press, pp. 129–152. Reprinted by The Blackburn Press, 2004.

Schneebeli, M. (1995). Development and stability of preferential flow paths in a layered snowpack. In *Proc. International Symposium on Biogeochemistry of Seasonally Snow-Covered Catchments*, Boulder, Colorado, July 1–14 (ed. Tonneson, K. A., *et al.*). IAHS Publication No. 228, Great Yarmouth, pp. 89–95.

Selker, J., Leclerq, P., Parlange, J-Y., and Steenhuis, T. (1992). Fingered flow in two dimensions. 1: Measurements of matric potential. *Water Resources Res.*, **28**(9), 2513–2521.

Sergent, C., Leroux, C., Pougatch, E., and Guirado, F. (1998). Hemispherical-directional reflectance measurements of natural snow in the 0.9- to 1.45-m spectral range: comparison with adding-doubling modelling. *Ann. Glaciol.*, **26**, 59–68.

Shapiro, L. H., Johnson, J. B., Sturm, M., and Blaisdell, G. L. (1997). *Snow Mechanics: review of the State of Knowledge and Applications*. U.S. Army Cold Regions Research and Engineering Laboratory, Hanover, NH, CRREL Report 97–03.

Shimizu, H. (1960). Determination of the resistance to air flow of snow layer. IV. Air permeability of deposited snow. *Low Temperature Science A*, **19**, 165–173 (in Japanese).

Shimizu, H. (1970). *Air Permeability of Deposited Snow*. Institute of Low Temperature Science, Sapporo, Japan, Contribution No. 1053 (English translation).

Singh, A. K. (1999). An investigation of the thermal conductivity of snow. *J. Glaciol.*, **45**(150), 346–351.

Sommerfeld, R. A. and Rocchio, J. (1989). The Darcy permeability of fine-grained compact snow. In *Proc. Eastern Snow Conference*, Quebec City, 8–9 June 1989, pp. 121–128.

Sturm, M. (1991). *The Role of Thermal Convection in Heat and Mass Transport in the Subarctic Snow Cover*. U.S. Army Cold Regions Research and Engineering Laboratory, Hanover, NH, CRREL Report 91–19.

Sturm, M. (1992). Snow distribution and heat flow in the taiga. *Arctic Alpine Res.*, **24**(2), 145–152.

Sturm, M. and Holmgren, J. (1998). Differences in compaction behaviour of three climate classes of snow. *Ann. Glaciol.*, **26**, 125–130.

Sturm, M., Holmgren, J., König, M., and Morris, K. (1997). The thermal conductivity of seasonal snow. *J. Glaciol.*, **43**, 26–41.

Sturm, M. and Johnson, J. B. (1992). Thermal conductivity measurements of depth hoar. *J. Geophys. Res.*, **97**(B2), 2129–2139.

Sturm, M., Perovich, D. K., and Holmgren, J. (2002). Thermal conductivity and heat transfer through the snow on the ice of the Beaufort Sea. *J. Geophys. Res.*, **107**(C10), 10.1029/2000JC00409.

Tuteja, N. K. and Cunnane, C. (1997). Modelling coupled transport of mass and energy into the snowpack: model development, validation and sensitivity analysis. *J. Hydrol.*, **195**, 232–255.

van Genuchten, M. T. A. (1980). A closed-form equation for predicting the hydraulic conductivity of unsaturated flow. *Soil Sci. Soc. Am. J.*, **44**, 892–898.

Waddington, E. D., Cunningham, J., and Harder, S. (1996). The effects of snow ventilation on chemical concentrations. In *Chemical Exchange Between the Atmosphere and Polar Snow* (ed. Wolff, E. W. and Bales, R. C.). Berlin: Springer-Verlag, pp. 403–451.

Wakahama, G. (1968). The metamorphism of wet snow. In *International Association of Scientific Hydrology Publication 79 (General Assembly of Bern 1967 – Snow and Ice)*, pp. 370–379.

Wang, P. K. and Ji, W. (2000). Collision efficiencies of ice crystals at low-intermediate Reynolds number colliding with supercooled cloud droplets: a numerical study. *J. Atmos. Sci.*, **57**(8), 1001–1009.

Wankiewicz, A. (1979). A review of water movement in snow. In *Proc. Meeting on Modelling of Snow Cover Runoff, 1978*, Hanover, NH (ed. Colbeck, S. C. and Ray,

M.). U.S. Army Cold Regions Research and Engineering Laboratory, Hanover, NH, pp. 222–252.

Warren, S. G. (1982). Optical properties of snow. *Rev. Geophys. Space Phys.*, **20**(1), 67–89.

Warren, S. G. and Wiscombe, W. J. (1980). A model for the spectral albedo of snow. II. Snow containing atmospheric aerosols. *J. Atmos. Sci.*, **37**(12), 2734–2745.

Wiscombe, W. J. and Warren, S. G. (1980). A model for the spectral albedo of snow. I. Pure snow. *J. Atmos. Sci.*, **37**(12), 2712–2733.

# 3

# Snow–atmosphere energy and mass balance

John C. King, John W. Pomeroy, Donald M. Gray,
Charles Fierz, Paul M. B. Föhn, Richard J. Harding, Rachel E. Jordan,
Eric Martin and Christian Plüss

## 3.1 Introduction

*John C. King, John W. Pomeroy, Donald M. Gray, and Charles Fierz*

Climates at the global, the regional, and the local scale determines the relative contributions of radiation, turbulent, and mass fluxes to the corresponding balance at the atmosphere–ground interface. In particular conditions, these fluxes induce the formation of a snowpack on the ground and predominantly effect its accumulation (depth), ablation (melt), sublimation, evaporation, as well as its structure (layering). Presence and transformation of the snow pack change the physical properties of the atmosphere–ground interface, which in turn affects the above fluxes, thereby influencing the properties of the lower atmosphere. For instance, the nearly black body emissivity of snow in the longwave range as well as surface temperature never exceeding 0 °C (273.15 K) promote an often persistent stable atmospheric boundary layer over the snowpack's surface, even in daytime. In addition, fluxes involved at the snow–atmosphere interface depend much on the ground vegetation, on the snowpack being continuous or patchy, as well as on topography.

These features, along with other particular physical properties of the snowpack such as its high albedo, its capability to store both frozen and liquid water and its low thermal conductivity lead to a strong feedback between the atmosphere and the snowpack, as well as to substantially altered climatic signals over seasonal snowpacks as compared with bare soil.

This chapter first presents general equations for both energy and mass balances (Section 3.2). Detailed discussions of each component of the energy balance follow (Section 3.3), including effects due to either vegetation or topography. Next, the influence of vegetation, blowing snow, as well as topography, on snow

*Snow and Climate: Physical Processes, Surface Energy Exchange and Modeling*, ed. Richard L. Armstrong and Eric Brun. Published by Cambridge University Press. 

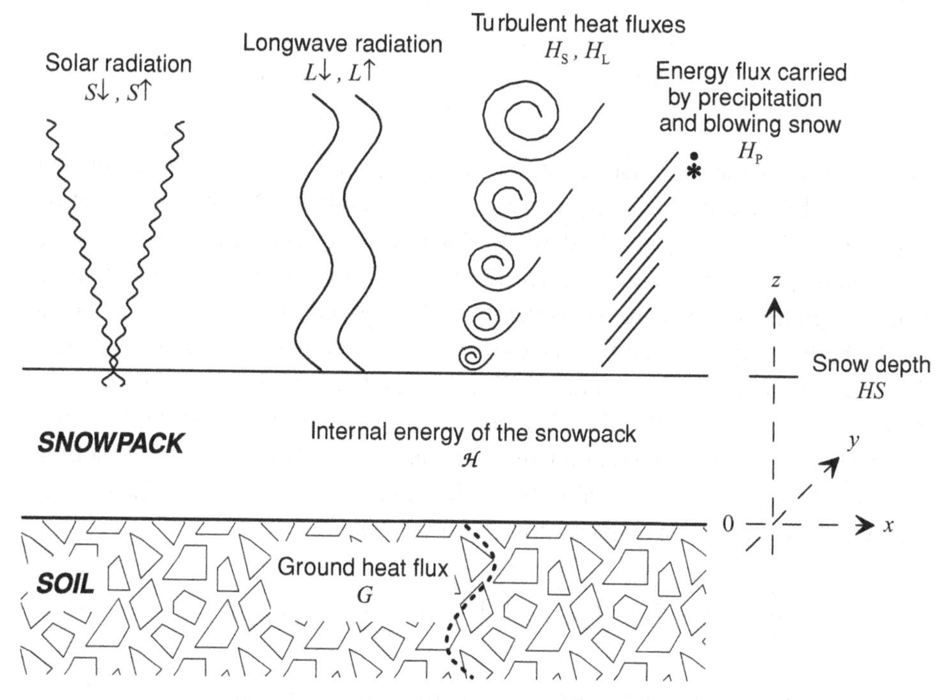


Figure 3.1. Energy balance for an open snowpack.

accumulation is treated (Section 3.4), including some considerations of the coupling of energy and mass balances. Finally, in order to give the reader a sense of the magnitude of the fluxes involved, some examples are presented (Section 3.5), considering either daily averages over typical periods (season) or the processes in detail.

## 3.2 Equations of energy and mass balance

*John C. King, John W. Pomeroy, Donald M. Gray, and Charles Fierz*

Energy balance is often formulated in terms of energy exchanges taking place at an interface (see e.g. Oke, 1987). Such an active surface has to be thought of as being infinitesimally thin and having neither mass nor specific heat. Penetration of shortwave radiation into the snowpack as well as mass movements and phase changes within the snowpack make this concept barely applicable to snow. A better approach is to consider a volume balance of the fluxes as shown in Fig. 3.1. Neglecting horizontal energy transfers as well as effects due to blowing snow or vegetation, the balance for an open and flat snow cover is given in units of W m$^{-2}$ by:[1]

$$-\frac{d\mathcal{H}}{dt} = S\downarrow + S\uparrow + L\downarrow + L\uparrow + H_S + H_L + H_P + G, \qquad (3.1)$$

where $d\mathcal{H}/dt$ is the net change rate of the snowpack's internal energy per unit area. $S\downarrow$ and $S\uparrow$ are the downward and reflected components of shortwave radiation respectively, $L\downarrow$ and $L\uparrow$ are the downward and upward components of longwave radiation respectively, $H_S$ and $H_L$ are the turbulent fluxes of sensible and latent heat through the atmosphere, $H_P$ is the flux of energy carried as sensible or latent heat by both precipitation and blowing snow. It is generally small, but may become significant in the case of warm rain falling onto and penetrating a cold snowpack. However, it will not be discussed further here. Finally, $G$ is the ground heat flux.

Since the change of internal energy is related to either warming and melting (positive change, i.e. energy gain) or cooling and freezing (negative change, i.e. energy loss) within the snowpack, it is also given by:

$$-\frac{d\mathcal{H}}{dt} = L_{\ell i}(R_F - R_M) - \int_{z=0}^{HS} \left[\frac{d}{dt}(\rho_s(z)c_{p,i}T_s(z))\right] dz, \qquad (3.2)$$

where $R_F$ and $R_M$ are the freezing and melting rate, respectively, $L_{\ell i}$ the latent heat of fusion of ice ($3.34 \times 10^5$ J kg$^{-1}$) and $c_{p,i}$ the specific heat capacity of ice ($2.1 \times 10^3$ J kg$^{-1}$ K$^{-1}$); $\rho_s$ is the snow density and $T_s$ is the snow temperature,[2] both at height $z$. The integral is over the snowpack depth $HS$ and is often referred to as the snowpack's cold content. Freezing and melting rates couple the energy balance through Equation (3.2) to the mass $\mathcal{M}$ per unit area of the snowpack which, neglecting the mass of air, is given by:

$$\mathcal{M} = \int_z (\theta_i(z)\rho_i(z) + \theta_\ell(z)\rho_\ell(z))\, dz, \qquad (3.3)$$

where $\theta_i$ and $\theta_\ell$ are the volumetric fractions of ice and water taken at height $z$ with densities $\rho_i$ and $\rho_\ell$, respectively. The mass balance of the snowpack is given in units of kg m$^{-2}$ s$^{-1}$ by (see Fig. 3.2):

$$\frac{d\mathcal{M}}{dt} = P \pm E - R_{\mathrm{runoff}}, \qquad (3.4)$$

where $d\mathcal{M}/dt$ is the snowpack mass change rate (positive in the case of accumulation); $P$ is the precipitation rate (accumulation) and $E = E_{\mathrm{subl}} + E_{\mathrm{evap}}$

[1] Here energy fluxes are the dot product of energy flux densities and the unit normal to the surface. Using the coordinate system shown in Fig. 3.1, energy flux densities directed away from a surface lead to a positive energy flux.

[2] Because 0 °C (273.15 K) is the melting point of ice, the Celsius scale is the natural choice for describing temperature conditions. However, the absolute Kelvin scale is used in the equations unless stated otherwise.

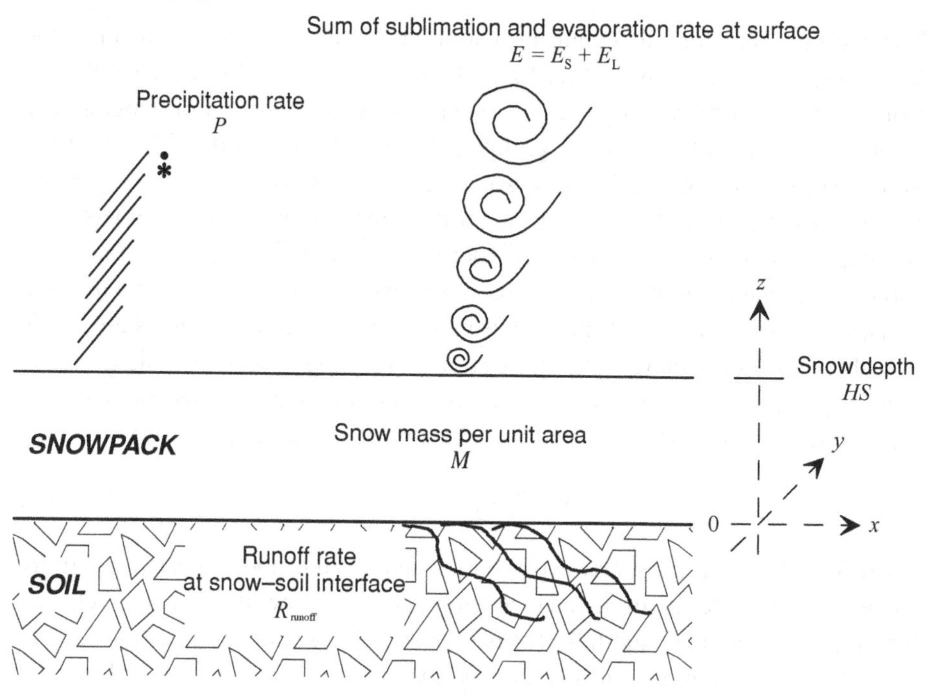


Figure 3.2. Mass balance for an open snowpack.

is the sum of sublimation and evaporation rates at the surface that may contribute either positively or negatively to the mass balance. The runoff rate, $R_{\mathrm{runoff}}$, is strongly coupled to the melting rate of so-called isothermal snowpacks ($T_{\mathrm{s}} = 0\,°\mathrm{C}$ throughout the snowpack) and contributes to ablation only.

A further coupling of energy and mass balances arises because the latent heat flux is simply related to both sublimation and evaporation rates through:

$$H_{\mathrm{L}} = L_{\mathrm{vi}} E_{\mathrm{subl}} + L_{\mathrm{v}\ell} E_{\mathrm{evap}} \approx L_{\mathrm{vi}} E, \tag{3.5}$$

where $L_{\mathrm{vi}}$ is the latent heat of sublimation for ice ($2.838 \times 10^6\ \mathrm{J\,kg^{-1}}$ at $0\,°\mathrm{C}$) and $L_{\mathrm{v}\ell}$ is the latent heat of evaporation for water ($2.505 \times 10^6\ \mathrm{J\,kg^{-1}}$ at $0\,°\mathrm{C}$).

## 3.3 The fluxes involved in the energy balance

*John C. King, John W. Pomeroy, Donald M. Gray, and Charles Fierz*

In the following sections, these fluxes will be discussed in greater detail and schemes for modeling them will be presented.

### 3.3.1 Shortwave radiation

The starting point for calculating insolation at the surface is the solar radiation flux at the top of the atmosphere, $S_{\text{toa}}$, 99.9% of which lies in the spectral band from 0.2 to 100 μm. $S_{\text{toa}}$ may easily be calculated as a function of geographical location, season, and time (see, e.g., Iqbal, 1983). The part of the solar spectrum up to 4.0 μm is usually called shortwave radiation, representing about 99.2% of $S_{\text{toa}}$. However, solar radiation is absorbed by some atmospheric constituents and is reflected and scattered by clouds and aerosol. Accurate calculation of incoming shortwave radiation at the earth's surface, $S\downarrow$, from $S_{\text{toa}}$ thus requires the use of a radiative transfer model, but the data required for driving such models are rarely available in practical snow-cover applications. Incoming shortwave radiation is the most important energy source for snow cover in most situations (see Section 3.5). The net flux of shortwave radiation, $S_N$, at the snow surface is given by:

$$S_{\text{N}} = S\downarrow + S\uparrow = S\downarrow (1 - \alpha), \tag{3.6}$$

where $S\uparrow$ is the reflected shortwave radiation and $\alpha$ is the albedo, i.e. the spectrally integrated reflectance as discussed in Section 2.5.2. As snow albedo ranges between 0.50 for old, wet snow and 0.95 for new snow, changes in snow albedo lead to substantially different amounts of energy absorbed by the snowpack. Moreover, the interaction between the surface and the atmosphere is substantially altered over seasonal snow covers because albedo is quite different over snow-covered ground as opposed to bare ground.

If measurements of $S\downarrow$ are not available, it is generally necessary to parameterize this flux in terms of solar zenith angle and easily observed quantities such as total cloud cover. Key *et al.* (1996) discuss and evaluate a number of such parameterizations. Generally, $S\downarrow$ is calculated first evaluating the insolation under clear sky conditions and then applying a correction for cloud cover. Clouds are reflective; hence, increasing cloud cover tends to *reduce* the magnitude of the downward component of shortwave radiation $S\downarrow$. Over high-albedo surfaces, such as snow, it is important to include the effect of multiple reflections between the snow surface and the cloud base in any parameterization used (Shine, 1984; Gardiner, 1987).

A non-negligible part of downward shortwave radiation $S\downarrow$ reaches the ground as diffuse, nearly isotropic radiation. Its percentage depends primarily on cloudiness and isotropy greatly simplifies its parameterization. As Varley *et al.* (1996) pointed out, many areas in heterogeneous terrain (e.g. alpine topography) receive little or no direct radiation in wintertime, making the separate modeling of direct and diffuse radiation an important issue there. In addition, diffuse radiation reflected from the surrounding topography has to be taken into account (e.g. Dozier, 1980; Plüss, 1997).

### 3.3.2 Longwave radiation

Longwave or terrestrial radiation encompasses wavelengths from approximately 4 to 100 μm. Downward longwave radiation at the surface, $L\downarrow$, results from thermal emission from both atmospheric gases (notably water vapor and carbon dioxide) and clouds while upward longwave radiation, $L\uparrow$, is thermally emitted from the surface.

While an accurate evaluation of $L\downarrow$ once again requires the use of a radiative transfer model, the relatively strong absorption of infrared radiation by water vapor means that $L\downarrow$ is largely determined by conditions in the lowest few hundred meters of the atmosphere (Ohmura, 2001) and, consequently, can be parameterized sufficiently well for many practical applications in terms of near-surface variables. In analogy with the Stefan–Boltzmann equation, parameterizations generally take the form:

$$L\downarrow = -\varepsilon_{\mathrm{eff}}\,\sigma_{\mathrm{SB}} T_{\mathrm{a}}^4, \tag{3.7}$$

where $\sigma_{\mathrm{SB}}$ is the Stefan–Boltzmann constant, $T_{\mathrm{a}}$ is a near-surface air temperature (in K) and $\varepsilon_{\mathrm{eff}}$ is an "effective" emissivity for the atmosphere. $\varepsilon_{\mathrm{eff}}$ is usually specified as a function of cloud cover only, cloud cover and near-surface humidity (Konzelmann *et al.*, 1994) or near-surface humidity only under clear sky (Brutsaert, 1975). Clouds are very efficient infrared emitters and have effective emissivities close to unity, while, under clear skies, $\varepsilon_{\mathrm{eff}}$ is typically around 0.75 but may reach values as low as 0.55 in alpine regions (Marty, 2000). Thus, for fixed $T_{\mathrm{a}}$, increasing cloud cover will *increase* the magnitude of the downward component of longwave radiation $L\downarrow$. Key *et al.* (1996) as well as König-Langlo and Augstein (1994) give further examples of such parameterizations.

Upward longwave radiation can be calculated from the snow surface temperature, $T_{\mathrm{o}}$ (in K), and snow infrared emissivity $\varepsilon$ (see Section 2.5.3) as:

$$L\uparrow = \varepsilon\sigma_{\mathrm{SB}} T_{\mathrm{o}}^4 - (1-\varepsilon)L\downarrow\,. \tag{3.8}$$

In the stably stratified conditions that often prevail in the surface boundary layer over snow-covered surfaces, $T_{\mathrm{o}}$ may be several degrees colder than $T_{\mathrm{a}}$ and the use of $T_{\mathrm{a}}$ instead of $T_{\mathrm{o}}$ in (3.8) may result in a significant overestimate of $L\uparrow$. Over a melting snow cover, $T_{\mathrm{o}}$ may be set to 273.15 K (0 °C). It should be further noted that both downward and upward longwave radiation are usually assumed to be isotropic, resulting in negligible errors only (Plüss, 1997).

To account for longwave radiation in heterogeneous terrain, it is necessary to calculate the incoming fluxes from the sky and the surrounding terrain separately (Plüss and Ohmura, 1997). They showed that in snow-covered environments, where the surface temperature is usually below the air temperature, neglecting the effects due to air temperature leads to an underestimation of the incoming longwave radiation flux on inclined slopes.

### 3.3.3 Net radiation

In summary, the net radiative flux at the surface is:

$$R_N = S_N + L_N = S\downarrow (1-\alpha) + (L\downarrow + L\uparrow). \tag{3.9}$$

The value of $\alpha$ is crucial to the sign of net radiation. Under clear sky, e.g., albedo needs often to be below 0.75 before $R_N$ becomes negative and hence represents a positive energy gain to the snowpack (see Equation 3.1). However, $R_N$ is mostly negative under overcast conditions, independent of the value of $\alpha$.

Furthermore, as we have seen above, cloud cover has opposite effects on $S\downarrow$ and $L\downarrow$. So it is not immediately clear how an increasing cloud cover will effect net radiation. Over a high-albedo surface such as snow, the increase in magnitude of $L\downarrow$ with increasing cloud cover can more than outweigh the reduction in net shortwave radiation, leading to an increased energy gain to the snowpack (Ambach, 1974). This situation of positive cloud radiative forcing is most likely to arise in the polar regions, where both surface albedo and cloud transmissivities in the shortwave region are high and the sun is low. However, the effect of cloud radiative forcing depends critically on surface and cloud properties and both positive and negative forcing has been observed over polar snow surfaces (Bintanja and van den Broeke, 1996).

### 3.3.4 Turbulent heat fluxes

Sensible and latent heat may be carried to or from a snow surface by the action of turbulent eddies in the surface boundary layer (Morris, 1989). Formally, these fluxes may be written as the covariance of fluctuations in vertical velocity, $w$, with those either in temperature, $T_a$, or specific humidity, $Q$, i.e.:

$$\begin{aligned} H_S &= \rho_a c_{p,a} \overline{w' T_a'}, \\ H_L &= \rho_a L_{vi} \overline{w' Q'}, \end{aligned} \tag{3.10}$$

where the overbar denotes a mean over time, primes denote deviations from time averaged values, $\rho_a$ is the air density and $c_{p,a}$ is the specific heat of air at constant pressure ($1.01 \times 10^3$ J kg$^{-1}$ K$^{-1}$). Given suitable fast-response instrumentation, it is possible to measure these covariances and hence obtain direct estimates of the fluxes. However, such measurements are rarely available and in many experimental applications or modeling studies it is necessary to parameterize these fluxes. Most commonly this is done using a bulk transfer formulation, in which fluxes are expressed in terms of differences between surface variables and the values of those variables at some reference height, $z_{\mathrm{ref}}$, in the surface boundary layer, i.e.:

$$\begin{aligned}\tau_{\mathrm{o}} &= \rho_{\mathrm{a}} C_D u^2(z_{\mathrm{ref}}),\\ H_{\mathrm{S}} &= -\rho_{\mathrm{a}} c_{\mathrm{p,a}} C_H u(z_{\mathrm{ref}})(T_{\mathrm{a}}(z_{\mathrm{ref}}) - T_{\mathrm{o}}),\\ H_{\mathrm{L}} &= -\rho_{\mathrm{a}} L_{\mathrm{vi}} C_Q u(z_{\mathrm{ref}})(Q(z_{\mathrm{ref}}) - Q_{\mathrm{o}}).\end{aligned} \tag{3.11}$$

Here, $T_{\mathrm{a}}(z_{\mathrm{ref}})$, $Q(z_{\mathrm{ref}})$ and $u(z_{\mathrm{ref}})$ are the air temperature, specific humidity, and wind speed respectively at the reference height $z_{\mathrm{ref}}$, while $T_{\mathrm{o}}$ and $Q_{\mathrm{o}}$ are the temperature and specific humidity at the snow surface. $Q_{\mathrm{o}}$ may be taken as the specific humidity of air saturated with respect to ice at temperature $T_{\mathrm{o}}$. $\tau_{\mathrm{o}}$ is the surface stress and $C_D$, $C_H$, and $C_Q$ are the bulk transfer coefficients for momentum, heat, and water vapor, respectively. In order to calculate fluxes using (3.11), it is necessary to determine these coefficients that, in general, will depend on both surface roughness and atmospheric stability. This is most satisfactorily accomplished through the framework of the Monin–Obukhov surface layer similarity theory (see, e.g. Garratt, 1992, pp. 49–58). This approach also takes into account that the reference height $\zeta_{\mathrm{ref}} = z_{\mathrm{ref}} - HS$ above a snowpack of depth $HS$ will change with time as $HS$ increases or decreases.

The bulk transfer coefficients are related to the integrated forms of the surface layer similarity functions by:

$$\begin{aligned}C_D &= \kappa^2[\ln(\zeta_{\mathrm{ref}}/z_{\mathrm{o}}) - \Psi_M(\zeta_{\mathrm{ref}}/L_{\mathrm{O}})]^{-2},\\ C_H &= \kappa^2[\ln(\zeta_{\mathrm{ref}}/z_{\mathrm{o}}) - \Psi_M(\zeta_{\mathrm{ref}}/L_{\mathrm{O}})]^{-1}[\ln(\zeta_{\mathrm{ref}}/z_H) - \Psi_H(\zeta_{\mathrm{ref}}/L_{\mathrm{O}})]^{-1},\\ C_Q &= \kappa^2[\ln(\zeta_{\mathrm{ref}}/z_{\mathrm{o}}) - \Psi_M(\zeta_{\mathrm{ref}}/L_{\mathrm{O}})]^{-1}[\ln(\zeta_{\mathrm{ref}}/z_Q) - \Psi_Q(\zeta_{\mathrm{ref}}/L_{\mathrm{O}})]^{-1},\end{aligned} \tag{3.12}$$

where $\kappa$ is von Kármán's constant (generally taken to be around 0.4), $z_{\mathrm{o}}$, $z_H$, and $z_Q$ are the roughness lengths for momentum, heat, and water vapor respectively and $\Psi_M$, $\Psi_H$, $\Psi_Q$ are the corresponding integrated forms of the surface layer similarity functions $\Phi_\xi$:

$$\Psi_\xi = \int_{\zeta_0}^{\zeta_{\mathrm{ref}}} (1 - \Phi_\xi(\zeta'/L_{\mathrm{O}}))\, d(\ln \zeta'); \quad \xi = M, H, Q; \quad \zeta_0 = z_{\mathrm{o}}, z_H, z_Q. \tag{3.13}$$

The surface layer similarity functions express how profiles of wind speed, temperature and humidity deviate from the logarithmic forms that are observed under neutral conditions ($\zeta_{\mathrm{ref}}/L_{\mathrm{O}} = 0$) as a result of stability effects. These functions depend solely on the dimensionless height $\zeta_{\mathrm{ref}}/L_{\mathrm{O}}$, where $L_{\mathrm{O}}$ is the Obukhov length defined as:

$$L_{\mathrm{O}} = -\frac{u_*^3 T_{\mathrm{mean}} \rho_{\mathrm{a}} c_{\mathrm{p,a}}}{\kappa g H_{\mathrm{S}}} \tag{3.14}$$

and

$$u_*^2 = \frac{\tau_o}{\rho_a}, \tag{3.15}$$

where $u_*$ is the friction velocity, $T_{mean}$ is the mean air temperature in the layer of depth $\zeta_{ref}$, and $g$ is the acceleration due to gravity. Note the negative sign on the right-hand side of Equation (3.14), which results from the chosen sign convention (see Stull, 1988). Since $L_O$ is a function of the fluxes to be calculated, the equation set (3.11, 3.12 and 3.14) must generally be solved iteratively. However, as shown below, if some simplifying assumptions are made concerning the form of the $\Psi$-functions, a direct solution is possible under some circumstances.

Experimental studies have determined the forms of the similarity functions in (3.12) even though little has been based on measurements over snow covers (Morris, 1989). Under stable conditions, where the sensible heat flux is directed towards the surface ($H_S < 0$), it is found that, for $0 \leq \zeta_{ref}/L_O \leq 1$:

$$\begin{aligned} \Psi_M &= \beta_{1M}\zeta_{ref}/L_O, \\ \Psi_H &= \Psi_Q = \beta_{1H}\zeta_{ref}/L_O, \end{aligned} \tag{3.16}$$

while for unstable conditions ($H_S > 0$), for $-5 \leq \zeta_{ref}/L_O \leq 0$:

$$\begin{aligned} \Psi_M &= 2\ln[(1 + x_M)/2] + \ln[(1 + x_{M^2})/2] - 2\tan^{-1} x_M + \pi/2, \\ \Psi_H &= \Psi_Q = 2\ln[(1 + y_H)/2], \end{aligned} \tag{3.17}$$

where:

$$\begin{aligned} x_M &= (1 + \gamma_1\zeta_{ref}/L_O)^{1/4}, \\ y_H &= (1 + \gamma_2\zeta_{ref}/L_O)^{1/2}. \end{aligned} \tag{3.18}$$

Measurements indicate $\beta_{1M} \approx \beta_{1H} \approx 5$ and $\gamma_1 \approx \gamma_2 \approx 16$, but there is a considerable range in experimentally determined values (see Garratt 1992, appendix 4).

In what follows, we shall concentrate on stable conditions, since this regime tends to prevail over snow covers. In winter, particularly in high latitudes, $R_N$ is positive for much of the time, leading to a downward heat flux, while advection of warm air over a melting snow cover also leads to the establishment of stable stratification since $T_o$ cannot rise above 0 °C. If we simplify (3.16) by assuming $\beta_{1M} = \beta_{1H} = \beta_1$, manipulation of (3.12) and (3.14) yields an explicit expression for the transfer coefficients in terms of the bulk Richardson number (Garratt, 1992):

$$Ri_B = gz(T_a(z_{ref}) - T_o)/(T_{mean}\, u(z_{ref})^2), \tag{3.19}$$

where $Ri_B \geq 0$ for stable conditions and $T_{mean}$ is the mean temperature in the layer of depth $\zeta_{ref} = z_{ref} - (HS + z_o)$.

For $0 \leq Ri_\mathrm{B} < \beta_1{}^{-1}$:

$$\begin{aligned} C_D &= \kappa^2[\ln(\zeta_\mathrm{ref}/z_\mathrm{o})]^{-2}\,(1-\beta_1 Ri_\mathrm{B})^2 = C_{DN}\,f(Ri_\mathrm{B}),\\ C_H &= \kappa^2[\ln(\zeta_\mathrm{ref}/z_\mathrm{o})\ln(\zeta_\mathrm{ref}/z_H)]^{-1}\,(1-\beta_1 Ri_\mathrm{B})^2 = C_{HN}\,f(Ri_\mathrm{B}),\\ C_Q &= \kappa^2[\ln(\zeta_\mathrm{ref}/z_\mathrm{o})\ln(\zeta_\mathrm{ref}/z_Q)]^{-1}\,(1-\beta_1 Ri_\mathrm{B})^2 = C_{QN}\,f(Ri_\mathrm{B}), \end{aligned} \tag{3.20}$$

and for $Ri_\mathrm{B} \geq \beta_1{}^{-1}$:

$$C_D = C_H = C_Q = 0. \tag{3.21}$$

In practice, there may be problems with using (3.20) and (3.21) as presented above. In runs of coupled surface boundary layers – snow-cover models that parameterize turbulent fluxes using (3.20) and (3.21), it has been found that, when strong radiative cooling is imposed, snow surface temperatures can drop to the point where $Ri_\mathrm{B} = \beta_1^{-1}$, at which point the fluxes are "switched off" and the snow surface becomes effectively decoupled from the atmosphere. This leads to further rapid (and unrealistic) cooling as the surface temperature evolves towards a pure radiative equilibrium (e.g. Morris *et al.*, 1994). Measurements (e.g. King, 1990) show that small but non-zero heat fluxes persist even when $Ri_\mathrm{B} \geq \beta_1^{-1}$ and this effect should be incorporated into any heat flux parameterization used in snow models. Theoretical understanding of turbulent transport in this high-stability regime is still developing but a number of practical alternative schemes to (3.20) and (3.21) have been proposed to avoid the problem of decoupling (Beljaars and Holtslag, 1991; King and Connolley, 1997).

The ratio of $H_\mathrm{S}/H_\mathrm{L}$ is known as the Bowen ratio, $B$. Over snow surfaces, the air is often close to saturation, so $Q(z)$ may be related to $T_\mathrm{a}(z)$ through the Clausius–Clapeyron equation. It can be demonstrated (Andreas, 1989) that, if supersaturation is forbidden and if both $H_\mathrm{S}$ and $H_\mathrm{L}$ are directed downwards, $B$ is limited by:

$$B \geq B_* = \left.\frac{\rho_\mathrm{a} c_\mathrm{p,a}}{L_\mathrm{vi}(\partial\rho_\mathrm{v,sat}/\partial T)}\right|_{T=T_\mathrm{o}}, \tag{3.22}$$

where $\rho_\mathrm{v,sat}$ is the saturation water vapor density. Andreas and Cash (1996) have extended this result to other combinations of $H_\mathrm{S}$ and $H_\mathrm{L}$ and have deduced general formulations for $B$ over saturated surfaces. Such relationships can be of value if estimates of $H_\mathrm{L}$ are required but no humidity measurements are available.

In order to calculate the turbulent fluxes, it is necessary to know the appropriate roughness lengths for momentum, heat and water vapor. The momentum (or aerodynamic) roughness length, $z_\mathrm{o}$, is related to the geometric roughness characteristics of the snow surface. Snow cover is one of the smoothest land surface types encountered in nature and, consequently, measurements of $z_\mathrm{o}$ over snow (Table 3.1) indicate small values, of the order $10^{-4}$ to $10^{-3}$ m. Usually, $z_Q$ and $z_H$ are assumed to be one

Table 3.1 *Aerodynamic roughness lengths measured over various snow and ice surfaces.*

| Surface | | Roughness length $z_o(m)$ | Location | Reference |
|---|---|---|---|---|
| Seasonal snow cover | | $2.3 \times 10^{-4}$ | Finse | Kondo and Yamazawa (1986) |
| | | $2.0 \times 10^{-4}$ to $4.0 \times 10^{-3}$ | | Harding (1986) |
| | | $2.0 \times 10^{-4}$ to $2.0 \times 10^{-2}$ | | Konstantinov (1966) |
| | | $2.5 \times 10^{-3}$ | Spitsbergen | Sverdrup (1936) |
| Antarctic ice shelves | | $5.6 \times 10^{-5}$ | | King and Anderson (1994) |
| | | $1.0 \times 10^{-4}$ | | Heinemann (1989) |
| | | $1.0 \times 10^{-4}$ | | König (1985) |
| Antarctic blue ice | | $2.8 \times 10^{-6}$ | | Bintanja and van den Broeke (1995) |
| Sea ice | Snow-covered | $3$ to $5 \times 10^{-4}$ | | Joffre (1982) |
| Subantarctic glacier | Fresh snow | $2.0 \times 10^{-4}$ | | Poggi (1976) |
| Alpine glacier | Accumulation zone | $2.3 \times 10^{-3}$ | | van den Broeke (1997) |
| | Ablation zone | $4.4 \times 10^{-3}$ | | van den Broeke (1997) |
| | Sastrugi | $(1.10 \pm 0.25) \times 10^{-2}$ | | Grainger and Lister (1966) |
| | Undulating wet snow | $(6.8 \pm 1.4) \times 10^{-3}$ | | Grainger and Lister (1966) |
| Icelandic glacier | Ablation zone | $2 \times 10^{-3}$ to $1 \times 10^{-1}$ | | Smeets *et al.* (1998) |

order of magnitude smaller, as suggested by Garrat (1992) and Morris (1989). With values this small, it is clear that roughness elements of the scale of individual snow grains must be making the greatest contribution to the surface drag, with larger micro-topographic features, such as sastrugi, making a lesser contribution (Kondo and Yamazawa, 1986; Inoue, 1989). Ablating glaciers can develop large roughness elements on their surfaces, leading to aerodynamic roughness lengths of up to 0.1 m. The roughness of such surfaces can change rapidly as surface features develop during the ablation season (Smeets *et al.*, 1998). Bare ice surfaces have particularly small roughness lengths (Bintanja and van den Broeke, 1995; see Table 3.1). Flow over such surfaces is "aerodynamically smooth," i.e. the surface Reynolds number:

$$Re = \frac{u_* z_*}{\mu_a}, \tag{3.23}$$

where $\mu_a$ is the kinematic viscosity of air and $z_*$, the scale of the roughness elements, is less than about 5. In this low Reynolds number regime, flow around individual roughness elements is laminar and the roughness length is given by:

$$z_o = 0.135 \frac{\mu_a}{u_*}. \tag{3.24}$$

If the surface stress is great enough to generate blowing snow (see Section 3.4), suspended snow grains may contribute to the momentum transfer from atmosphere to surface and may thus cause an apparent change in $z_o$. Owen (1964), Tabler (1980), Chamberlain (1983), and others have suggested that, under these conditions, the apparent roughness length is increased by drag exerted on saltating snow and will therefore be proportional to the surface stress, i.e.

$$z_o = c_1 \frac{u_*^2}{2g}. \tag{3.25}$$

Experimental evidence for such a relationship is variable. Tabler (1980), Tabler and Schmidt (1986) as well as Pomeroy and Gray (1990) present extensive quantitative measurements that support such behavior with values of $c_1$ of 0.1203 over continuous snowfields (Pomeroy and Gray, 1990) and $c_1$ of 0.026 48 over a mixture of snow and lake ice (Tabler, 1980).

Bintanja and van den Broeke (1995), however, suggest that, in some cases, the alteration of surface characteristics by the transport of fresh snow onto a smoother underlying snow or ice surface may be the dominant process leading to an apparent increase in $z_o$ with wind speed. More observations are needed to resolve this issue.

Fewer measurements exist for the scalar roughness lengths, $z_H$ and $z_Q$, largely because of the difficulty of defining $T_o$ and $Q_o$ other than over a melting snow surface, when $T_o$ may be taken as 0 °C. Heat and water vapor transfer at the snow surface must ultimately be accomplished purely by molecular diffusion, since there is no equivalent to the form drag of roughness elements that is responsible for the majority of the momentum transport. Andreas (1989) developed a theory that predicts the ratio of $z_o$ to the scalar roughness lengths as a function of the surface Reynolds number (see Equation 3.23). In the aerodynamically smooth regime, $z_H/z_o = 3.49$ and $z_Q/z_o = 5.00$. As $Re$ increases, the ratio of scalar roughness length to momentum roughness length decreases rapidly and, for moderate wind speeds over a typical snow cover, the scalar roughness lengths will be one or two orders of magnitude smaller than $z_o$. Measurements (Kondo and Yamazawa, 1986; Bintanja and van den Broeke, 1995) generally support this functional dependence. However, in many modeling applications, the scalar roughness lengths are set equal to $z_o$ for simplicity, on the grounds that the ensuing errors in surface fluxes will be no greater than the uncertainties resulting from other parts of the flux computation procedure.

### 3.3.5 *Heat fluxes over a non-uniform snow cover*

The results of Section 3.3.4 are strictly applicable only over an extensive and uniform snow cover, where the atmospheric conditions at the reference height used for the flux computation are in equilibrium with the underlying surface. This will not, in general, be the case over a patchy snow cover. Areas of bare ground will have different roughness and albedo characteristics from snow-covered areas. Advection of air warmed over bare ground onto a snow-covered area will lead to an enhanced downward heat flux at the upwind edge of the snow patch, with heat fluxes decreasing with increased fetch over the snow as the air comes into a new equilibrium with the snow surface. For instance, Weisman (1977) showed that the dimensionless sensible heat flux $H'_S$ at any point downwind of the leading edge of a snow patch varies with snow-covered fetch distance as

$$H'_S = -a_4 X^{-0.125}, \tag{3.26}$$

where $X$ is a dimensionless distance downwind of the leading edge of the snow patch (influenced by roughness length) and $a_4$ is controlled by a temperature difference stability parameter (Weisman, 1977).

Marsh and Pomeroy (1996) proposed that the additional sensible heat advected to a snow patch is a function of the snow-free fraction of the upwind domain,

$$H_S - H_{S,s} = (-H_{S,b}(1 - \mathrm{SCA})/\mathrm{SCA})\, h_{S,b}, \tag{3.27}$$

where $H_{S,s}$ is the sensible heat flux to snow over a completely snow-covered fetch, $H_{S,b}$ is sensible heat over bare ground, SCA is the proportion of snow-covered area and $h_{S,b}$ expresses the portion of bare ground sensible heat that is advected to the snow patch. Neumann and Marsh (1998) showed that $h_{S,b}$ declines from 0.3 to 0.001 as SCA decreases from 1 to 0.01 and that for SCA of 0.5, $h_{S,b}$ ranges from 0.02 to 0.2 depending on wind speed and snow patch size. Even more extreme effects may occur over a sea-ice cover interspersed with leads of open water (Claussen, 1991). Interactions with the vegetation will also affect the heat fluxes. They will be discussed in more detail in Sections 3.5.4 and 3.5.5.

The computation of surface fluxes over such non-uniform surfaces is very important for climate simulation (see Section 4.3) and it is a developing subject that requires the application of numerical or analytic models of varying degrees of complexity. A detailed description of such techniques is beyond the scope of the present work and the reader is referred to Weisman (1977), Liston (1995), Essery (1997), and Marsh *et al.* (1997) for further information.

To conclude, few authors have investigated the variation of the turbulent fluxes in heterogeneous terrain although it was termed a priority research topic by Garratt (1992). Based on such an investigation, a simplified approach is presented in Plüss (1997).

## 3.4 Snow accumulation

*John C. King, John W. Pomeroy, Donald M. Gray, and Charles Fierz*

The surface mass balance described in Equation (3.4) is affected by distinctive snow accumulation fluxes that are influenced by the air mass characteristics, surface snow conditions, vegetation, and topography. Interception and unloading must be considered in forested environments where evergreen canopies intercept substantial portions of snowfall. Blowing snow fluxes are significant in exposed, poorly vegetated regions such as alpine terrain, steppes, prairies, tundra, and ice sheets.

### 3.4.1 Interception by vegetation

Vegetation intercepts snow as a function of its winter leaf and stem area and the size of the snowfall event. Because sublimation and melt remove snow from canopies, interception (the snowfall trapped in the canopy, normally event-based) is distinguished from snow load (the snow held in the canopy at a particular time). Hedstrom and Pomeroy (1998) developed and field tested a model of snow interception and snow unloading of the following form (see Fig. 3.3):

$$I = I_1 e^{-U_I t}, \quad I_1 = (\mathcal{L}_{\mathrm{L}}^* - \mathcal{L}_{\mathrm{L,o}})\left(1 - e^{c_{\mathrm{can}}\mathcal{P}/\mathcal{L}_{\mathrm{L}}^*}\right), \tag{3.28}$$

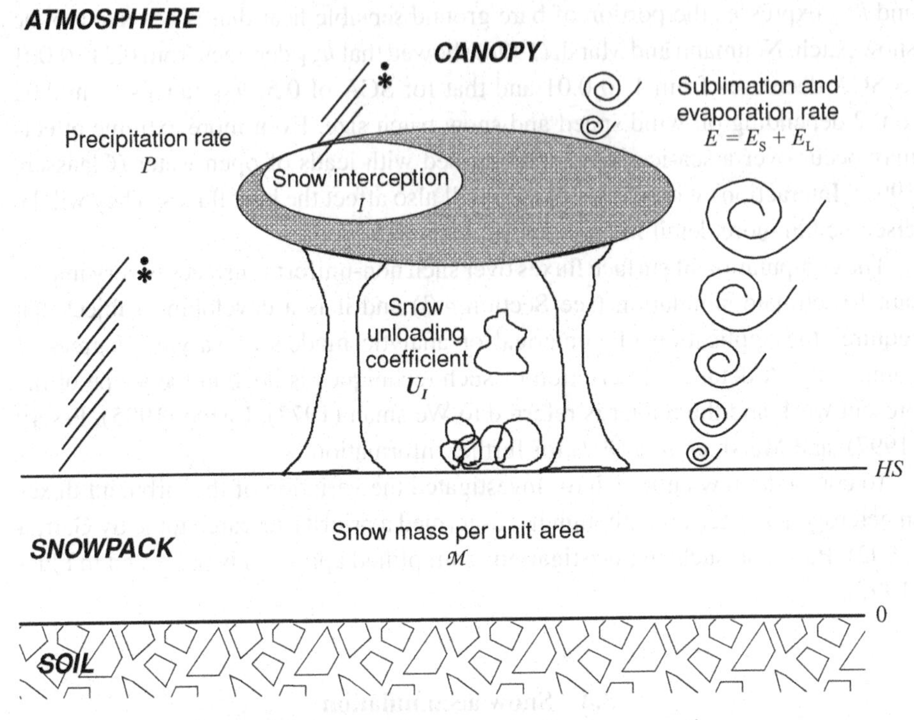


Figure 3.3. Interception and unloading by vegetation.

where $I$ is snow interception after unloading has occurred (kg m$^{-2}$), $I_1$ is interception at the start of unloading, $U_I$ is a snow unloading coefficient, $t$ is time (s), $\mathcal{L}_{\mathrm{L}}^{*}$ is the maximum canopy snow load, $\mathcal{L}_{\mathrm{L,o}}$ is the initial snow load, $\mathcal{P}$ is the accumulated snowfall (kg m$^{-2}$) and $c_{\mathrm{can}}$ is the canopy closure.

In a southern boreal forest, Hedstrom and Pomeroy (1998) found that $e^{-U_I t} = 0.678$ using empirical data with a weekly time resolution. $\mathcal{L}_{\mathrm{L}}^{*}$ can be determined using an empirical relationship developed by Schmidt and Gluns (1991), where:

$$\mathcal{L}_{\mathrm{L}}^{*} = \mathcal{L}_{\mathrm{L,b}}\left(0.27 + \frac{46}{\rho_{\mathrm{s}}}\right) LAI \tag{3.29}$$

and $\mathcal{L}_{\mathrm{L,b}}$ is the maximum snow load per unit area of branch (kg m$^{-2}$), $\rho_{\mathrm{s}}$ is the density of fresh snow (kg m$^{-3}$) and $LAI$ is winter leaf and stem area index (m$^2$ m$^{-2}$). $\mathcal{L}_{\mathrm{L,b}}$ ranges from 5.9 kg m$^{-2}$ for spruce to 6.6 kg m$^{-2}$ for pine. Equation (3.28) is an extension of the expression proposed by Calder (1990), which is based on formulations used for the interception of rainfall. Important differences from earlier formulations are the consideration of canopy snow load in reducing interception efficiency and the substantially high maximum canopy snow loads that are found, compared to maximum rainfall interception. Interception efficiency ($I/\mathcal{P}$) varies from small values up to 0.6 for dense conifer canopies and declines with increasing storm snowfall amounts and initial canopy snow loads.

Snow in the canopy represents a "snow surface" with very different characteristics from snow on the ground (Harding and Pomeroy, 1996). For instance, it is relatively well-exposed to the atmosphere and subject to high net radiation because snow-filled canopies retain their low snow-free albedo (Pomeroy and Dion, 1996; Yamazaki *et al.*, 1996), are aerodynamically rough (Lundberg *et al.*, 1998), and are usually associated with slightly unstable surface boundary layers (Nakai *et al.*, 1999). The varying snow load and degree of its exposure in the canopy means that snow-filled canopies do not behave as a continuous, saturated surface; this affects the free availability of moisture in the canopy. Nakai *et al.* (1999) accounted for this effect by varying the ratio of bulk transfer coefficients $C_Q/C_H$ (see Equation 3.12) from 0.1 for the low moisture case (low snow load) to 1.0 for the high moisture case (high snow load). Lundberg *et al.* (1998) used a Penman–Monteith combination model with a resistance parameterization to estimate sublimation from snow-covered canopies, and found that the resistance of a snow-covered canopy had to be set at 10 times that of a rain-covered canopy to provide results that matched measurements. Parviainen and Pomeroy (2000) use a coupled calculation scheme where turbulent transfer from snow clumps is calculated (presuming no evaporation from the snow-free canopy) to provide a within-canopy humidity and temperature field that is then matched with the bulk transfer formulation for the whole canopy as shown in Equation (3.11). Both sensible and latent heat transfer from snow clumps are calculated from variants of Equation (3.11) where the terms $C_H$ and $C_Q$ are replaced with $C_{hq}$ and $u(z)$ is replaced with the Sherwood and Nusselt numbers for turbulent transfer of heat and water vapor from particles respectively and solved with the assumption that sublimating snow clumps are in thermodynamic equilibrium at the ice-bulb surface temperature (Schmidt, 1991). $C_{hq}$ is then found as:

$$C_{hq} = \frac{3\mathcal{L}_{\mathrm{L}} k_{\mathrm{cl}}}{2r_{\mathrm{pt,n}}^2 \rho_{\mathrm{i}}} \left(\frac{\mathcal{L}_{\mathrm{L}}}{\mathcal{L}_{\mathrm{L}}^*}\right)^{-\chi_{\mathrm{I}}}, \tag{3.30}$$

where $\mathcal{L}_{\mathrm{L}}$ is snow load (equal to initial interception, $I_1$, less any sublimation and unloading), $k_{\mathrm{cl}}$ is a dimensionless snow clump shape coefficient, $r_{\mathrm{pt,n}}$ is a nominal snow particle radius (0.0005 m), $\rho_{\mathrm{i}}$ is the density of ice (kg m$^{-3}$), and $\chi_{\mathrm{I}}$ is 1.0 less the fractal dimension of intercepted snow (Pomeroy and Schmidt, 1993). Pomeroy *et al.* (1998a) found $\chi_{\mathrm{I}}$ is normally 0.4 for mature evergreens and Parviainen and Pomeroy (2000) report empirically derived values of $k_{\mathrm{cl}}$ (given the nominal $r_{\mathrm{pt,n}}$) of 0.0114 for a mature conifer forest and 0.0105 for a young conifer plantation in western Canada.

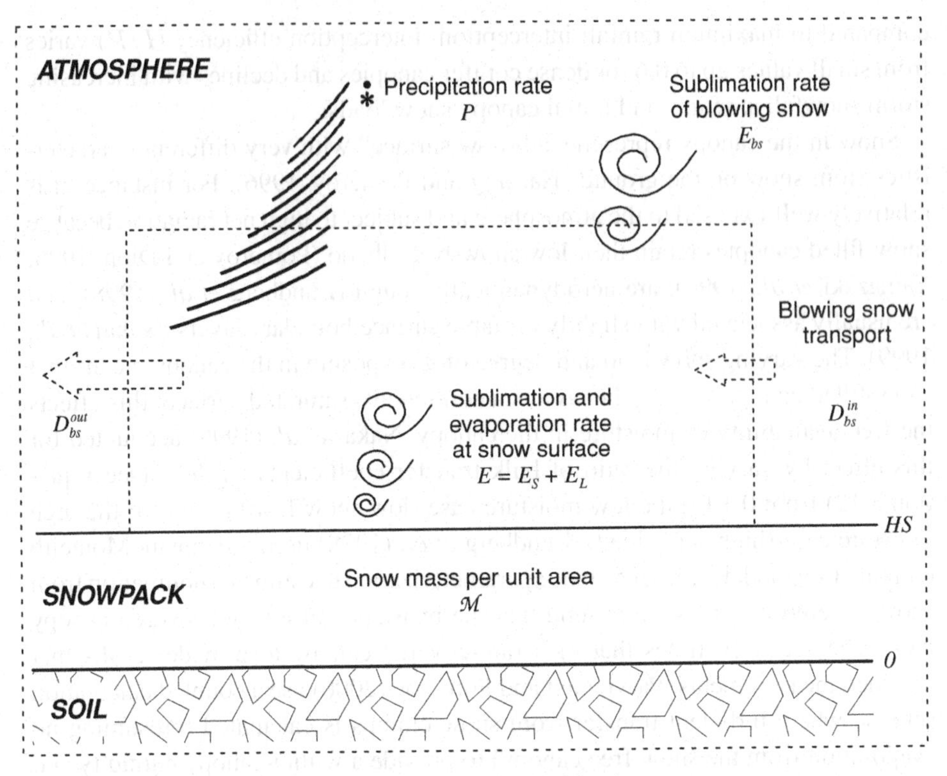


Figure 3.4. Mass balance with blowing snow.

Experimentally derived sublimation rates from mature evergreen forests range from 3 to about 12 kg m$^{-2}$ d$^{-1}$ (Pomeroy *et al.*, 1998a; Nakai *et al.*, 1999) and can return 13–40% of seasonal snowfall to the atmosphere as sublimation from northern forests (Pomeroy and Gray, 1995; Pomeroy *et al.*, 1998a).

### 3.4.2 *Blowing snow*

Blowing snow fluxes occur over open snow surfaces with good exposure to wind and a supply of erodible snow (see Fig. 3.4). On large, relatively uniform surfaces such as ice sheets steady-state conditions may develop. However, on seasonal snow surfaces, exposed vegetation and larger scale topographic features disrupt steady-state conditions. At small scales (e.g. grass stalks, boulders) the features act as an additional surface roughness, and impede erosion or sometimes induce small depositions of wind-blown snow. At large scales (e.g. woodlands, valleys, mountain crests) the features can cause flow separation in the boundary layer and significantly reduce the shear stress on downwind surfaces, which may lead to the formation of large snowdrifts. The result is a strong heterogeneity to blowing snow fluxes and a need to consider the landscape in calculating these fluxes. One important aspect of blowing snow is that it can dramatically increase latent heat transfer to the atmosphere via sublimation and that blowing snow transport and sublimation can control the mass balance of non-melting snow covers in open areas. Including blowing snow in the mass balance, Equation (3.4) may be rewritten as:

$$\frac{d\mathcal{M}}{dt} = P - \nabla \cdot D_{\text{bs}} - E_{\text{bs}} \pm E - R_{\text{runoff}}, \tag{3.31}$$

where $D_{\text{bs}}$ is the horizontal blowing snow transport (with dimensions of mass per unit length per unit time) and $E_{\text{bs}}$ is the rate of sublimation of blowing snow. Horizontal inhomogeneities in either the surface wind field or surface characteristics can generate corresponding inhomogeneities in snow transport leading to a divergence of $D_{\text{bs}}$ and a redistribution of surface snow.

Blowing snow transport, $D_{\text{bs}}$, primarily occurs in two modes: saltation, $D_{\text{salt}}$ and suspension, $D_{\text{susp}}$, thus:

$$D_{\text{bs}} = D_{\text{salt}} + D_{\text{susp}}. \tag{3.32}$$

Saltation is the mechanism by which snow particles are eroded from or deposited onto the snow surface and transported near to the surface. The process involves particles bouncing downstream and shattering new particles from the snow surface. Suspension occurs above the saltation layer and involves smaller particles lofted by turbulence to heights several tens of meters above the snow surface. At very low wind speeds saltation is the primary mode of transport, while at higher wind speeds suspension becomes predominant.

Steady-state saltation is generally considered to operate in a dynamic balance with atmospheric driving forces, where the atmospheric shear stress, $\tau_{\text{atm}}$, is partitioned into that applied to the non-erodible surface, $\tau_{\text{n}}$, that applied to the stationary erodible surface, $\tau_{\text{s}}$, and that applied to the saltating bed of snow particles, $\tau_{\text{p}}$. Thus:

$$\tau_{\text{atm}} = \tau_{\text{n}} + \tau_{\text{s}} + \tau_{\text{p}}. \tag{3.33}$$

Most blowing snow models (Pomeroy *et al.*, 1993; Pomeroy and Gray, 1995; Bintanja, 1998; Dery *et al.*, 1998; Liston and Sturm, 1998; Essery *et al.*, 1999; Pomeroy and Li, 2000) use a solution for the saltating mass flux, $D_{\text{salt}}$, derived and field-tested by Pomeroy and Gray (1990) where:

$$D_{\text{salt}} = \frac{a_5 u_{*t} \rho_a}{u_* g} \left(u_*^2 - u_{*n}^2 - u_{*t}^2\right) \tag{3.34}$$

and $a_5$ is an experimentally determined "saltation efficiency" coefficient ($0.68\ \text{m}\,\text{s}^{-1}$), $u_*$ is the friction velocity, $u_{*t}$ is the threshold friction velocity (that at the initiation or cessation of saltation), $u_{*n}$ is the friction velocity applied to small-scale non-erodible elements. Li and Pomeroy (1997a) outline operational techniques to calculate the threshold friction velocity from the meteorological history of the snowpack, while Pomeroy and Li (2000) discuss the influence of vegetation on $u_{*n}$. Doorschot (2002) and Doorschot and Lehning (2002) address the relative importance of aerodynamic entrainment compared to rebound and ejection during steady-state saltation, a topic which is still not fully resolved.

Suspended blowing snow particles are presumed to move downwind with the same speed as fluid points in the atmosphere, but are not lifted as effectively due to inertia and a significant terminal fall velocity. The downwind mass flux of suspended snow, $D_{\mathrm{susp}}$, may be described as,

$$D_{\mathrm{susp}} = \frac{u_*}{\kappa} \int_{h_*}^{z_{\mathrm{b}}} c_{\mathrm{susp}}(\zeta) \ln(\zeta / z_{\mathrm{o}})\, d\zeta, \tag{3.35}$$

where $\kappa$ is von Kármán's constant, $h_*$ is the lower boundary for suspended snow (upper boundary for saltating snow), $z_{\mathrm{b}}$ is the upper boundary for suspended snow, $c_{\mathrm{susp}}(\zeta)$ is the mass concentration of suspended snow at height $\zeta = z - HS$ above the snow surface and $z_{\mathrm{o}}$ is the aerodynamic roughness length. Presuming steady-state, one-dimensional diffusion and in analogy to Fick's law, the instantaneous mass concentration may be found as

$$(w c_{\mathrm{susp}})' = -\kappa_{\mathrm{s}}(\zeta) \frac{\partial c_{\mathrm{susp}}}{\partial \zeta}, \tag{3.36}$$

where $w$ is the vertical particle velocity at height $\zeta$ and $\kappa_{\mathrm{s}}(\zeta)$ is the turbulent diffusion coefficient for snow particles at height $\zeta$. For steady-state, non-advective conditions, this formulation can be solved to find a mass concentration profile as:

$$c_{\mathrm{susp}}(\zeta + d\zeta) = c_{\mathrm{susp}}(\zeta) \left( \frac{\zeta + d\zeta}{\zeta} \right)^{w_*(\zeta)}, \tag{3.37}$$

where $w_*$ is a dimensionless vertical velocity. Pomeroy and Male (1992) experimentally derived expressions for $h_*$ and $w_*$. Kind (1992) proposed an improved theoretical solution for $c_{\mathrm{susp}}(\zeta)$ assuming steady-state turbulent diffusion. This solution has been used in the Liston and Sturm (1998) model. Dery *et al.* (1998) considered non-steady-state diffusion formulations involving explicit calculation over the particle size spectrum, and the effects of sublimation in setting the upper boundary, $z_{\mathrm{b}}$. Unfortunately, measurements that can describe the individual behaviors of $\kappa_{\mathrm{s}}$ and $w$ are lacking, and hence there is considerable uncertainty in derivations of $w_*$.

Determination of the amount of sublimation during blowing snow is based on the work of Thorpe and Mason (1966) with modifications by Schmidt (1972) and Pomeroy (1989). The sublimation rate within a column of blowing snow, $E_{\mathrm{bs}}$, is calculated with reference to the sublimation rate $dm_{\mathrm{p}}/dt$ of a single blowing snow particle of mass $m_p$, the variation in mass concentration with height, $c_{susp}(r_{pt}, z)$, and the change in particle radius, $r_{pt}(m_p)$, as

$$E_{bs} = \int_0^{z_b} \int \frac{1}{m_p(r_{pt}, z)} \frac{dm_p}{dt} c_{susp}(r_{pt}, z)\, dz\, dr_{pt}. \tag{3.38}$$

The sublimation rate of a single blowing snow particle is,

$$\frac{dm_p}{dt} = 2r_{pt} D_a Sh(\rho_{v,a} - \rho_{v,pt}), \tag{3.39}$$

which is balanced by the convective heat transferred to the particle,

$$L_{vi}\left(\frac{dm_p}{dt}\right) = -2\pi r_{pt} k_a Nu(T_a - T_{o,sp}), \tag{3.40}$$

where $D_a$ is the diffusion coefficient of water vapor in air ($m^2\ s^{-1}$), $Sh$ is the Sherwood number, $\rho_{v,a}$ is the water vapor density in ambient air and $\rho_{v,pt}$ is that at the particle surface and $k_a$ is the thermal conductivity of air ($W\ m^{-1}\ K^{-1}$), $Nu$ is the Nusselt number, $T_{o,sp}$ is the ice sphere surface temperature (K) and $T_a$ is the ambient air temperature (K). Lee (1975) found that both the Nusselt and Sherwood numbers can be related to the particle Reynolds number, $N_{Re}$, as

$$Nu = Sh = 1.79 + 0.606 N_{Re}^{0.5}. \tag{3.41}$$

Presuming that blowing snow particles are in thermodynamic equilibrium, Schmidt (1972) combined Equations (3.39) and (3.40) using the Clausius–Clapeyron equation to solve for $dm/dt$. The combined equation can be found in several later publications (Schmidt, 1991; Pomeroy *et al.*, 1993; Dery *et al.*, 1998).

One problem in applying sublimation equations is the difficulty in determining the fields of air temperature and humidity in the lowest meter of the atmosphere where the mass concentration of blowing snow is greatest. Thermodynamic models such as that of Dery *et al.* (1998) suggest that the atmosphere cools and approaches saturation near the snow surface during blowing snow. Measurements of these fields in extremely stable surface boundary layers over ice shelves support cold, humid conditions in this layer (King *et al.*, 1996); however, measurements above terrestrial snow surfaces during blowing snow often show a notable undersaturation of water vapor with respect to ice (Pomeroy and Li, 2000). Pomeroy and Essery (1999) measured sublimation fluxes during light blowing snow of up to 60 W $m^{-2}$; these fluxes were far larger than those predicted from simple surface sublimation (approx 10 W $m^{-2}$) and were well matched by the model predictions using the Prairie Blowing Snow Model driven by surface observations (Pomeroy and Li, 2000).

Attempts to scale up the calculations of blowing snow fluxes to heterogeneous terrain have been made by Pomeroy *et al.* (1993, 1997), Gauer (1998), Liston and Sturm (1998), Essery *et al.* (1999), Lehning *et al.* (2004), Bowling *et al.* (2004), and Essery and Pomeroy (2004). In all cases considerable difficulties are encountered in these attempts because of the development of non-steady-state conditions and flow convergence and divergence within the surface boundary layer over complex terrain. Typical small-scale features such as flow separation, recirculation and speed-up effects are indeed the driving mechanisms behind inhomogeneous snow distribution. For the simple case of scaling up blowing snow fluxes over heterogeneous snow covers on level, uniform terrain (Li and Pomeroy, 1997b; Pomeroy and Li, 2000), a cumulative normal probability distribution is used to weight fluxes by the probability of a blowing snow occurrence, $P_{bs}$. The formulation is based on an observed association between the probability of occurrence of blowing snow, the average hourly 10 m wind speed, $u_{10}$, average hourly air temperature, $T_a$, in °C and snow surface age, $t_h$, in hours where:

$$P_{bs}(u_{10}, T_a, t_h) = \frac{1}{\sqrt{2\pi}\,\delta}\int_0^{u_{10}} \exp\left[-\frac{(\overline{u}-u)^2}{2\delta^2}\right] du \tag{3.42}$$

and for dry snow, the mean wind speed is found as

$$\overline{u} = 11.2 + 0.365T_a + 0.00706T_a^2 + 0.9\ln(t_h) \tag{3.43}$$

and the standard deviation of wind speed is found as

$$\delta = 4.3 + 0.145T_a + 0.00196T_a^2. \tag{3.44}$$

Note that $\overline{u}$ and $\delta$ are properties of the snow surface and affected by the meteorological history of the snowpack, whilst $u$ depends on current wind speed. If the snow surface is wet or icy, then values can be assigned as $\overline{u} = 21$ m s$^{-1}$ and $\delta =$ 7 m s$^{-1}$ based on observations in the Canadian Prairies (Li and Pomeroy, 1997b).

There are different conclusions among experimental and modeling studies regarding the importance of blowing snow fluxes to regional snow mass balance and surface fluxes. Bintanja (1998) suggests that blowing snow sublimation in the Antarctic varies from negligible in the interior to 170 kg m$^{-2}$ near the coast and is therefore a major term in the mass balance of the Antarctic ice sheet. Dery and Yau (2002) used the re-analysis (ERA15) data of the European centre for medium-range weather forecasts (ECMWF) at a resolution of 2.5° to study the effects of surface sublimation and blowing snow on the surface mass balance of the largely forested Mackenzie River basin, Canada. They found that surface sublimation removes

29 kg m$^{-2}$ SWE per year, or about 7% of the annual precipitation of the largely subarctic Mackenzie Basin.

From blowing snow model results and measured areal snow mass balances, Pomeroy *et al.* (1997) as well as Liston and Sturm (1998) suggest that the percentage of annual snowfall that is sublimated via blowing snow from arctic surfaces is 20% in northern Canada and 9% to 22% in northern Alaska. The model of Essery *et al.* (1999) suggests annual sublimation losses from 25% to 47% of annual snowfall for northern Canada, depending upon surface vegetation and topography. In removing these large sublimation estimates from their blowing snow model, Essery *et al.* (1999) found that snow accumulation was grossly overestimated in deposition zones, whereas it was well estimated when sublimation was included. Despite the deficiencies in these relatively simple blowing snow models, it appears that sublimation estimates can reasonably match seasonal losses and that, in field situations, the humidity feedback effects in the lower surface boundary layer proposed by Dery *et al.* (1998) are not normally sufficient to significantly lower seasonal sublimation rates.

### 3.4.3 Accumulation in alpine topography

Both orography and wind strongly affect the spatial distribution of snow depth in mountainous regions. Föhn (1977, 1985) addressed the question of representative new snow depth measurements in mountainous terrain. Clearly, crests, wind- and leeward slopes as well as depressions have to be avoided. Even at the most suited level sites, the catch efficiency of precipitation gauges (60–80%) does not closely match the reliability of a snow board laid on the snow surface. Even so, Föhn (1977, 1985) states that precipitation gauges located in surrounding valley floors may help to extrapolate new snow depths over larger areas in wintertime; very often one has to rely on available weather forecast outputs as well. By combining the latter with statistical and climatic approaches, promising results have been obtained (Durand *et al.*, 1993; Raderschall, 1999).

Yet for many applications redistribution by wind on leeward slopes has to be taken into account. Föhn and Hächler (1978) give an empirical relation for a 35° steep leeward slope as:

$$HN_{\mathrm{w}} = c_2 \overline{u}_{\mathrm{crest}}^3, \qquad \overline{u} \leq 20\,\mathrm{m\,s^{-1}}, \tag{3.45}$$

where $HN_{\mathrm{w}}$ is the mean additional new snow (in m) deposited on the leeward slope over a time period of 24 h, $c_2 = 0.8 \times 10^{-4}$ (s$^3$ m$^{-2}$) is an empirical coefficient and $\overline{u}_{\mathrm{crest}}$ is the mean wind speed measured on the crest. Despite the complexity of both modeling and measurements, blowing and drifting snow in heterogeneous terrain is currently a topic of strong interest (Pomeroy, 1991; Liston and Sturm, 1998; Pomeroy *et al.*, 1999c; Durand *et al.*, 2001; Michaux *et al.*, 2001). Modeling blowing snow over a steep alpine ridge, Gauer (2001) obtained a power relation for the simulated steady-state flux, decreasing from approximately 4 to 2 with increasing wind speed. The mass flux decrease with increasing wind speed further indicates that a saturation of snow transport occurs. Combining a saltation model (Doorschot and Lehning, 2002) with an analytical wind profile taking into account speed-up effects over a ridge, Doorschot *et al.* (2001) observed a similar trend. The latter authors also included preferential deposition of precipitation as a third mode of snow transport to the lee slope. Preferential deposition does not require any threshold value to be exceeded to occur and may thus be predominant under certain conditions in steep Alpine terrain.

Finally, on the scale of a few hundred to one thousand square kilometers, Föhn (1992) as well as Martin *et al.* (1994) studied the impact of climate on snow depth in alpine regions. Such compilations provide valuable information for governmental agencies, hydropower facilities, winter tourism, and flood mitigation.

## 3.5 Examples of energy and mass balances

The relative contribution of the different fluxes involved in the energy and mass balance of the snow cover strongly depends on the prevailing meteorological and topographical conditions as well as on vegetation. The examples presented are "typical" snow covers although they do not all formally correspond to a classification scheme as proposed by Sturm *et al.* (1995); neither are they exhaustive.

Sections 3.5.1–3.5.3 deal with daily averages over continuous snow covers, stressing the relative contribution of each flux as the season proceeds. Prevailing meteorological conditions are also shown. Table 3.2 emphasizes this time average approach summarizing means for chosen time periods. Sections 3.5.4 and 3.5.5 are process oriented, showing and discussing detailed balances over a few days' time. Where conditions influencing the energy balance vary strongly, this latter approach is best suited to describing short time variations that are smeared out by averaging.

Note that the radiative components of the energy balance as well as the mass balance are usually measured *in situ*. The turbulent fluxes, however, are parameterized using a bulk transfer formulation (see e.g. Section 3.3, Equation 3.11). Remember that according to the sign convention introduced in Section 3.2, negative fluxes are associated with positive changes of the snowpack's internal energy, i.e. with an energy gain. In Sections 3.5.1–3.5.3, mass fluxes are integrated over a time period of one day, with precipitations (accumulation) and runoff (ablation) having opposite signs.

Table 3.2 *Detailed investigations of the energy balance over alpine snow covers and glaciers.*

| Surface | Location | Period | Surface fluxes[a] | | | $d\mathcal{H}/dt$[b] | Source |
|---|---|---|---|---|---|---|---|
| | | | $R_N$[c] | $H_S$ | $H_L$ | | |
| Melting snow cover | Weissfluhjoch, Switzerland (47° N) | May to Jun | −53.0[d] | −0.8 | 0.1 | 53.7 | 1 |
| | Finse, Norway (61° N) | 15 d in May | −25.0 | −21.4 | −0.3 | 46.7 | 2 |
| | Southern Alps, New Zealand (43° S) | Oct 28 to Nov 9 | −20.0 | −69.0 | −31.0 | 120.0 | 3 |
| | Sierra Nevada, USA. (37° N) | May | −44.0 | −50.0 | 54.0 | 40.0 | 4 |
| | | Jun | −93.0 | −92.0 | 71.0 | 114.0 | |
| Melting snow cover over glacier | Peyto Glacier, Canada (52° N) | 14 d in Jul | −79.8 | −87.0 | −14.5 | 181.3 | 5 |
| | Mt. Blanc area, French Alps (46° N) | 23 d in Jul | −20.8 | −5.0 | 4.9 | 20.9 | 6 |
| Over glacier under melting condition | Urumqi Glacier N° 1, Tien Shan (42° N) | Jun to Aug | −54.0 | −12.0 | 14.0 | 52.0 | 7 |

[a] Average surface fluxes in W $m^{-2}$ over the given period.
[b] ($d\mathcal{H}/dt$) is the net change rate of the snowpack's internal energy per unit area, which is the negative sum of net radiative ($R_N$), sensible heat ($H_S$) and latent heat ($H_L$) fluxes, but neglecting advective and ground heat fluxes (see Equation 3.1).
[c] The net radiative flux ($R_N$) is the sum of net shortwave radiation flux ($S_N$) and net longwave radiation flux ($L_N$).
[d] $S_N = -79.0$ W $m^{-2}$; $L_N = 26.0$ W $m^{-2}$

References: 1, Plüss (1997); 2, Harding (1986); 3, Moore and Owens (1984); 4, Marks and Dozier (1992); 5, Föhn (1973); 6, de la Casinière (1974); 7, Calanca and Heuberger (1990).

### 3.5.1 *High-elevation alpine snow cover*

*Christian Plüss, Charles Fierz, and Paul M. B. Föhn*

*Relevance and characteristics*

High alpine seasonal snow covers are present in mountainous areas around the world. The duration and spatial distribution of alpine snow covers is extremely variable and depends mainly on the geographical location, the climatic conditions, and the elevation of the mountain range. Alpine snow covers are of large economic and social importance in many areas, for example as a water resource for hydropower or as a base for tourism. In all alpine areas the snow cover is an important climate element because of its high albedo and low surface temperature. However, seasonal alpine snow covers may also cause natural hazards such as avalanches or flooding. The snow cover of the Alps – situated in the heart of densely populated Europe – meets all the above-mentioned issues and therefore considerable research effort has been spent towards its investigation.

The central part of the Alps is seasonally snow covered from about December to April above an altitude of 1000 m. The snow cover is highly variable in time and space because of both the complex topography and large differences in elevation within the mountain range. Wind influence as well as variable elevation, slope angles, and surface conditions lead to a highly structured snowpack being spatially inhomogeneous at scales as small as a slope, contrary to the Arctic, Antarctic, or prairie snowpack. Accordingly, the distribution of snow accumulation is very difficult to investigate (Elder *et al.*, 1989; Sturm *et al.*, 1995).

*Site*

The site of Weissfluhjoch lies in the eastern Swiss Alps near the town of Davos. The measurement site is located at a horizontal site in a southeasterly slope at an altitude of 2540 m a.s.l. (46.83° N, 9.81° E). On this well-equipped site, daily manual observations have been performed since 1936 and today most relevant nivo-meteorological parameters are measured automatically at a half-hourly time step (down- and up-welling shortwave and longwave radiation, air temperature, humidity, wind speed, snow surface temperature, snow depth). The snow cover lasts on average nine months (269 days), from mid-October to mid-July, and the mean snow depth reaches its maximum around mid-April (221 cm). The mean maximum snow water equivalent amounts to 857 mm w.e.

*Energy balance*

Energy balance investigations over an alpine snow cover were performed mainly in the European Alps and the North American Rocky Mountains. Many of these studies were performed over snow-covered glaciers under melt conditions (see Tables 3.2 and 3.3). All these investigations show that, on a daily basis, net radiation is the primary energy source, while turbulent fluxes are generally of minor importance.

Along with the daily means of air temperature, wind, and albedo, the daily means of energy fluxes at the surface and mass balance at Weissfluhjoch (eastern Swiss Alps) are presented in Fig. 3.5. Using the aforementioned measured forcing data, the snow-cover model SNOWPACK (Bartelt and Lehning, 2002; Lehning *et al.*, 2002a, 2002b) computes the energy balance, allowing for the extracting of parameterized turbulent fluxes.

Despite the high albedo of the snow cover, the shortwave net radiation flux $S_N$ is the dominant energy source for the snow cover during most of the investigation period. Longwave net radiation flux $L_N$ is, in general, an energy loss and depends mainly on cloud conditions. During the ablation period, net radiative flux $R_N$ is by far the dominant energy source for snowmelt. The magnitude of the turbulent fluxes of sensible and latent heat, $H_S$ and $H_L$, respectively, are very small on average, but the daily mean values may exceed the magnitude of net radiative flux and are therefore not negligible for the investigation of short-term processes. At Weissfluhjoch, the small average magnitude of the turbulent fluxes is attributed firstly to the relatively small wind speed at this site and secondly to the frequent changes of the weather patterns, which lead to changes of the sign of these fluxes.

### *Modeling aspects*

The high variability of the alpine snow cover in space and time proves to be very difficult to model. Accumulation depends largely on wind influence and small-scale precipitation differences may lead to very inhomogeneous conditions. The energy fluxes at the snow surface are highly variable due to topographic influence on the radiation fluxes and due to high variability of the turbulent fluxes. Dozier (1980) presented a spatially distributed model for shortwave radiation, Plüss and Ohmura (1997) proposed a modeling approach for longwave radiation. For the turbulent fluxes only very simple models have been proposed so far, despite the fact that over melting snow, the turbulent fluxes may be of considerable importance (Olyphant and Isard, 1988).

For ablation estimation, hydrologic models (see Kirnbauer *et al.*, 1994) have proven to successfully model the spatial distribution of ablation. Fierz *et al.* (1997) showed that snow temperature profiles could be modeled on several aspects using a distributed energy balance model to drive a point snow-cover model.

Table 3.3 *Energy balance over snow covers for selected time periods.*

| Type of snow cover | Location | Period | Surface fluxes[a] | | | | | |
|---|---|---|---|---|---|---|---|---|
| | | | $S_N$ | $L_N$ | $R_N$[c] | $H_S$ | $H_L$ | $d\mathcal{H}/dt$[b] |
| High Alpine | Weissfluhjoch, Switzerland (47° N) | Nov to Dec 95 | −13.7 | 31.9 | 18.2 | −19.7 | 5.0 | −3.5 |
| | | Jan to Apr 96 | −29.7 | 46.8 | 17.1 | −19.4 | 7.5 | −5.2 |
| | | May to Jun 96 | −86.7 | 40.8 | −45.8 | −28.0 | −2.5 | 76.3 |
| Middle Alpine | Col de Porte, France (61° N) | Jan 95 | −4.9 | 11.2 | 6.3 | −6.1 | −2.0 | 1.8 |
| | | Apr 95 | −46.9 | 12.5 | −34.4 | −11.0 | −4.1 | 49.5 |
| | | Jan to Apr 95 | −21.8 | 14.7 | −7.1 | −7.7 | −2.1 | 16.9 |
| | | Jan 94 | −6.6 | 16.0 | 9.4 | −7.0 | −2.4 | 0.0 |
| | | Apr 94 | −25.3 | 13.8 | −11.5 | −8.4 | −2.5 | 22.4 |
| | | Jan to Apr 94 | −22.0 | 17.4 | −4.6 | −9.9 | −2.0 | 16.5 |
| Snow-covered sea ice | High Arctic ice flow, North Pole 4, (>85° N) | May to Jun 56 | −48.0 | 26.3 | −21.7 | −11.6 | 12.3 | −2.2 |
| | | Jul to Aug 56 | −36.7 | −14.6 | −22.1 | 2.9 | 8.4 | 10.8 |
| | | Sep to Oct 56 | −2.5 | 13.4 | 10.9 | −1.6 | 1.0 | −10.3 |
| | | Nov to Dec 56 | 0.0 | 11.9 | 11.9 | −9.3 | 1.3 | −3.9 |
| | | Jan to Mar 57 | −0.4 | 16.5 | −16.1 | −10.7 | 0.8 | −6.2 |

[a] Average surface fluxes in W $m^{-2}$ over the given period.
[b] $(d\mathcal{H}/dt)$ is the net change rate of the snowpack's internal energy per unit area, which is the negative sum of net radiative ($R_N$), sensible heat ($H_S$) and latent heat ($H_L$) fluxes, but neglecting advective and ground heat fluxes (see Equation 3.1).
[c] The net radiative flux ($R_N$) is the sum of net shortwave radiation flux ($S_N$) and net longwave radiation flux ($L_N$).

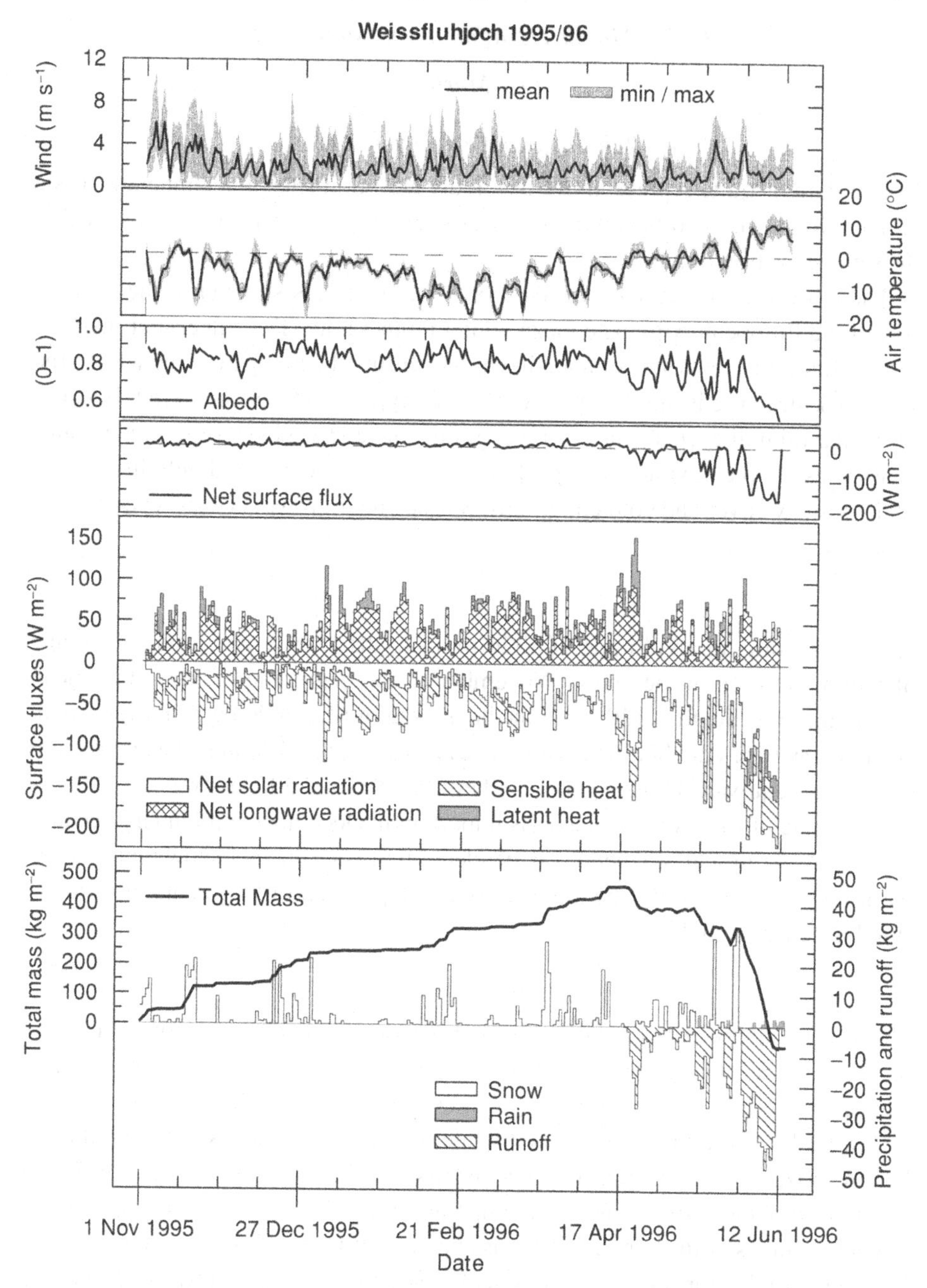


Figure 3.5. Daily means and, where appropriate, daily minima and maxima (shaded band) at Weissfluhjoch during the winter 1995/96. The snow cover was continuous from November 2, 1995 to June 11, 1996. Net surface flux is the sum of surface fluxes, which corresponds to the net negative change rate of the snowpack's internal energy per unit area ($-d\mathcal{H}/dt$) neglecting advective and ground heat fluxes (cf. Equation 3.1). Total mass is the accumulated difference of precipitations (snow and rain) to runoff, neglecting sublimation and evaporation (cf. Equation 3.4).

### 3.5.2 *Middle elevation alpine snow cover*

*Eric Martin*

#### *Relevance and characteristics*

Col de Porte is quite representative of middle elevation sites in a temperate alpine climate. Many ski resorts and hydropower stations in the Alps are located within the same elevation range.

The climate is wet (2000 mm w.e. $yr^{-1}$) because it is situated in the western part of the European Alps. The mean winter temperature (December–February) is −1 °C. Rainfall events are common in winter. During the first part of the winter, the snowpack structure is highly variable; all grain types can be encountered. Because of regular rainfall events, a layer of wet grains topped by a layer of rounded grains constitutes a typical winter snowpack. Depth hoar is encountered only in cases of shallow snow cover and cold conditions. Surface hoar growth occurs several times in winter.

#### *Site*

At Col de Porte, at an altitude of 1320 m a.s.l., the snow cover lasts an average of five months, from the end of November to the beginning of May. The mean maximum snow depth is 130 cm at the beginning of March. Snowmelt events may occur at any time but ablation usually takes place after mid-March. The maximum snow water equivalent varies generally between 200 and 500 mm w.e.

Screening of shortwave radiation is important in December and January because of trees. At large scales, albedo is dependent on the presence of snow on branches but it may sometimes also be affected by the deposition of fragments of needles. Wind is generally light.

#### *Energy balance*

The site is equipped with shortwave and longwave radiation sensors. Air temperature, humidity, and wind speed are also measured at hourly time steps. Energy balance investigations are made using the snow model CROCUS (Brun *et al.*, 1989, 1992). Radiation terms are measured while the model calculates turbulent fluxes. Parameterization of the latter fluxes is discussed in Martin and Lejeune (1997).

Along with the daily means of air temperature, wind and albedo, the daily means of energy fluxes at the surface and mass balance at Col de Porte are presented in Figs. 3.6 and 3.7 for the winter seasons 93/94 and 94/95, respectively. Turbulent fluxes (sensible and latent heat) transfer heat from the atmosphere to the snow cover as the atmosphere is usually stable. March 1994 was very warm and snowmelt occurred throughout this month. On the contrary, the first part of April was cold and snowy before the final melting period. In 94/95, melting only occurred in April.

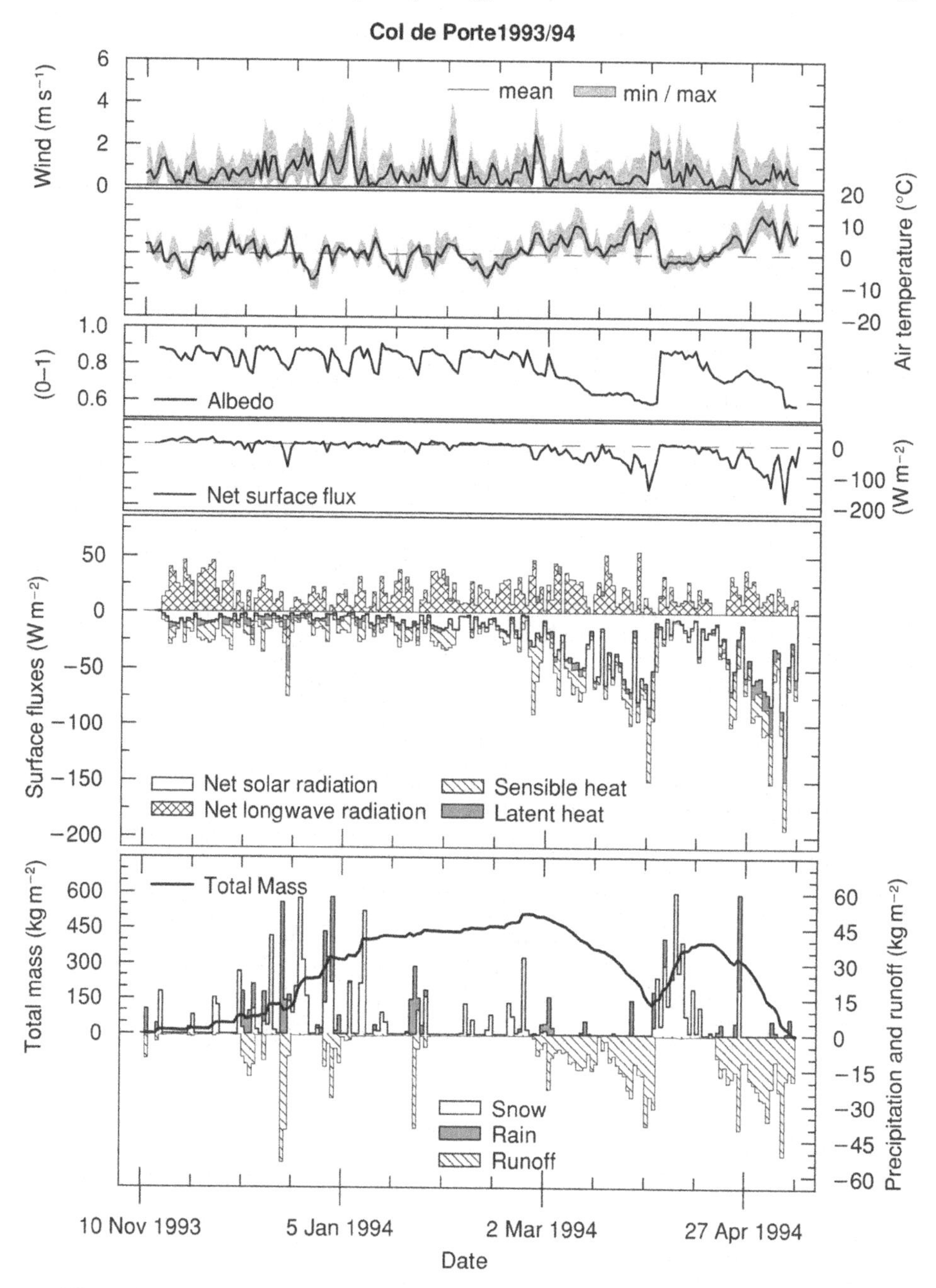


Figure 3.6. Daily means and, where appropriate, daily minima and maxima (shaded band) at Col de Porte during the winter 93/94. The snow cover was continuous from November 13, 1993 to May 5, 1994. Net surface flux is the sum of surface fluxes, which corresponds to the net negative change rate of the snowpack's internal energy per unit area ($-d\mathcal{H}/dt$) neglecting advective and ground heat fluxes (cf. Equation 3.1). Total mass is the accumulated difference of precipitations (snow and rain) to runoff, neglecting sublimation and evaporation (cf. Equation 3.4).

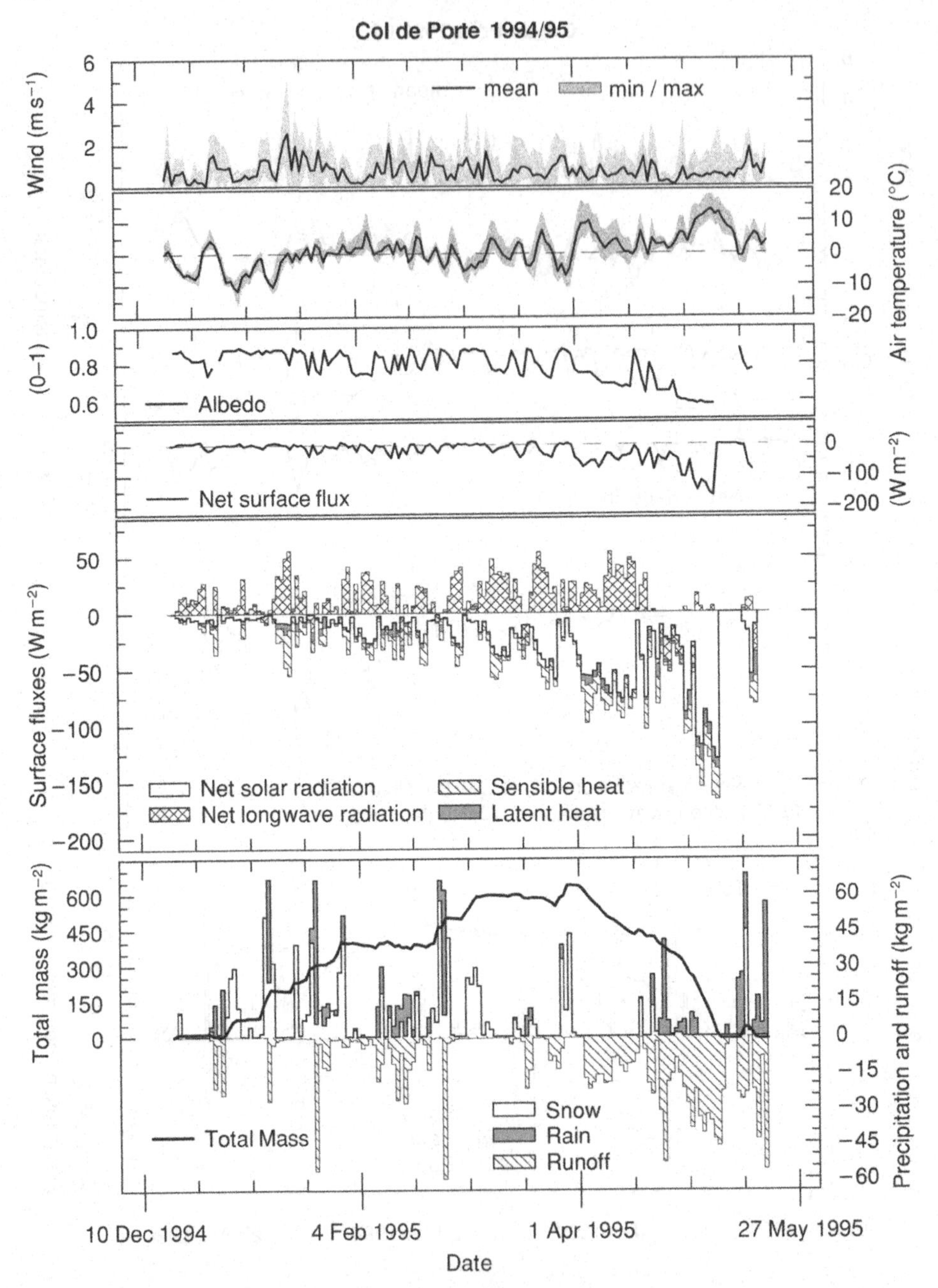


Figure 3.7. Daily means and, where appropriate, daily minima and maxima (shaded band) at Col de Porte during the winter 1994/95. The snow cover was continuous from December 31, 1994 to May 8, 1995. Net surface flux is the sum of surface fluxes, which corresponds to the net negative change rate of the snowpack's internal energy per unit area ($-d\mathcal{H}/dt$) neglecting advective and ground heat fluxes (cf. Equation 3.1). Total mass is the accumulated difference of precipitations (snow and rain) to runoff, neglecting sublimation and evaporation (cf. Equation 3.4).

The magnitude of net shortwave radiation is large during the melting period (March 94 and April 95). The variations of the net longwave radiation are more complex.

*Modeling aspects*

Turbulent fluxes are probably the most difficult factor to take into account at this site because of the complex topography. All types of snowpack can be encountered at this site: cold or completely wet even in winter, which induces difficulties in modeling the structure and layer texture of the snowpack.

### 3.5.3 Snow-covered sea ice on a high arctic ice flow

*Rachel E. Jordan*

*Relevance and characteristics*

Between 1937 and 1991, Russian scientists embarked on 31 "North Pole" field experiments within the Arctic Ocean. With the recent relaxing of East–West relations, the extensive set of meteorological, oceanographic and ice flow data from these expeditions has been made available to Western scientists on a CD-ROM (National Snow and Ice Data Center, 1996). There is now a general consensus that energy exchange over the polar ice caps is of critical importance to long-term climatological change. A comparison of historical data with that from the recent SHEBA expedition (Persson *et al.*, 2002; Uttal *et al.*, 2002) may reveal emerging trends in the surface energy balance of the Arctic Ocean.

*Site*

The Russian drifting station North Pole 4 (NP-4) was within 5° latitude of the North Pole from April 1956 until April 1957. Instrumentation at the NP-4 site is described by Kucherov and Sternzat (1959), Marshunova and Mishin (1994), and Jordan *et al.* (1999). The NP-4 site was characterized by high winds, fine-grained, dense snow, and a relatively shallow snowpack. Spring, summer, and winter seasons had distinct characteristics. Much of the snow was lost through melting during the summer and melt ponds formed in the sea ice, causing extreme variability in albedo. Cloud cover was heavy over the melting snow and the air temperature remained near 0 °C. In contrast, the amplitude of temperature swings associated with the passing of synoptic systems was up to 40 °C in winter (see Fig. 3.8).

*Energy balance*

Energy balance studies over arctic ice flows include works by Nazintsev (1963, 1964), Maykut (1982), Ebert and Curry (1993), Radionov *et al.* (1996), Lindsay (1998), and Uttal *et al.* (2002). Jordan *et al.* (1999) describe in detail the NP-4 data presented here in Fig. 3.8.

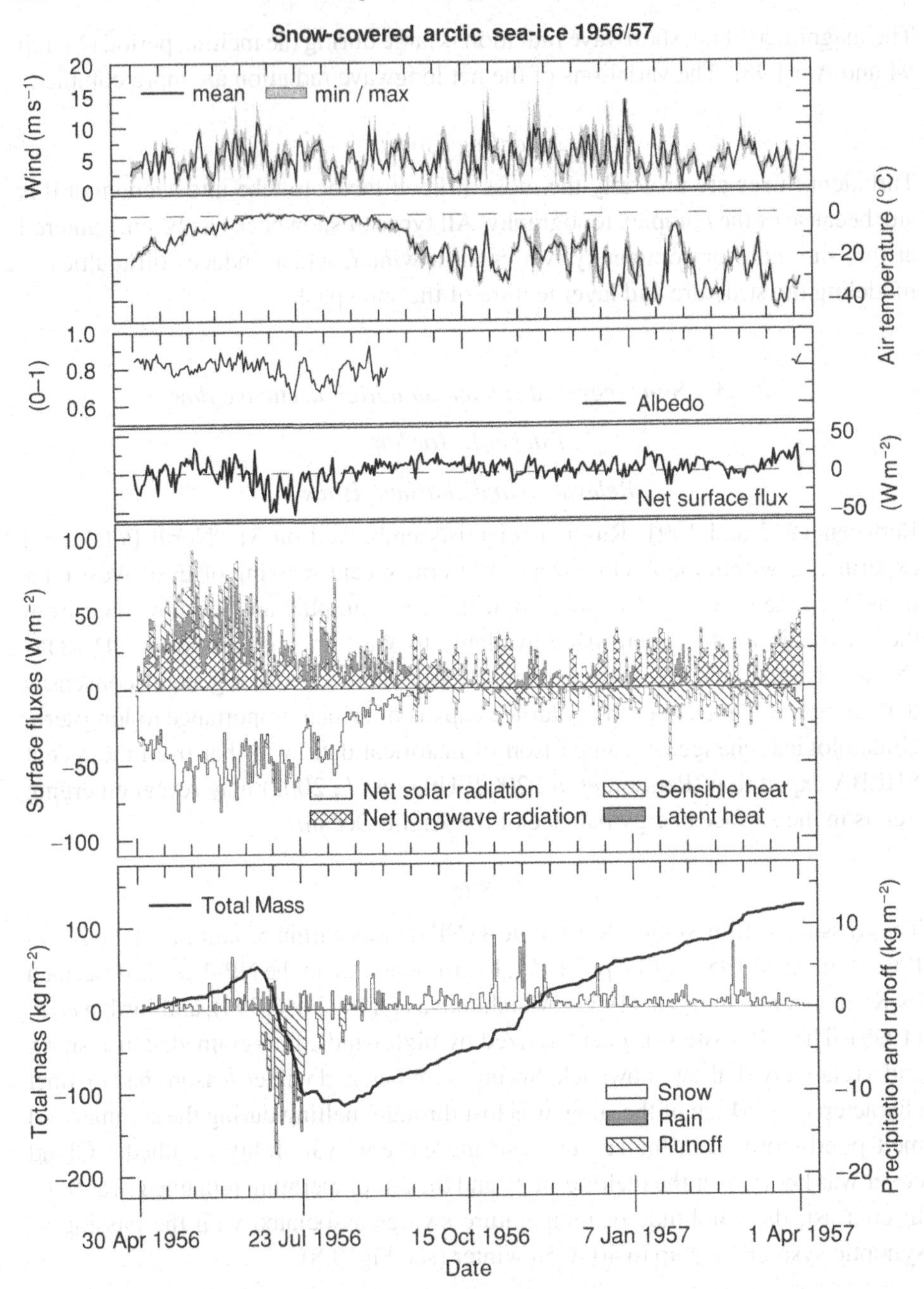


Figure 3.8. Daily means and, where appropriate, daily minima and maxima (shaded band) on snow-covered sea ice on a high arctic ice flow from April 30, 1956 to April 3, 1957. Net surface flux is the sum of surface fluxes, which corresponds to the net negative change rate of the snowpack's internal energy per unit area ($-d\mathcal{H}/dt$) neglecting advective fluxes and heat flux from the sea ice (cf. Equation 3.1). Total mass is the accumulated difference of precipitations (snow and rain) to runoff, neglecting sublimation and evaporation (cf. Equation 3.4).

Net shortwave radiation was the dominant energy flux to the snowpack during the late spring and summer. During these seasons at NP-4, the turbulent fluxes were close in magnitude and predominantly upwards, thus cooling the surface. Figure 3.8 shows a positive radiation balance after mid-September and simultaneously the appearance of downward spikes in the sensible heat flux, which increase in magnitude over the winter. The largest longwave radiation losses occurred under cloudless skies and were primarily compensated by sensible heat exchange. Latent heat fluxes were predominantly upward or evaporative. The magnitude of latent heat was small during the winter because saturation vapor pressure decreases exponentially with temperature.

Figure 3.8 shows a high snow albedo between 0.8–0.9 for new snow and a low albedo of 0.65 for older snow, near the value for bare ice. Snow cover persisted throughout the summer at this location, with depths ranging between 3 and 8 cm. About one-half of summer precipitation fell as snow and the increases in albedo reflect this in Fig. 3.8. The 1997–1998 SHEBA expedition, by contrast, reported loss of snow cover by early August. The remaining surface mix of bare ice and melt ponds had a much lower albedo than snow. Such dramatic alterations in albedo lead to an ice albedo feedback (Perovich *et al.*, 2002), which plays a key role in the energy budget of the high Arctic.

### *Modeling aspects*

High arctic sites are subject to blowing and drifting snow and exhibit considerable variability in snow depth. Thus, the wind transport of snow must be considered to build realistic snow covers. Precipitation measurements reported here are corrected for wind effects, evaporation, and wetting error (Yang, 1995; Jordan *et al.*, 1999). Because arctic snow is fine-grained and packed by wind, polar snowpacks are much denser than their temperate counterparts. The density of newly fallen snow should therefore be around 150–300 kg m$^{-3}$ and increase with wind speed.

Sturm *et al.* (2002) infer an effective thermal conductivity of 0.33 W m$^{-1}$ K$^{-1}$ from snow temperature profiles at SHEBA, which is higher than that for temperate snow and also higher than recent field measurements by Sturm *et al.* (1997, 2002). Sturm *et al.* (2002) conclude that mechanisms other than one-dimensional conduction may enhance the heat exchange within the polar snowpack. Jordan *et al.* (2003) suggest that high winds may induce wind ventilation in the upper 10–20 cm of the snowpack and thereby increase its effective thermal conductivity.

Persistent radiative losses in winter lead to periods when the standard stability correction for stable atmospheres can shut down sensible heat exchange. To avoid unrealistically low surface temperature predictions, Jordan *et al.* (1999) therefore replaced the usual log–linear stability function with one that maintains a minimal turbulent exchange.

### 3.5.4 Canadian prairies

*John W. Pomeroy and Donald M. Gray*

#### *Relevance and characteristics*

"Prairie" snow seasonally covers northern continental grain-growing and grassland regions of North America, Europe, and Asia. These snow covers impact both the social and the economic aspects of the region because: a large proportion of the population lives in rural districts, grain and livestock exports are globally important food supplies, the region experiences long and severe winters and summer water deficits, and agriculture and industry rely on long distance transport by road and rail (Steppuhn, 1981).

Although snowfall only comprises about one-third to one-half of annual precipitation, snowmelt runoff often exceeds 90% of annual streamflow (Gray, 1970). Spring floods caused by prairie snowmelt water are the most economically destructive natural phenomena in the U.S.A. and Canada and defensive measures such as large "ring-dikes" are used to protect major cities such as Winnipeg, Manitoba.

On the prairies, blowing snow erosion and sublimation may result in increased water deficits, termed "northern desertification" in Russia (Dyunin *et al.*, 1991), which may be enhanced by the suppression and removal of natural vegetation. In many areas, blowing snow storms result in winter restrictions on transportation and a specialized design of infrastructure to minimize snow removal costs and snow load damage (Tabler and Schmidt, 1986.)

Prairie snow covers persist from November to April. They are generally cold, dry, and wind-packed. Blowing snow causes redistribution several times during a winter season and results in dense (generally greater than 250 kg $m^{-3}$), crusted and variable (coefficient of variation of SWE 0.3–0.58) snow covers (Pomeroy *et al.*, 1998b). Snow accumulation is very sensitive to vegetation cover and topography, and the depths in sheltered sites can be five or six times that on exposed sites (Steppuhn and Dyck, 1974; Pomeroy *et al.*, 1993). In early spring the energy for melting snow is largely derived from shortwave radiation and as melt progresses the magnitude of net radiative flux generally increases due to the increase in magnitude of incoming shortwave flux and the decrease in areal albedo due to the decreases in snow depth and in snow-covered area (O'Neill and Gray, 1973). Advection of sensible heat from bare ground to snow has been shown to play an important role in the melting of a patchy snow cover (Shook and Gray, 1997). As a result, areal melt rates are greatest when the snow-covered area is between 40% and 60% (Shook, 1995). Ground heat flux is negligible during melt because of the infiltration of meltwater into frozen soils, which leads to the release of latent heat upon freezing and very small temperature gradients near the soil surface (Zhao *et al.*, 1997; Pomeroy *et al.*, 1998b).

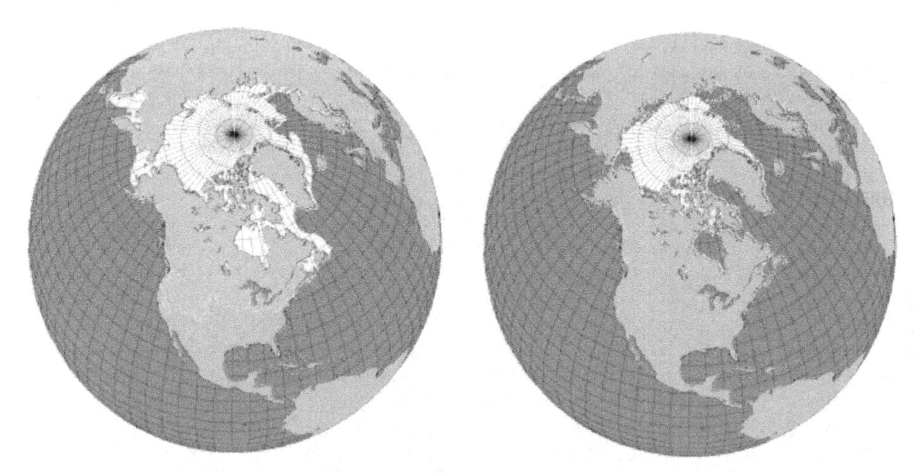

Plate 1.2. Mean seasonal variation in snow (gray) and sea-ice cover (white) between February (left) and August (right) as derived from satellite data. Data from NSIDC "Weekly Snow Cover and Sea Ice Extent," CD-ROM, NSIDC, 1996.

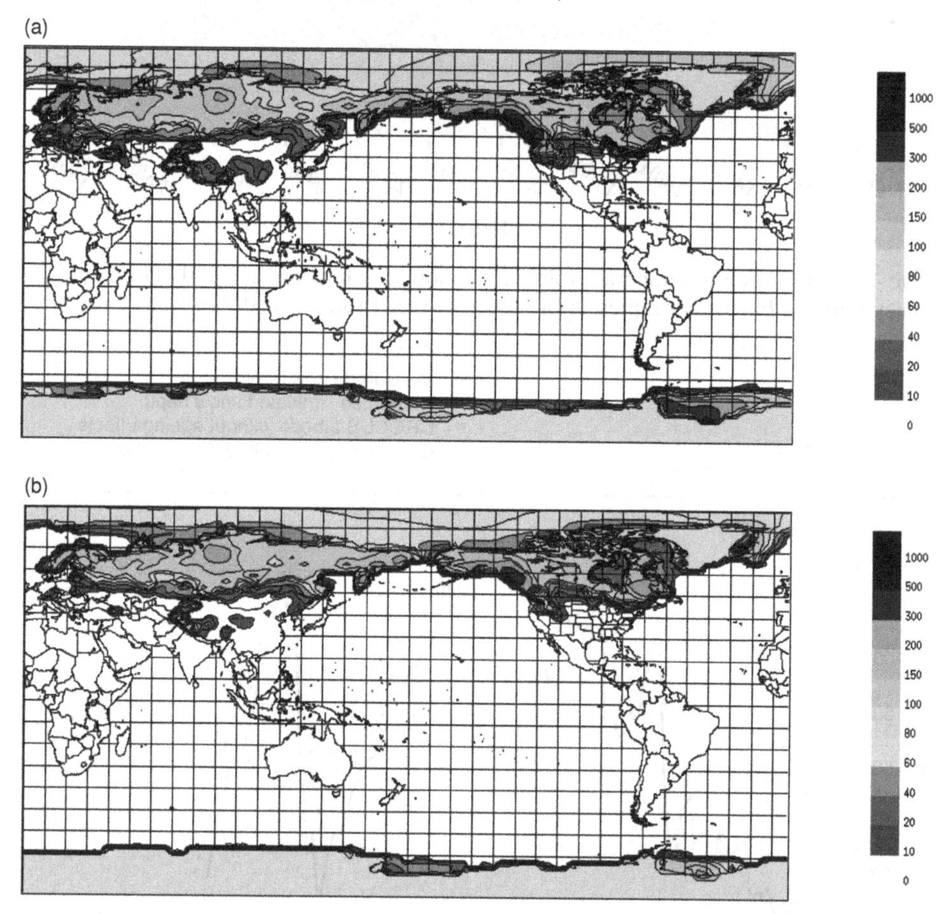


Plate 1.4. Comparison of mean March winter SWE (mm) simulated by the Canadian coupled global climate model (CGCM3) for the 1981–2000 "current climate" period (a) with simulated mean SWE for the 2081–2100 period (b) based on the SRES A2 emission scenario. Data courtesy of the Canadian centre for climate modeling and analysis.

A colour version of these plates is available for download from www.cambridge.org/9780521130653

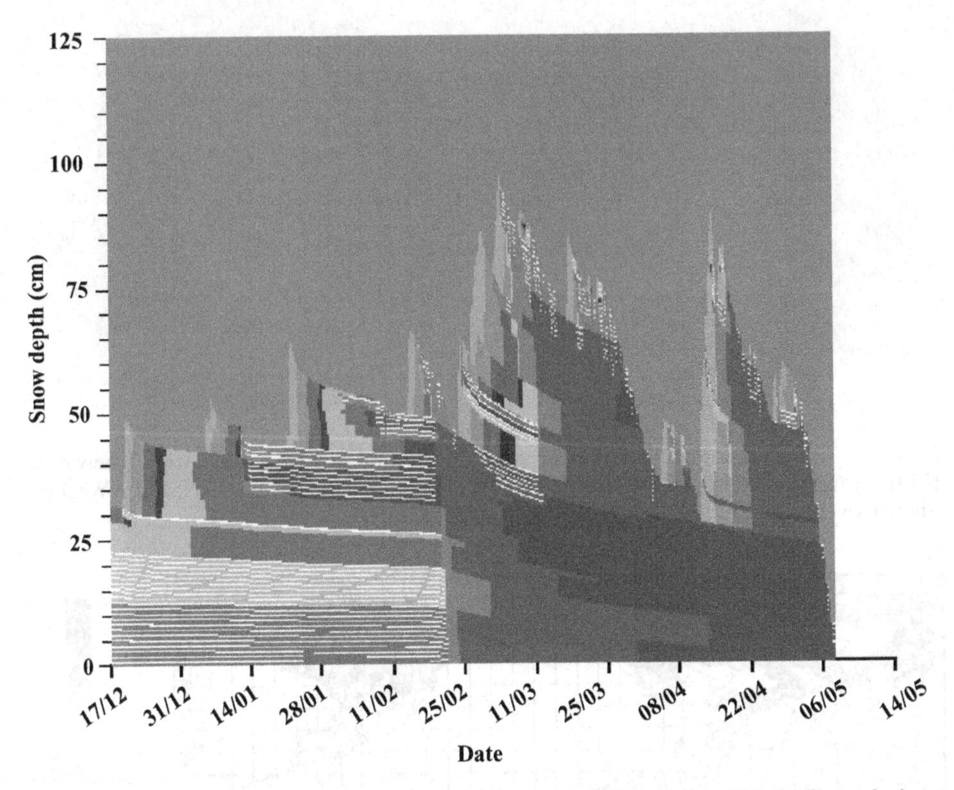


Plate 4.1. Simulation of temporal evolution of snowpack layering at Col de Porte during winter 1998/99. Each color represents a snow type (see Brun *et al.*, 1992).

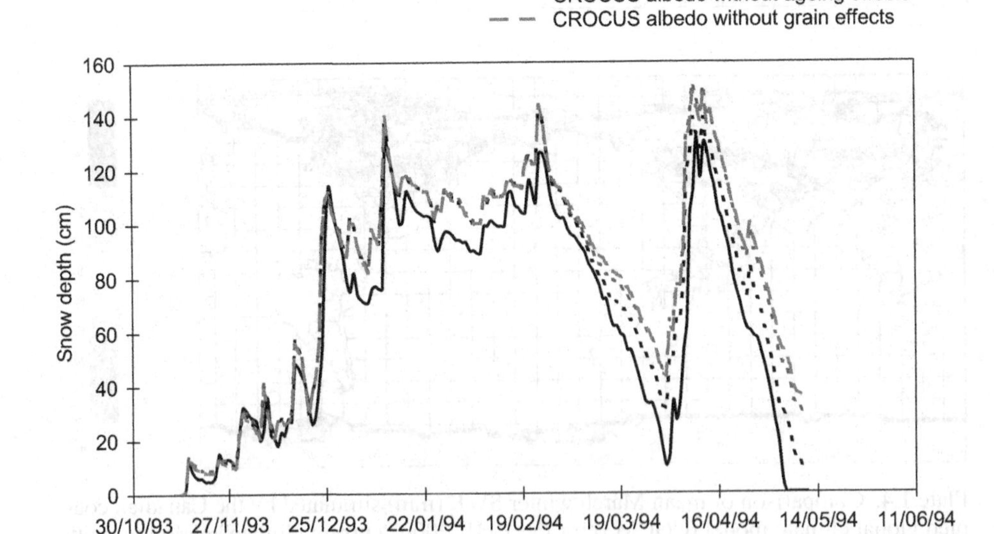


Plate 4.2. Sensibility to ageing and grain size in albedo calculation on snow depth simulations.

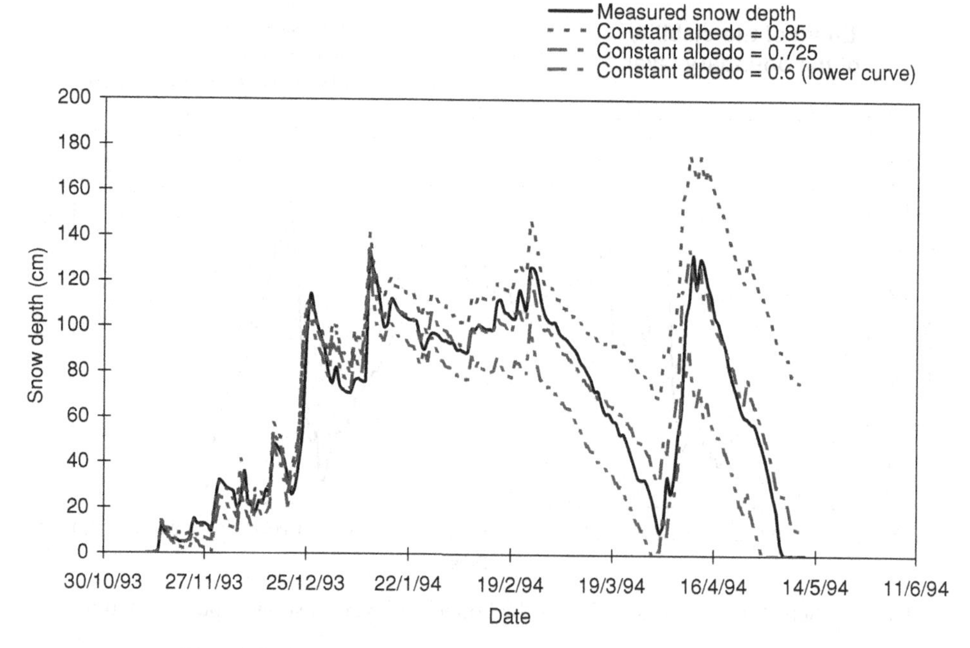


Plate 4.3. Sensibility to albedo on snow depth simulations.

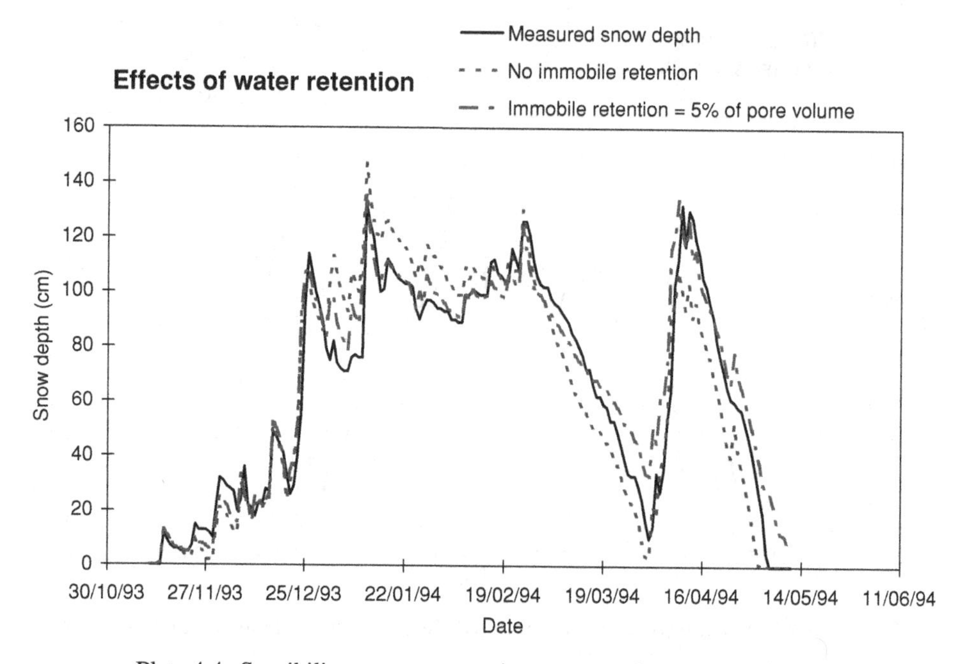


Plate 4.4. Sensibility to water retention on snow depth simulations.

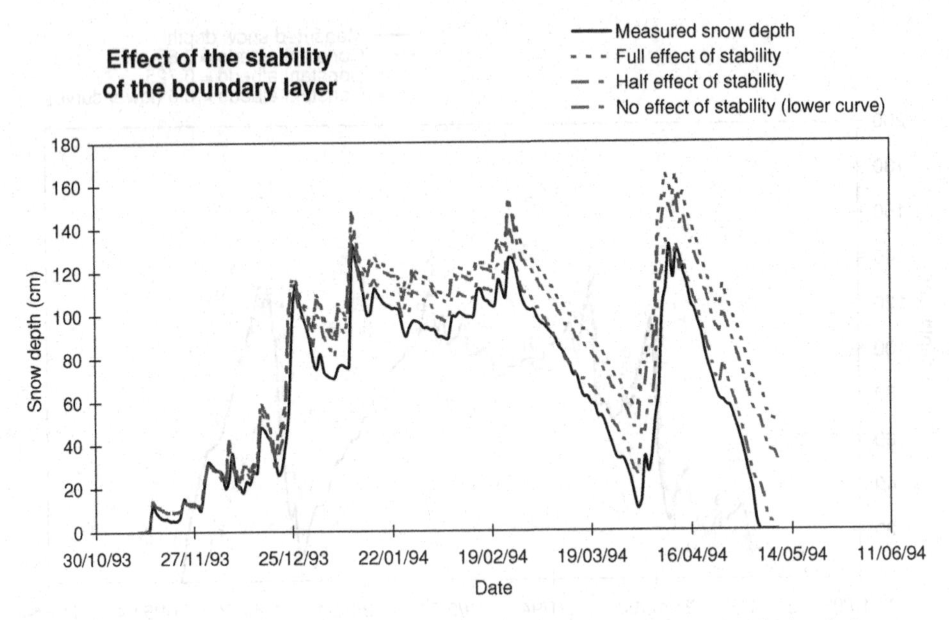


Plate 4.5. Sensibility to the stability of the boundary layer on snow depth simulations.

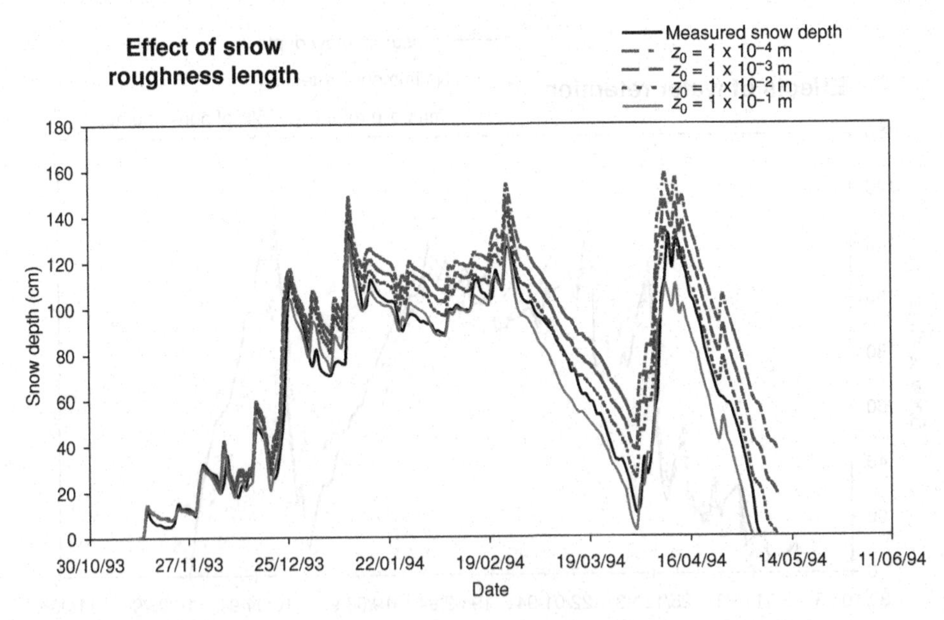


Plate 4.6. Sensibility to snow length roughness on snow depth simulations.

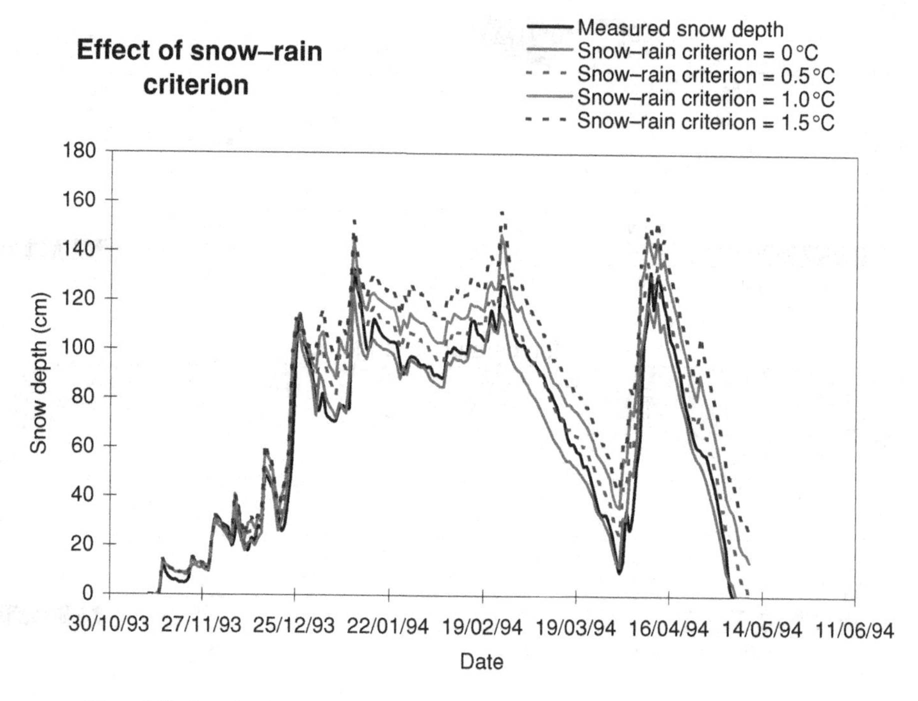


Plate 4.7. Sensibility to snow-rain criterion on snow depth simulations.

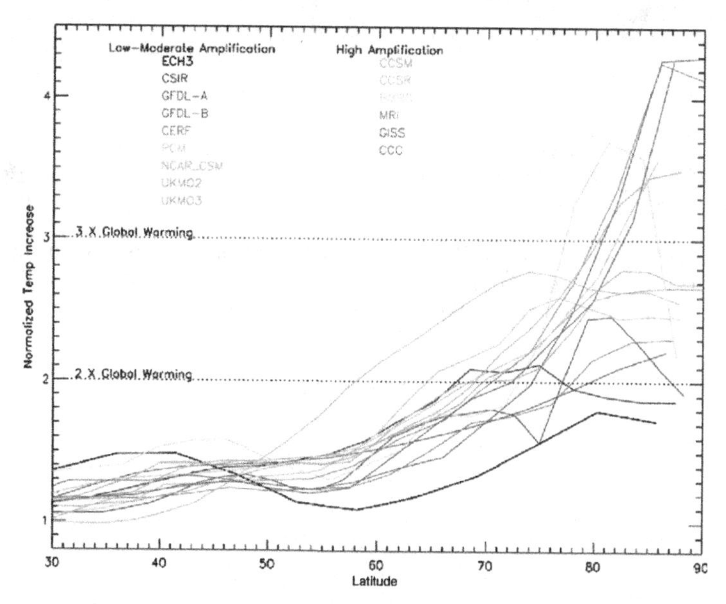


Plate 4.14. Surface air temperature change in 15 coupled GCMs that have doubled atmospheric $CO_2$ concentrations. All models show "polar amplification" and enhanced warming in the Arctic compared to tropical latitudes. Taken from Holland and Bitz (2003).

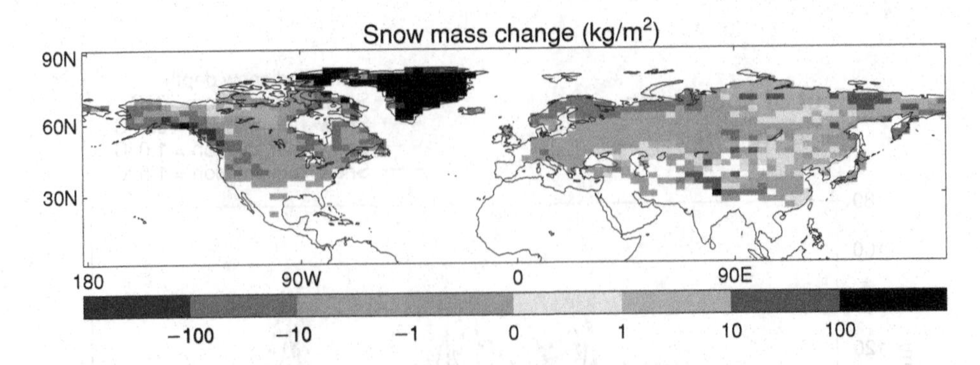


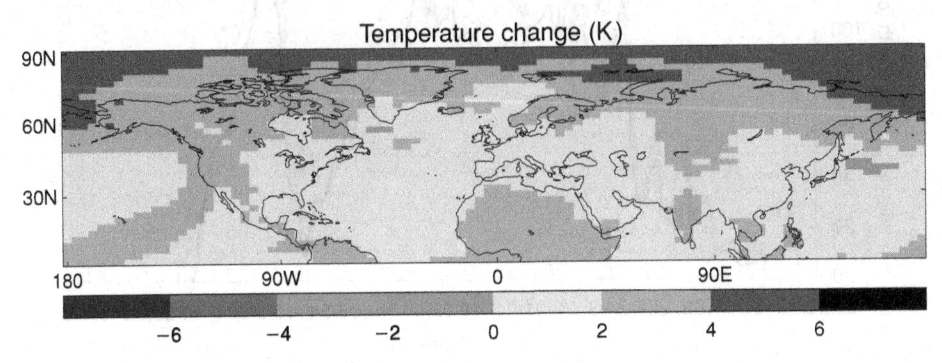


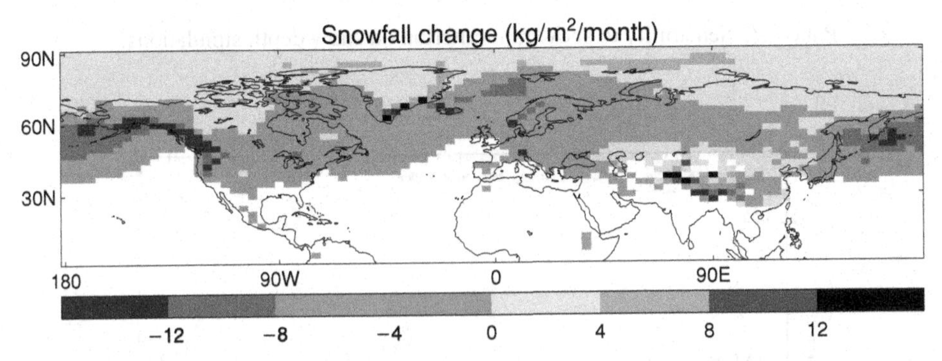


Plate 4.19. Differences between 2030–2050 averages from the climate-change simulation and 130-year averages from the control of the Hadley GCM for (a) snow mass (b) temperature and (c) snowfall. Taken from Essery (1997).

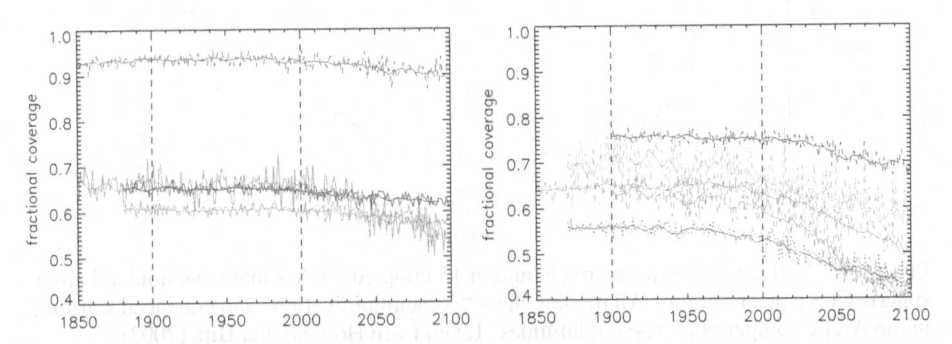


Plate 4.20. Annual time series (thin line), overlaid with nine-year running means (thick line), of ensemble-mean January North American snow cover extent, including both twentieth century and twenty-first century scenarios, for nine available coupled atmosphere-ocean GCMs. Taken from Frei and Gong (2005).

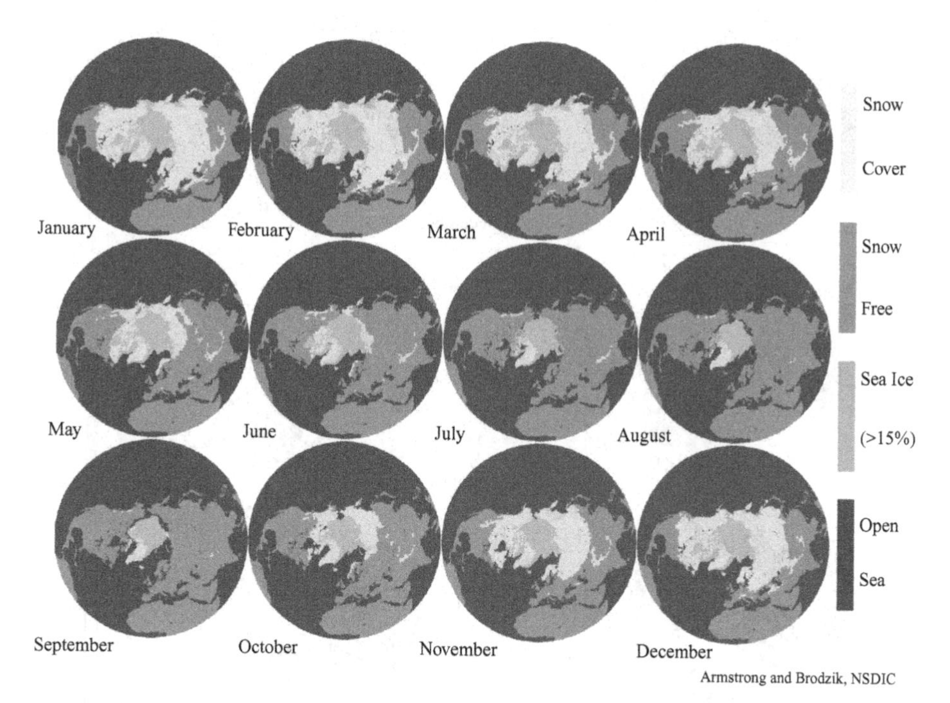


Plate 5.2. Monthly Northern Hemisphere snow cover (1966–2005) and sea ice extent (1978–2005) climatologies (Source: NSIDC *Northern Hemisphere EASE-Grid Weekly Snow Cover and Sea Ice Extent Version 3, 2005*).

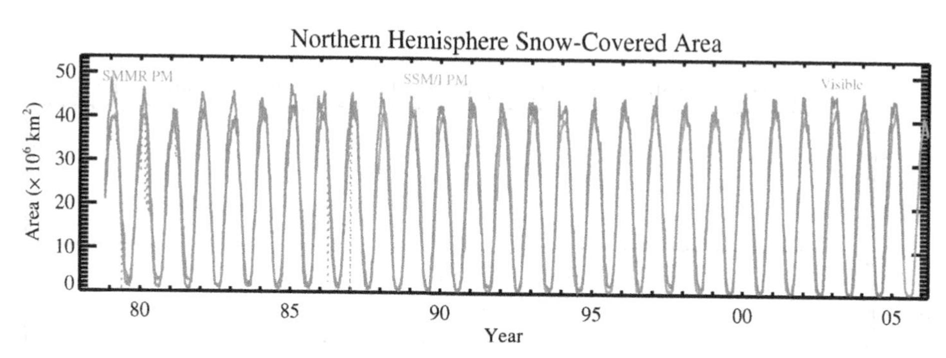


Plate 5.3. Northern Hemisphere monthly SCA, 1978–2005, from NOAA snow charts (orange) and microwave satellite (purple/green) data sets.

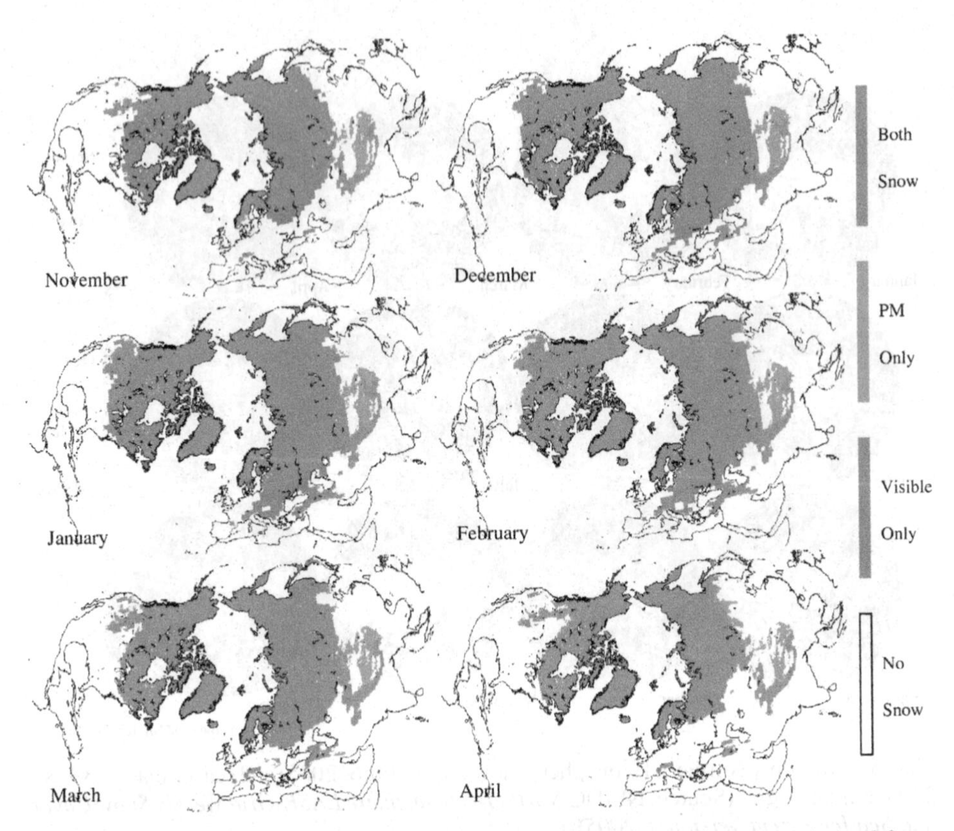


Plate 5.4. Comparison of mean monthly Northern Hemisphere snow extent derived from visible and passive microwave satellite data, 1978–2005 (50% or more of the weeks in the particular month over the total time period classified as snow covered).

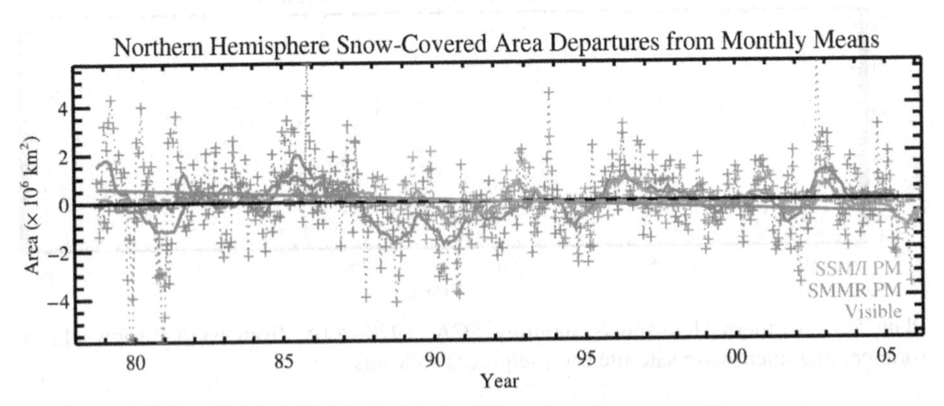


Plate 5.5. Northern Hemisphere SCA departures from monthly means, 1978–2005, from NOAA snow charts (orange) and microwave satellite (purple/green) data sets. The NOAA time series for this period exhibits a significant decreasing trend of −2.0% per decade (solid orange line); the microwave snow-cover time series exhibits a decreasing trend of −0.7% per decade that is not significant at a 90% level (dashed green line).

*Site*

Kernen Farm (500 m a.s.l.) is east of the City of Saskatoon (52° N, 107° W) in the central southern half of the Province of Saskatchewan, Canada. The farm is situated on an open, flat, lacustrine plain, which is cropped to cereal grains and pulse crops under the practice of dryland farming (Shook and Gray, 1996). Trees are limited to farmyards, which are located several kilometers distant from the site. The climate is subhumid and typical of northern prairies with cold winters and continuous snow cover from late November to early April.

Experiments were conducted in December 1998 and March 1999 at level sites with continuous snow cover on fields of uniform short vegetation or fallow. Energy balance and related parameters were measured and recorded half-hourly using eddy correlation equipment (Gill Instruments "Solent" or Campbell Scientific "CSAT" sonic anemometers, Campbell Scientific "Krypton" hygrometers, fine wire thermocouple controlled by a datalogger for covariance calculation), Radiation and Energy Balance System "REBS" net radiometers and ground heat flux plates, "NRG40" cup anemometer, "Everest" infrared thermometer, "Vaisala" HMP35CF platinum resistance thermometers, Campbell Scientific "SR-50" ultrasonic snow depth gauge and University of Saskatchewan blowing snow particle counters (Brown and Pomeroy, 1989; Shook and Gray, 1997; Pomeroy *et al.*, 1998b, 1999b). During the measurement periods, the sites were frequently manned, which provided a high confidence in the observations.

*Energy balance*

Two energy balances and related measurements are shown in Figs. 3.9 and 3.10. As stated previously, negative values indicate downward fluxes. Figure 3.9 shows fluxes during an early winter snow accumulation period with blowing snow, Fig. 3.10 shows fluxes during a snow warming and melt sequence in spring. The snow accumulation period (Fig. 3.9) shows a characteristic prairie weather pattern of highly variable meteorology associated with the passage of frontal systems.

On 17 December the air temperature warmed to slightly above freezing, then dropped dramatically with strong winds to below −20 °C in about 12 hours. During this cooling period snowfall and blowing snow were recorded and the lower surface boundary layer remained well mixed. Over the next two days the temperature dropped below −30 °C (by 20 December) with lower wind speeds and a stable lower surface boundary layer forming. During short periods around mid-day, the magnitude of net radiation was small but negative (peak −20 to −90 W m$^{-2}$), however at other times it was positive, reaching 40 W m$^{-2}$, while staying smaller than 10 W m$^{-2}$ during cloudy nights. Turbulent fluxes are enhanced over that expected from smooth snow covers because of exposed vegetation (snow depth <10 cm,

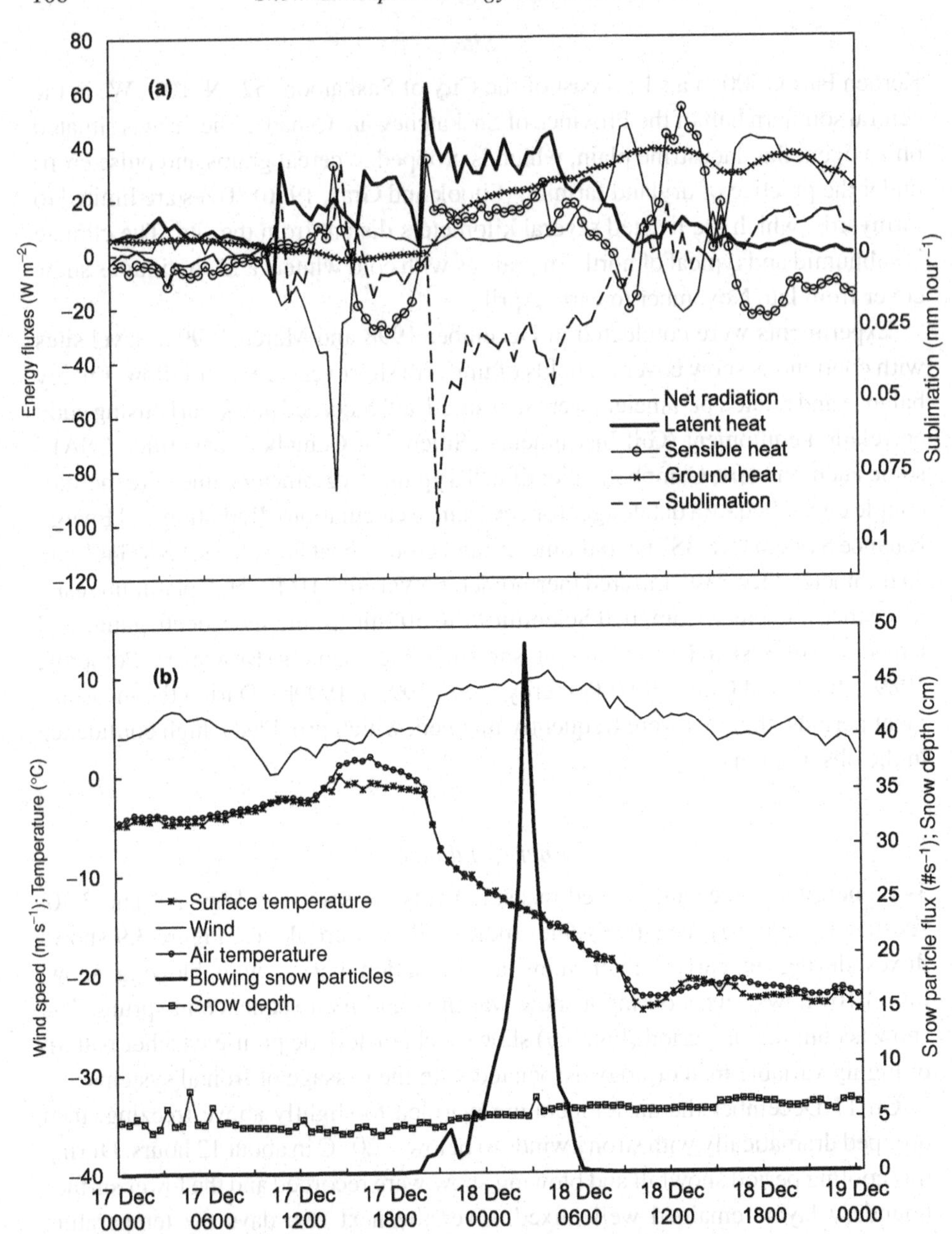


Figure 3.9. Direct measurements made during a snow accumulation period with blowing snow in December, 1998, Kernen Farm, Saskatoon, Saskatchewan, Canada. (a) Fluxes of net radiation, latent heat, and sensible heat, measured 1 m (net radiation) and 2 m (turbulent fluxes) above the snow cover, as well as ground heat flux measured 5 cm into the soil and (b) snow surface temperature, air temperature, and wind speed measured 1.3 m above the snow surface as well as blowing snow particle flux (measured 0.2 m above the snow surface) and snow depth.

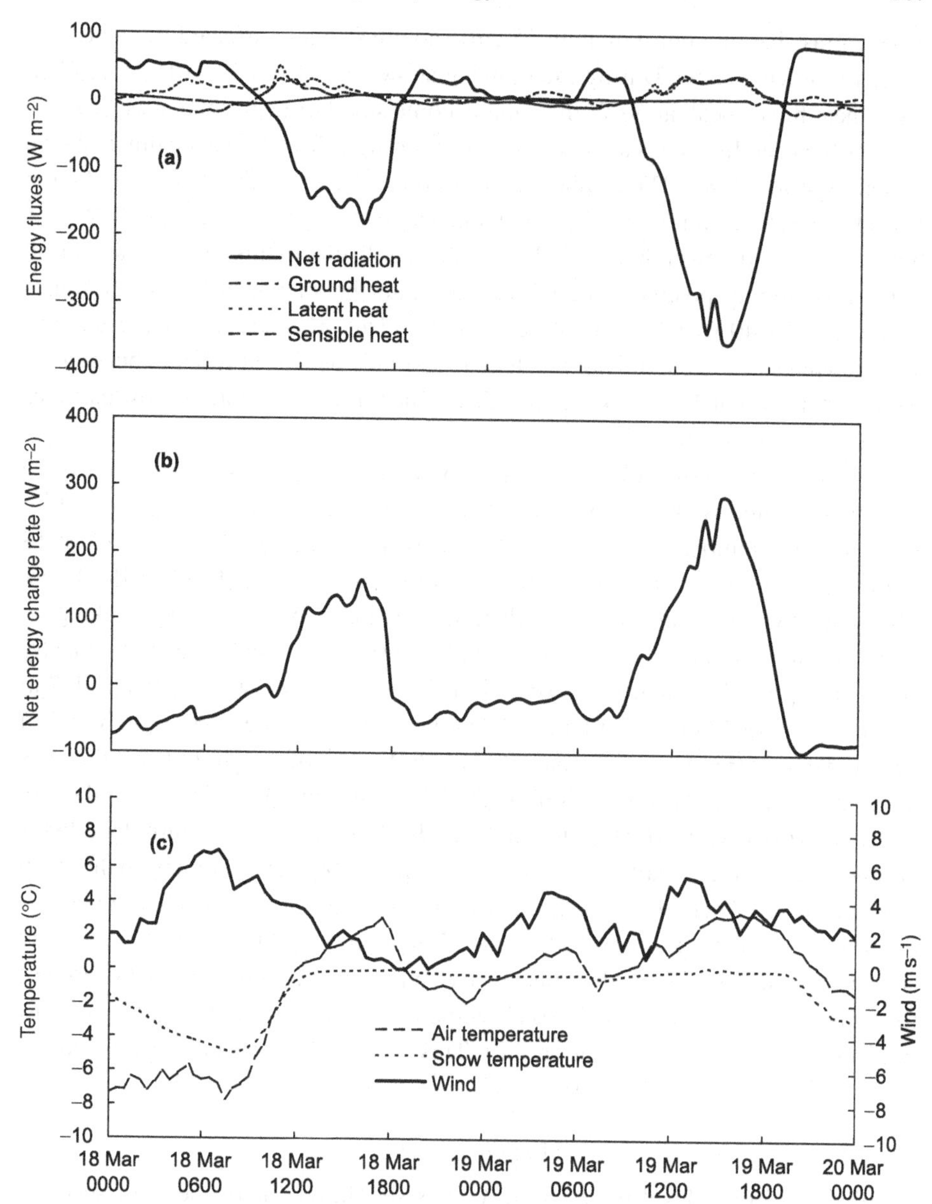


Figure 3.10. Direct measurements made during a snowmelt period in March, 1999, Kernen Farm, Saskatoon, Saskatchewan, Canada. (a) Fluxes of net radiation, latent heat, and sensible heat, measured 1 m (net radiation) and 2 m (turbulent fluxes) above the snow cover, as well as ground heat flux measured 5 cm into the soil, (b) net change rate of the snowpack's internal energy per unit area ($d\mathcal{H}/dt$). Positive values correspond to either warming or melting (cf. Equation 3.2), and (c) snow temperature at mid-pack depth as well as air temperature and wind speed measured 1.3 m above the snow surface.

patchy grass height approximately 25 cm) and are strongly affected by the occurrence of blowing snow. During the blowing snow event from 17 to 18 December, latent heat fluxes peak at 60 W $m^{-2}$ and become small or negligible after that.

Sensible heat flux peaks at up to 50 W $m^{-2}$ during cold, relatively calm, negative radiation periods (18–20 December). During cold, relatively calm, positive radiation periods, sensible heat is generally slightly negative, with peak values at −30 W $m^{-2}$ but most often not lower than −20 W $m^{-2}$. A notable flux in this early winter period is the consistently positive ground heat flux into the snowpack, which becomes prominent (30–40 W $m^{-2}$) as cooling of snow and air proceeds (18–20 December). The blowing snow event did not result in a large increase in snow depth but density increased from near 100 to 140 kg $m^{-3}$ due to the impact of saltating snow particles and subsequent sintering.

An exemplary snowmelt period (Fig. 3.10) occurred in March 1999 over a smooth, continuous snow cover (average depth = 14.5 cm, density = 330 kg $m^{-3}$). Internal temperatures show that the snowpack warmed from −5 °C on 18 March to isothermal conditions as temperatures reached nearly 0 °C on 19 March. The magnitude of daytime net radiation was large but negative, peaking at −182 W $m^{-2}$ on 18 March (cloudy) and −360 W $m^{-2}$ on 19 March (mostly clear). The preceding melt was a small blowing snow event (wind speed peak 7 m $s^{-1}$) early on 18 March. The largest latent heat fluxes of the period (50 W $m^{-2}$) occurred during this event. The magnitude (<6 W $m^{-2}$) of the ground heat flux was quite small on both days, becoming slightly negative and thus directed to the soil as the snowpack warmed. Despite wind speeds of up to 5.5 m $s^{-1}$, turbulent fluxes during the melt were small and similar in size (peak 40 W $m^{-2}$), much smaller than the magnitude of net radiation. The net energy change rate shows a large positive input to the snowpack on both days (see Equation 3.1). On 18 March warming snow temperatures suggest that this positive input increased the internal energy of the snowpack while constant internal temperatures on 19 March imply that most went into snowmelt (see Equation 3.2). This melt period was ephemeral and the snowpack returned to subfreezing conditions at the end of 19 March.

### *Modeling aspects*

Relatively uniform, level prairie snow covers should be one of the most successful types of snow for physical models; however, the complexity of prairie snow phenomena have resulted in several modeling challenges towards which long-term process research has been directed (Male and Gray, 1975, 1981; Pomeroy *et al.*, 1998b). Shook and Gray (1994) described the variation of depth and density and the influence of covariance between these properties in determining areal snow mass. Granger and Male (1978) measured the turbulent exchange over melting prairie snow and derived stability corrections for eddy diffusivities for water vapor and heat with respect to momentum transfer; these corrections dampened the turbulent exchange from the normal log–linear formulations (e.g. Webb, 1970). Shook (1993, 1995) studied the depletion of the snow-covered area during melt and showed that changes in areal albedo can be adequately explained by the decrease in the snow-covered area and the assumption that albedo does not change significantly during melt. Shook (1995) found that the advection of sensible heat could contribute substantial melt energy when snow cover was incomplete. Pomeroy *et al.* (1998b) examined the performance of certain land surface schemes for prairie snowmelt and found that turbulent fluxes and ground heat flux were generally overestimated.

### 3.5.5 Boreal forest

*John W. Pomeroy and Richard J. Harding*

*Relevance and characteristics*

The boreal forest in winter is a complex mosaic of land surface types varying from closed coniferous canopies, mixed-wood deciduous forests, sparsely vegetated areas (clearings, wetlands, clear-cuts, burns) to ice and snow-covered lakes. At roughly 20% of the earth's land area, the boreal forest is the largest terrestrial type of land cover and extends in a circumpolar band across North America, Europe, and Asia. Canada, Russia, and the Scandinavian countries are dominated by boreal forest, although in most cases it lies to the north of major population centers. Economic activities in the boreal forest center on forest harvesting, tourism, and mining, yet many boreal forests retain an indigenous aboriginal population whose members conduct aspects of a traditional hunting, fishing, and gathering lifestyle. The large number of lakes and rivers (up to 40% of some boreal regions are covered by water) promote fishing and water transport that has been fundamental to the development of northern Canada and Siberia. Recent environmental concerns focus on the extensive clearcutting of boreal forest in Canada and Russia, episodic acidic precipitation (including snow) from anthropogenic pollution, and the apprehension that climate warming will result in a major northward shift in the boreal climate zone and loss of forest lands in the southern boreal forest.

One distinction of the boreal forest in comparison to more temperate forests is its long snow-covered period and cold winter temperatures (Harding and Pomeroy, 1996). The depth, density, and duration of snow cover is ecologically important to mammals and various microbial life-forms in this forest; in some cases the snow cover provides a thermally moderated habitat, in others a means of avoiding predators (Jones *et al.*, 2001). Boreal forest productivity and carbon cycling are strongly influenced by the supply of available nitrogen and soil moisture. Snow influences productivity and carbon cycling by providing, upon melt and infiltration, a significant portion of the annual water and inorganic nitrogen input (Pomeroy *et al.*, 1999a). The global boreal forest exerts a strong control on climate and because of its low winter albedo, its removal and the resulting higher albedo in spring might result in a cooling of the Northern Hemisphere (Thomas and Rowntree, 1992). Snowmelt provides 40–60% of annual streamflow from boreal forests, with increases in snowmelt runoff of 24–75% when forest cover is removed (Hetherington, 1987).

Boreal forest snow covers are strongly influenced by the forest canopy, its interception of snow and radiation and dampening of wind speed and mixing above the snowpack's surface. Pomeroy *et al.* (1998a) observed in a mixed range of boreal forest cover types that 20–65% of cumulative snowfall was intercepted in early winter, and 10–45% of snowfall sublimated over the season. Leaf area strongly controls interception efficiency (Hedstrom and Pomeroy, 1998) and clearcutting or conversion of coniferous stands to deciduous species reduces interception to insignificant levels (Pomeroy and Granger, 1997). The energetics of intercepted snow in the boreal forest have been studied by Nakai *et al.* (1993, 1994, 1999), Lundberg and Halldin (1994), Harding and Pomeroy (1996), Pomeroy and Dion (1996), Pomeroy *et al.* (1998a) and Parviainen and Pomeroy (2000). These studies show that the albedo of snow-covered forest canopies is low ($<0.2$), that sublimation rates up to 3 kg $m^{-2}$ $d^{-1}$ are possible from snow-covered canopies and that the direction and magnitude of sensible and latent heat fluxes are influenced by the presence of snow in the canopy because it represents a "wetter," cooler surface than a snow-free canopy. Parviainen and Pomeroy (2000) suggest that sublimation is driven by local-scale advection of sensible heat from exposed branches heated in the sun to intercepted snow clumps and that the efficiency of this advection is related to the fractal geometry of intercepted snow clumps (Pomeroy and Schmidt, 1993).

Studies of snow under the canopy have been directed towards snow accumulation and melt prediction. Boreal forest snow covers have relatively low coefficients of variation of snow water equivalent (0.04–0.14) and maximum densities near 200 kg $m^{-3}$ (Pomeroy *et al.*, 1998b). At small scales, the snow water equivalent generally decreases with distance from coniferous tree stems (Woo and Steer, 1986; Jones, 1987; Sturm, 1992) and at stand scales it decreases with increasing canopy density (Kuz'min, 1960; Pomeroy and Gray, 1995). The forest cover attenuates the magnitude of incoming shortwave radiation and large-scale advection of warm air, reducing the "connectivity" between subcanopy snow and the atmosphere. Ni *et al.* (1997) and Pomeroy and Dion (1996) have measured and modeled winter subcanopy radiation and found its magnitude greatly reduced from the above canopy values and strongly dependent on solar zenith angle, leaf area, and needle orientation. Typically, subcanopy net radiation in a mature conifer stand is one-tenth that above the canopy at the time of snowmelt. Davis *et al.* (1997), Hardy *et al.* (1997a, b), and Metcalfe and Buttle (1995, 1998) have measured and modeled snow ablation under boreal forest canopies and conclude that though the magnitude of net radiation is strongly reduced in forests and decreases with increasing canopy density, it still comprises the largest component of the energy balance because the subcanopy snow albedo drops substantially during melt as forest leaf litter and debris in the snowpack are exposed and because subcanopy turbulent fluxes are extremely small in magnitude and usually of the opposite direction. Faria *et al.* (2000) suggest that because subcanopy snowmelt energy and the pre-melt snow water equivalent have a spatial covariance, depletion of the snow-covered area is accelerated as the covariance increases. Pomeroy and Granger (1997) compared melt rates in various forest types and found that melt timing was accelerated three-fold in a clear-cut compared to under a mature boreal forest canopy because the net melt energy was up to four times greater in the clear-cut.

### *Site*

Beartrap Creek (550 m a.s.l.) is a research basin, located at 54° N, 106° W, near the village of Waskesiu Lake, in Prince Albert National Park, Saskatchewan, Canada. The site has been the subject of intense investigations of boreal forest hydrology and climate under the Mackenzie global energy and water cycle experiment (MAGS), Prince Albert model forest hydrology study, and boreal ecosystem–atmosphere study (BOREAS). The region has a subhumid continental climate with six months of snow cover during a cold dry winter that experiences few melt events until April. Mean annual precipitation is 463 mm w.e. of which 33% occurs as winter snowfall. Topography is rolling with 700 m of local relief. Forest cover is typical of mature southern boreal forest: pine and mixed stands of aspen and white spruce on uplands, spruce, larch, and open muskeg in lowlands and about 15% covered by lakes.

The site studied is a mature, slightly open jack pine (*Pinus banksiana*) stand, 16–22 m tall with a winter leaf and stem area index of 2.2 $m^2$ $m^{-2}$ and canopy coverage of 82%. The fetch is level and uniform for about 100 m. Experiments were conducted in March, 1994 and 1996, using a canopy access tower (27 m). At the tower top two eddy correlation flux systems were installed, an Institute of Hydrology "Hydra" system for sensible and latent heat and a Gill Instruments "Solent" 3-axis sonic anemometer (Harding and Pomeroy, 1996). Above canopy net radiation and ground heat flux were measured using radiation and energy balance systems "REBS" net radiometers and heat flux plates and below canopy net radiation was measured using a Delta "T" tube net radiometer. Above canopy wind speed was measured using an RM Young propeller anemometer and temperature using a Vaisala HMP35CF hygro-thermometer. The intercepted snow load was measured using a suspended full size pine tree, which was weighed with an in-line force transducer (Hedstrom and Pomeroy, 1998). Weight of snow on the tree (kg) was converted to an areal mass (kg $m^{-2}$) using an empirical conversion developed from comparing event-based snow interception on the single tree to areal interception determined from above canopy snowfall measurements and changes in snow accumulation along a line of 25 snow survey points in subfreezing conditions.

*Energy balance*

Two sets of energy balance and related surface conditions are shown. Figure 3.11 shows sensible and latent heat flux measurements made with a Hydra and checked against a Solent sonic anemometer along with net radiation above the canopy over a five day period in late March 1994. Fresh snowfall resulted in an initial intercepted load of about 4.5 kg $m^{-2}$ on 27 March which then sublimated in temperatures ranging from −13 to 0 °C until the end of 29 March, when above freezing temperatures (5 °C) resulted in melt and the unloading of any remaining snow.

Daily maximum temperatures then increased dramatically to 17 °C on 31 March resulting in some early melt under the canopy. Four days had high net radiation inputs, i.e. negative peaks ranging from −450 to −500 W $m^{-2}$, while 28 March was overcast with a peak net radiation of only −100 W $m^{-2}$. When snow was in the canopy (27–29 March), daytime latent heat fluxes were directed away from the surface at approximately one-half the magnitude of net radiation. Sensible heat fluxes over the snow-covered canopy were similar in magnitude and direction to latent fluxes on the high-insolation days (27 and 29 March) but negligible on 28 March when low insolation resulted in minimal canopy heating. When the canopy snow load ablated (30 March), the daytime magnitude of sensible heat flux remained half that of net radiation but latent heat became negligible early in the day and directed downward later in the day. On the 31 March sensible heat behavior was unchanged but latent heat became directed upward at one-half the magnitude of sensible heat. This may reflect evaporation from melting snow beneath the canopy or, more likely given the magnitude, transpiration from the pine canopy induced by extraordinarily warm temperatures. The weighed tree did show some weight loss in this period reflecting desiccation due to evapotranspiration.

Figure 3.12 shows a consistently subfreezing sequence from the same site on 16–18 March 1996. A Solent sonic anemometer measured sensible heat fluxes (not latent) and the ablation rate of intercepted snow was measured using the weighed tree. The ablation rate was converted to equivalent energy units (flux) as if all the energy was consumed for phase change to vapor (a reasonable assumption given the −15 to −1 °C air temperatures).

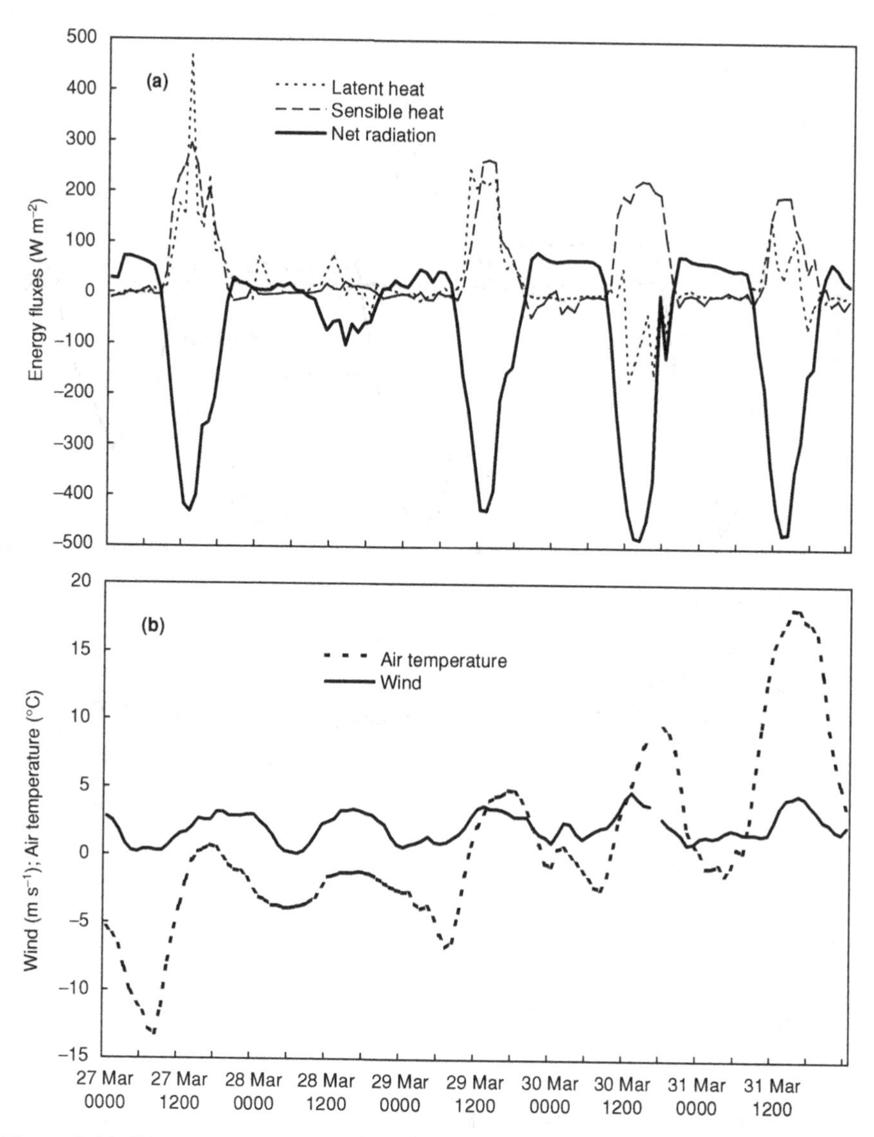


Figure 3.11. Fluxes and climate measured above a jack pine stand in the southern boreal forest of Saskatchewan, Canada, March, 1994. (a) Latent heat, sensible heat, and net radiation fluxes measured five meters above an initially snow-covered canopy and (b) air temperature and wind speed measured five meters above the canopy.

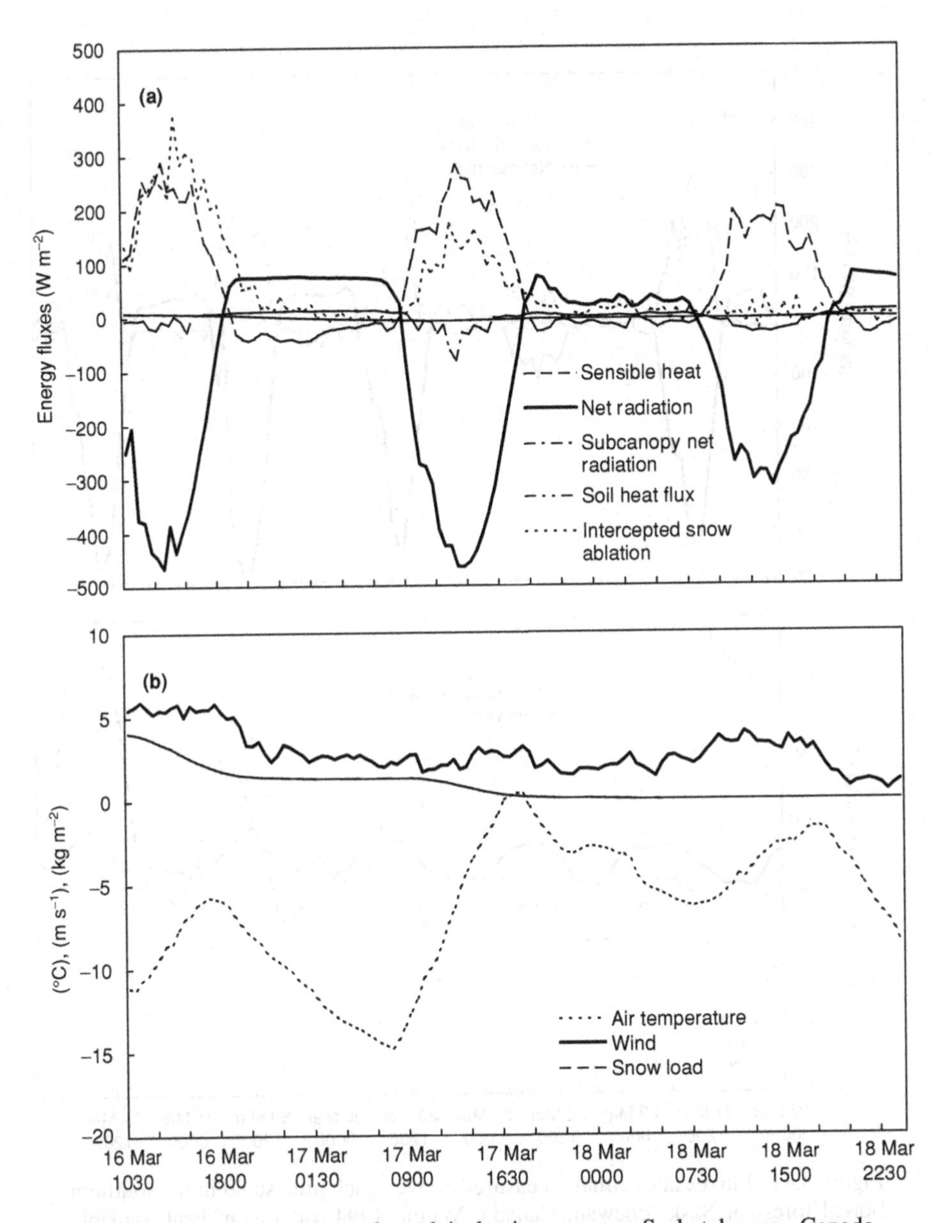


Figure 3.12. Fluxes about a boreal jack pine canopy, Saskatchewan, Canada, March, 1996. (a) Sensible heat, net radiation, subcanopy net radiation, ground heat flux, and estimated latent heat flux from intercepted snow ablation. Above canopy fluxes were measured 5 m above the canopy, below canopy radiation 1 m above the snow cover and ground heat flux 5 cm into the soil. Intercepted snow ablation was measured using a weighed, suspended full-size jack pine tree. (b) Air temperature and wind speed measured 5 m above the pine canopy. Intercepted snow load measured using a weighed, suspended pine tree.

Subcanopy net radiation and ground heat flux were also measured. An initial snow load of 4.2 kg $m^{-2}$ ablated to 1.5 kg $m^{-2}$ at the end of 16 March (strong winds, cold temperatures, and high insolation) and completely ablated by the end of 17 March. In this case sensible heat flux showed similar behavior to that in Fig. 3.11, at about one-half the magnitude of net radiation when the canopy is snow covered, increasing to three-quarters when snow free. Latent heat flux estimated from ablation equaled the sensible heat magnitude on the first (most snow-covered) day, then dropped to one-half the sensible magnitude on the second day and became negligible on the third day (snow-free canopy). Subcanopy net radiation was never more than one-tenth that of above canopy values, but remained slightly positive on 16 March when fresh snow covered the canopy and, along with a cool air mass, suppressed canopy temperatures and therefore downward longwave radiation. Ground heat fluxes were extremely small in magnitude and slowly fluctuated around zero.

### *Modeling aspects*

The snow-covered canopy represents a separate snow layer that warrants its own mass and energy balance but is not represented by many land surface schemes (Essery, 1997). For instance, the ECMWF model recently used a routine that set the boreal forest canopy albedo to a high value (0.8) after a snowfall. When corrected to a much lower and appropriate value, air temperature predictions improved dramatically over the boreal region (Betts and Ball, 1997). CLASS and SiB are exceptions that do consider canopy snow, but calculate the snow interception process in a similar manner to rainfall and therefore underestimate intercepted load by an order of magnitude for large snowfalls (Pomeroy *et al.*, 1998b). The interception models of Calder (1990) and Hedstrom and Pomeroy (1998) provide possible corrections to these schemes. Turbulent fluxes above a snow-covered canopy can be calculated using a resistance scheme with resistance set at 10 times the value of a rain-wetted canopy (Lundberg *et al.*, 1998) or by varying the ratio of bulk transfer coefficients with snow load (Nakai *et al.*, 1999). Parviainen and Pomeroy (2000) modeled the Beartrap Creek pine site using a nested control volume approach in which an energy and mass balance was conducted for an intercepted snow control volume, using the Reynolds number to calculate turbulent transfer between the intercepted snow and the atmosphere. The CLASS land surface scheme was then coupled to this small-scale calculation to calculate turbulent transfer between the canopy and atmosphere. Using a geometric radiative transfer model (GORT) to calculate subcanopy radiation (Ni *et al.*, 1997) and SNTHERM to calculate snowpack heat fluxes, snowmelt modeling was successful when snow surface albedo was reduced during melt and turbulent heat fluxes were given small values (Hardy *et al.*, 1997b).

## References


Ambach, W. (1974). The influence of cloudiness on the net radiation balance of a snow surface with high albedo. *J. Glaciol.*, **13**(67), 73–84.

Andreas, E L. (1989). A physical bound on the Bowen ratio. *J. Appl. Meteorol.*, **28**(11), 1252–1254.

Andreas, E. L and Cash, B. A. (1996). A new formulation for the Bowen ratio over saturated surfaces. *J. Appl. Meteorol.*, **35**(8), 1279–1289.

Bartelt, P. and Lehning, M. (2002). A physical SNOWPACK model for the Swiss avalanche warning; Part I: numerical model. *Cold Reg. Sci. Technol.*, **35**(3), 123–145.

Beljaars, A. C. M. and Holtslag, A. A. M. (1991). Flux parametrization and land surfaces in atmospheric models. *J. Appl. Meteorol.*, **30**, 327–341.

Betts, A. K. and Ball, J. H. (1997). Albedo over the boreal forest. *J. Geophys. Res.*, **102**(D24), 28 901–28 909.

Bintanja, R. (1998). The contribution of snowdrift sublimation to the surface mass balance of Antarctica. *Ann. Glaciol.*, **27**, 251–259.

Bintanja, R. and van den Broeke, M. R. (1995). Momentum and scalar transfer coefficients over aerodynamically smooth Antarctic surfaces. *Bound.-Lay. Meteorol.*, **74**, 89–111.

Bintanja, R. and van den Broeke, M. R. (1996). The influence of clouds on the radiation budget of ice and snow surfaces in Antarctica and Greenland in summer. *Int. J. Climatol.*, **16**, 1281–1296.

Bowling, L. C., Pomeroy, J. W., and Lettenmaier, D. P. (2004). Parameterisation of the sublimation of blowing snow in a macroscale hydrology model. *J. Hydrometeor.*, **5**, 745–762.

Brown, T. and Pomeroy, J. W. (1989). A blowing snow particle detector. *Cold Reg. Sci. Technol.*, **16**, 167–174.

Brun, E., David, P., Sudul, M., and Brunot, G. (1992). A numerical model to simulate snow-cover stratigraphy for operational avalanche forecasting, *J. Glaciol.*, **38**(128), 13–22.

Brun, E., Martin, E., Simon, V., Gendre, C., and Coléou, C. (1989). An energy and mass model of snow cover suitable for operational avalanche forecasting. *J. Glaciol.*, **35**(121), 333–342.

Brutsaert, W. (1975). On a derivable formula for long wave radiation from clear skies. *Water Resources Res.*, **11**, 742–744.

Calanca, P. and Heuberger, R. (1990). *Glacial Climate Research in the Tianshan* (ed. Ohmura, A. *et al.*). Zürcher Geographische Schriften (ZGS), Heft 38. Zürich: Swiss Federal Institute of Technology ETHZ, pp. 60–72.

Calder, I. R. (1990). *Evaporation in the Uplands*. Chichester: Wiley, p. 144.

Chamberlain, A. C. (1983). Roughness length of sea, sand and snow. *Bound.-Lay. Meteorol.*, **25**, 405–409.

Claussen, M. (1991). Local advection processes in the surface layer of the marginal ice zone. *Bound.-Lay. Meteorol.*, **54**, 1–27.

Davis, R. E., Hardy, J. P., Ni, W., *et al.* (1997). Variation of snow cover ablation in the boreal forest: a sensitivity study on the effects of conifer canopy. *J Geophys. Res.*, **102**(D24), 29 389–29 398.

de la Casinière, A. C. (1974). Heat exchange over a melting snow surface. *J. Glaciol.*, **13**, 55–72.

Dery, S. J. and Yau, M. K. (2002). Large-scale mass balance effects of blowing snow and surface sublimation. *J. Geophys. Res.*, **107**(D23), 4679.

Dery, S. J., Taylor, P. A., and Xiao, J. (1998). The thermodynamic effects of sublimating blowing snow in the atmospheric boundary layer. *Bound.-Lay. Meteorol.*, **89**, 251–283.

Doorschot, J. (2002). Mass transport of drifting snow in high alpine environments. Ph.D. Thesis, Swiss Federal Institute of Technology ETHZ, Zürich. http://e-collection.ethbib.ethz.ch/show?type=diss&nr=14515

Doorschot, J. and Lehning, M. (2002). Equilibrium saltation: mass fluxes, aerodynamic entrainment, and dependence on grain properties. *Bound.-Lay. Meteorol.*, **104**, 111–130.

Doorschot, J., Raderschall, N., and Lehning, M. (2001). Measurement and one-dimensional model calculations of snow transport over a mountain ridge. *Ann. Glaciol.*, **32**, 153–158.

Dozier, J. (1980). A clear sky spectral solar radiation model for snow-covered mountainous terrain. *Water Resources Res.*, **16**, 709–718.

Durand, Y., Brun, E., Mérindol, L., *et al.* (1993). A meteorological estimation of relevant parameters for snow models. *Ann. Glaciol.*, **18**, 65–71.

Durand, Y., Guyomarc'h, G., and Mérindol, L. (2001). Numerical experiments of wind transport over a mountainous instrumented site: I. Regional scale. *Ann. Glaciol.*, **32**, 187–194.

Dyunin, A. K., Kvon, Ya. D., Zhilin A. M., and Komorov, A. A. (1991). Effect of snow drifting on large-scale aridization. In *Glaciers–Ocean–Atmosphere Interactions* (ed. Kotlyakov, V. M., Ushakov, A., and Glasovsky, A.). IAHS Publication No. 208. Wallingford: IAHS Press, pp. 489–494.

Ebert, E. E. and Curry, J. A. (1993). An intermediate one-dimensional thermodynamic sea ice model for investigating ice–atmosphere interactions. *J. Geophys. Res.*, **98**(C6), 10 085–10 109.

Elder, K., Dozier, J., and Michaelsen, J. (1989). Spatial and temporal variation of net snow accumulation in a small alpine watershed. Emerald Lake basin, Sierra Nevada, California, U.S.A. *Ann. Glaciol.*, **13**, 56–63.

Essery, R. (1997). Modelling fluxes of momentum, sensible heat and latent heat over heterogeneous snow cover. *Q. J. Roy. Meteorol. Soc.*, **123**, 1867–1883.

Essery, R., Li, L. and Pomeroy, J. W. (1999). A distributed model of blowing snow over complex terrain. *Hydrol. Process.*, **13**(14–15), 2423–2438.

Essery, R. and Pomeroy, J. W. (2004). Vegetation and topographic control of wind-blown snow distributions in distributed and aggregated simulations for an Arctic tundra basin. *J. Hydrometeorol.*, **5**, 734–744.

Faria, D. A., Pomeroy, J. W., and Essery, R. L. H. (2000). Effect of covariance between ablation and snow water equivalent on depletion of snow-covered area in a forest. *Hydrol. Process.*, **14**(15), 2683–2695.

Fierz, C., Plüss, C., and Martin, E. (1997). Modelling the snow cover in complex alpine topography. *Ann. Glaciol.*, **25**, 312–316.

Föhn, P. M. B. (1973). Short term snow melt and ablation derived from heat- and mass-balance measurements. *J. Glaciol.*, **12**(65), 275–289.

Föhn, P. M. B. (1977). Representativeness of precipitation measurements in mountainous areas. In *Proc. Joint Scientific Meeting on Mountain Meteorology and Biometeorology AMS, SGBB, SSG*, Interlaken, Switzerland, 10–14 June 1976 (ed. Primault, B.). Geneva: Blanc et Wittwer, pp. 61–71.

Föhn, P. M. B. (1985). Besonderheiten des Schneeniederschlages. In *Der Niederschlag in der Schweiz*. Beitr. Geol. Schweiz – Hydrol., vol. 31, Bern: Kümmerly und Frey, pp. 87–96.

Föhn, P. M. B. (1992). Climatic change, snow cover and avalanches. In *Greenhouse-Impact on Cold-Climate Ecosystems and Landscape* (ed. Boer, M. and Koster, E.). Catena supplement 22. Cremlingen-Destedt: Catena, pp. 11–21.

Föhn, P. and Hächler, P. (1978). Prévision de grosses avalanches au moyen d'un modèle déterministe-statistique. In *Comptes Rendues de la Deuxième Rencontre Internationale sur la Neige et les Avalanches*, Grenoble, France, 12–14 avril 1978. Grenoble: Association Nationale pour l'Etude de la Neige et des Avalanches (ANENA), pp. 151–165.

Gardiner, B. G. (1987). Solar radiation transmitted to the ground through cloud in relation to surface albedo. *J. Geophys. Res.*, **92**(D4), 4010–4018.

Garratt, J. R. (1992). *The Atmospheric Boundary Layer*. Cambridge: Cambridge University Press.

Gauer, P. (1998). Blowing and drifting snow in alpine terrain: Numerical simulation and related field measurements. *Ann. Glaciol.*, **26**, 174–178.

Gauer, P. (2001). Numerical modeling of blowing and drifting snow in Alpine terrain. *J. Glaciol.*, **47**(156), 97–110.

Grainger, M. E. and Lister, H. (1966). Wind speed, stability and eddy viscosity over melting ice surfaces. *J. Glaciol.*, **6**(43), 101–127.

Granger, R. J. and Male, D. H. (1978). Melting of a prairie snowpack. *J. Appl. Meteorol.*, **17**, 1833–1842.

Gray, D. M. (ed.). (1970). *Handbook on the Principles of Hydrology*. Ottawa: Canadian National Committee for the International Hydrological Decade.

Harding, R. J. (1986). Exchanges of energy and mass associated with a melting snow pack. In *Modelling Snowmelt-Induced Processes* (ed. Morris, E. M.). IAHS Publication No. 155. Wallingford: IAHS Press, pp. 3–15.

Harding, R. J. and Pomeroy, J. W. (1996). The energy balance of the winter boreal landscape. *J. Climate*, **9**, 2778–2787.

Hardy, J. P., Davis, R. E., Jordan, R., *et al.* (1997a). Snow ablation modelling at the stand scale in a boreal jack pine forest. *J. Geophys. Res.*, **102**(D24), 29 397–29 406.

Hardy, J. P., Davis, R. E., Jordan, R., Li, X., and Woodcock, C. (1997b). Snow ablation modelling in conifer and deciduous stands of the boreal forest. *Proc. Western Snow Conf.*, **65**, 114–124.

Hedstrom, N. R. and Pomeroy, J. W. (1998). Accumulation of intercepted snow in the boreal forest: measurements and modelling. *Hydrol. Process.*, **12**, 1611–1625.

Heinemann, G. (1989). Über die Rauhigkeitslänge $z_0$ der Schneeoberfläche des Filchner-Ronne Schelfeises. *Polarforschung*, **59**, 17–24.

Hetherington, E. D. (1987). The importance of forests in the hydrological regime. *Can. Bull. Fish. Aquatic Sci.*, **215**, 179–211.

Inoue, J. (1989). Surface drag over the snow surface of the Antarctic Plateau. 1. Factors controlling surface drag over the katabatic wind region. *J. Geophys. Res.*, **94**(D2), 2207–2217.

Iqbal, M. (1983). *An Introduction to Solar Radiation*. Toronto: Academic Press.

Joffre, S. M. (1982). Momentum and heat transfers in the surface layer over a frozen sea. *Bound.-Lay. Meteorol.*, **24**, 211–229.

Jones, H. G. (1987). Chemical dynamics of snowcover and snowmelt in a boreal forest. In *Seasonal Snowcovers: Physics, Chemistry, Hydrology* (ed. Jones, H. G. and Orville-Thomas, W. J.). *NATO ASI Series C*, vol. 211. Dordrecht: Reidel Publishing, pp. 531–574.

Jones, H. G., Pomeroy, J. W., Walker, D. A., and Hoham, R. (eds.). (2001). *Snow Ecology: an Interdisciplinary Examination of Snow-Covered Ecosystems.* Cambridge: Cambridge University Press.

Jordan, R. E., Andreas, E. L., Fairall, C. W., *et al.* (2003). Modeling surface exchange and heat transfer for the shallow snow cover at SHEBA. In *Seventh Conference on Polar Meteorology and Oceanography*, Hyannis, MA (CD-ROM of preprints). Washington, DC: American Meteorological Society.

Jordan, R. E., Andreas, E. L., and Makshtas, A. S. (1999). The heat-budget of snow-covered sea ice at North-pole 4. *J. Geophys. Res.*, **104**(D4), 7785–7806.

Key, J. R., Silcox, R. A., and Stone, R. S. (1996). Evaluation of surface radiative flux parameterizations for use in sea ice models. *J. Geophys. Res.*, **101**(C2), 3839–3849.

Kind, R. J. 1992. One-dimensional aeolian suspension above beds of loose particles – a new concentration-profile equation. *Atmos. Environ.*, **26A**, 927–931.

King, J. C. 1990. Some measurements of turbulence over an Antarctic ice shelf. *Q. J. Roy. Meteor. Soc.*, **116**, 379–400.

King, J. C. and Anderson, P. S. (1994). Heat and water vapour fluxes and scalar roughness lengths over an Antarctic ice shelf. *Bound.-Lay. Meteorol.*, **69**, 101–121.

King, J. C., Anderson, P. S., Smith, M. C., and Mobbs, S. D. (1996). The surface energy and mass balance at Halley, Antarctica, during winter. *J. Geophys. Res.*, **101**(D14), 19 119–19 128.

King, J. C. and Connolley, W. M. (1997). Validataion of the surface energy balance over the Antarctic ice sheets in the U.K. Meteorological Office Unified Climate Model. *J. Climate*, **10**, 1273–1287.

Kirnbauer, R., Blöschl, G., and Gutknecht, D. (1994). Entering the era of distributed snow models. *Nordic Hydrol.*, **25**, 1–24.

Kondo, J. and Yamazawa, H. (1986). Bulk transfer coefficient over a snow surface. *Bound.-Lay. Meteorol.*, **34**, 123–135.

König, G. (1985). Roughness length of an Antarctic ice shelf. *Polarforschung*, **55**, 27–32.

König-Langlo, G. and Augstein, E. (1994). Parameterization of the downward longwave radiation at the Earth's surface in polar regions. *Meteorol. Z.*, **3**, 343–347.

Konstantinov, A. R. (1966). *Isparenie v Prirode*. Leningrad: Gidrometeoizdat. Published 1968 as *Evaporation in Nature*. (English translation by Israel Programme for Scientific Translation, Jerusalem.)

Konzelmann, T., van de Wal, R., Greuell, W., *et al.* (1994). Parameterization of global and longwave incoming radiation for the Greenland ice sheet. *Global Planet. Change*, **9**, 143–164.

Kucherov, N. V. and Sternzat, M. S. (1959). The apparatus and method of investigations at stations North Pole 4 and North Pole 5. *Trudy. Arkt. Antarkt. Nauchno-Issl. Inst.*, **226**, 5–18. (In Russian; English translation available from the CRREL Library.)

Kuz'min, P. P. (1960). *Formirovanie Snezhnogo Pokrova I Metody Opredeleniya Snegozapasov*, Leningrad. Published 1963 as *Snow Cover and Snow Reserves*. Washington, DC: National Science Foundation. (English translation by Israel Programme for Scientific Translation, Jerusalem.)

Lee, L. W. (1975). Sublimation of snow in a turbulent atmosphere. Ph.D. Thesis, University of Wyoming, Laramie, WY.

Lehning, M., Bartelt, P., Bethke, S., *et al.* (2004). Review of SNOWPACK and Alpine3D applications. In *Snow Engineering*, vol. V. (ed. Bartelt, P., Adams, E. E., Christen, M., Sack, R. L., and Sato, A.). Leiden: Balkema Publishers, pp. 299–307.

Lehning, M., Bartelt, P., Brown, B., and Fierz, C. (2002a). A physical SNOWPACK model for the Swiss avalanche warning; Part III: Meteorological forcing, thin layer formation and evaluation. *Cold Reg. Sci. Technol.*, **35**(3), 169–184.

Lehning, M., Bartelt, P., Brown, B., Fierz, C., and Satyawali, P. (2002b). A physical SNOWPACK model for the Swiss avalanche warning; Part II: Snow microstructure. *Cold Reg. Sci. Technol.*, **35**(3), 147–167.

Li, L. and Pomeroy, J. W. (1997a). Estimates of threshold wind speeds for snow transport using meteorological data. *J. Appl. Meteorol.*, **36**, 205–213.

Li, L. and Pomeroy, J. W. (1997b). Probability of occurrence of blowing snow. *J. Geophys. Res.*, **102**(D18), 21 955–21 964.

Lindsay, R. W. (1998). Temporal variability of the energy balance of thick Arctic pack ice. *J. Climate*, **11**, 313–333.

Liston, G. E. (1995). Local advection of momentum, heat and moisture during the melt of patchy snow covers. *J. Appl. Meteorol.*, **34**, 1705–1715.

Liston, G. E. and Sturm, M. (1998). A snow-transport model for complex terrain. *J. Glaciol.*, **44**(148), 498–516.

Lundberg, A., Calder, I., and Harding, R. (1998). Evaporation of intercepted snow: measurements and modelling. *J. Hydrol.*, **206**, 151–163.

Lundberg, A. and Halldin, S. (1994). Evaporation of intercepted snow – an analysis of governing factors. *Water Resources Res.*, **30**, 2587–2598.

Male, D. H. and Gray, D. M. (1975). Problems in developing a physically-based snowmelt model. *Can. J. Civil Eng.*, **2**, 474–488.

Male, D. H. and Gray, D. M. (1981). Snowcover ablation and runoff. In *Handbook of Snow: Principles, Processes, Management and Use* (ed. Gray, D. M. and Male, D. H.). Toronto: Pergamon Press, pp. 360–436.

Marks, D. and Dozier, J. (1992). Climate and energy exchange at the snow surface in the Alpine region of the Sierra Nevada. 2. Snow cover energy balance. *Water Resources Res.*, **28**, 3042–3054.

Marsh, P. and Pomeroy, J. W. (1996). Meltwater fluxes at an Arctic forest tundra site. *Hydrol. Process.*, **10**, 1383–1400.

Marsh, P., Pomeroy, J. W., and Neumann, N. (1997). Sensible heat flux and local advection over a heterogeneous landscape at an Arctic tundra site during snowmelt. *Ann. Glaciol.*, **25**, 132–136.

Marshunova, M. S. and Mishin, A. A. (1994). Handbook of the radiation regime of the Arctic Basin (Results from the drift stations). *Technical Report APL-UW TR 9413*. Seattle, WA: Applied Physics Laboratory, University of Washington.

Martin, E., Brun, E., and Durand, Y. (1994). Sensitivity of the French Alps snow cover to the variation of climatic variables. *Ann. Geophys.*, **12**, 469–477.

Martin, E. and Lejeune, Y. (1997). Investigations on turbulent fluxes above the snow surface. *Ann. Glaciol.*, **26**, 179–183.

Marty, C. (2000). Surface radiation, cloud forcing and greenhouse effect in the Alps. Ph.D. Thesis, Swiss Federal Institute of Technology ETHZ, Zürich. http://e-collection.ethbib.ethz.ch/cgi-bin/show.pl?type=diss&nr=13609

Maykut, G. A. (1982). Large-scale heat exchange and ice production in the central Arctic. *J. Geophys. Res.*, **87**(C10), 7971–7984.

Metcalfe, R. A. and Buttle J. M. (1995). Controls of canopy structure on snowmelt rates in the boreal forest. *Proc. Eastern Snow Conf.*, **52**, 249–257.

Metcalfe, R. A. and Buttle, J. M. (1998). A statistical model of spatially distributed snowmelt rates in a boreal forest basin. *Hydrol. Process.*, **12**, 1701–1722.

Michaux, J. L., Naaim-Bouvet, F., and Naaim, M. (2001). Drifting snow studies over a mountainous instrumented site: measurements and numerical model. *Ann. Glaciol.*, **32**, 175–181.

Moore, R. D. and Owens, I. F. (1984). Controls on advective snowmelt in a maritime alpine basin. *J. Appl. Meteorol.*, **23**, 135–142.

Morris, E. M. (1989). Turbulent transfer over snow and ice. *J. Hydrol.*, **105**, 205–223.

Morris, E. M., Anderson, P. S., Bader, H.-P., Weilenman, P., and Blight, C. (1994). Modelling mass and energy exchange over polar snow using the DAISY model. In *Snow and Ice Covers: Interactions with the Atmosphere and Ecosystems* (ed. Jones, H. G., Davies, T. D., Ohmura, A., and Morris, E. M.). IAHS Publication No. 223. Wallingford: IAHS Press, pp. 53–60.

Nakai, Y., Kitahara, H., Sakamoto, T., Saito, T., and Terajima, T. (1993). Evaporation of snow intercepted by forest canopies. *J. Jpn. Forest Soc.*, **75**, 191–200.

Nakai, Y., Sakamoto, T., Terajima, T., Kitahara, H., and Saito, T. (1994). Snow interception by forest canopies: weighing a conifer tree with meteorological observation and analysis with Penman–Monteith formula. In *Snow and Ice Covers: Interactions with the Atmosphere and Ecosystems* (ed. Jones, H. G., Davies, T. D. Ohmura, A., and Morris, E. M.). IAHS Publication No. 223. Wallingford: IAHS Press, pp. 227–236.

Nakai, Y., Sakamoto, T., Terajima, T., Kitamura, K., and Shirai, T. (1999). Energy balance above a boreal coniferous forest: a difference in turbulent fluxes between snow-covered and snow-free canopies. *Hydrol. Process.*, **13**, 515–529.

National Snow and Ice Data Center (NSIDC). (1996). *Arctic Ocean Snow and Meteorological Observations from Drifting Stations: 1937, 1950–1991*, CD-ROM Version 1.0. Boulder, CO: University of Colorado.

Nazintsev, Yu. L. (1963). On the role of thermal processes in sea ice melting and in the transformation of the relief of multiyear ice floes in the central Arctic (in Russian). *Prob. Arkt. Antarkt.*, **12**, 69–75. (In Russian; English translation available from the CRREL Library).

Nazintsev, Yu. L. (1964). Thermal balance of the surface of the perennial ice cover in the central Arctic. *Trudy, Arkt. Antarkt. Nauchno-Issl. Inst.*, **267**, 110–126. (In Russian; English translation available from the CRREL Library).

Neumann, N. and Marsh, P. (1998). Local advection of sensible heat in the snowmelt landscape of Arctic tundra. *Hydrol. Process.*, **12**, 1547–1560.

Ni, W., Li, X., Woodstock, C. E., Roujean, J-L., and Davis, R. E. (1997). Transmission of solar radiation in boreal conifer forests: measurements and models. *J. Geophys. Res.*, **102**(D24), 29 555–29 566.

Ohmura, A. 2001. Physical basis for the temperature-based melt-index method. *J. Appl. Meteorol.*, **40**, 753–761.

Oke, T. R. (1987). *Boundary Layer Climates*, 2nd edn. London: Routledge.

Olyphant, G. and Isard, S. (1988). The role of advection in the energy balance of late-lying snowfields: Niwot Ridge, Front Range, Colorado. *Water Resources Res.*, **24**, 1962–1968.

O'Neill, A. D. J. and Gray, D. M. (1973). Spatial and temporal variations of the albedo of a prairie snowpack. In *The Role of Snow and Ice in Hydrology: Proc., Banff Symposium*, vol. 1. Geneva-Budapest-Paris: UNESCO-WMO-IAHS, pp. 176–186.

Owen, P. R. (1964). Saltation of uniform grains in air. *J. Fluid Mech.*, **20**, 225–242.

Parviainen, J. and Pomeroy, J. W. (2000). Multiple-scale modelling of forest snow sublimation: initial findings. *Hydrol. Process.*, **14**(15), 2669–2681.

Perovich, D. K., Grenfell, T. C., Light, B., and Hobbs, P. V. (2002). Seasonal evolution of the albedo of multiyear Arctic sea ice. *J. Geophys. Res.*, **107**(C10), 8044, doi:10.1029/2000JC000438.

Persson, P. O. G., Fairall, C. W., Andreas, E. L., Guest, P. S., and Perovich, D. K. (2002). Measurements near the Atmospheric Surface Flux Group tower at SHEBA: near-surface conditions and surface energy budget. *J. Geophys. Res.*, **107**(C10), 8043, doi:10.1029/2000JC000705.

Plüss, C. (1997). *The Energy Balance over an Alpine Snowcover – Point Measurements and Areal Distribution*. Zürcher Geographische Schriften (ZGS), Heft vol. 65. Zürich: Swiss Federal Institute of Technology ETHZ.

Plüss, C. and Ohmura, A. (1997). Longwave radiation on snow-covered mountainous surfaces. *J. Appl. Meteorol.*, **36**, 818–824.

Poggi, A. (1976). Heat balance in the ablation area of the Ampere Glacier (Kerguelen Islands). *J. Appl. Meteorol.*, **16**, 48–55.

Pomeroy, J. W. (1989). A process-based model of snow drifting. *Ann. Glaciol.*, **13**, 237–240.

Pomeroy, J. W. (1991). Transport and sublimation of snow in wind-scoured alpine terrain. In *Snow, Hydrology and Forests in High Alpine Areas* (ed. Bergmann, H., Lang, H., Frey, W., Issler, D., and Salm, B.). IAHS Publication No. 205. Wallingford: IAHS Press, pp. 131–140.

Pomeroy, J. W., Davies, T. D., Jones, H. G., *et al.* (1999a). Transformations of snow chemistry in the boreal forest: accumulation and volatilization. *Hydrol. Process.*, **13**, 2257–2273.

Pomeroy, J. W. and Dion, K. (1996). Winter radiation extinction and reflection in a boreal pine canopy: measurements and modelling. *Hydrol. Process.*, **10** 1591–1608.

Pomeroy, J. W. and Essery, R. (1999). Turbulent fluxes during blowing snow: field tests of model sublimation predictions. *Hydrol. Process.*, **13**, 2963–2975.

Pomeroy, J. W., Essery, R. L. H., Gray, D. M., *et al.* (1999b). Modelling snow–atmosphere interactions in cold continental environments. In *Interactions Between the Cryosphere, Climate and Greenhouse Gases* (ed. Tranter, M., Armstrong, R., Brun, E., *et al.*). IAHS Publication No. 256. Wallingford: IAHS Press, pp. 91–101.

Pomeroy, J. W. and Granger, R. J. (1997). Sustainability of the western Canadian boreal forest under changing hydrological conditions – I – Snow accumulation and ablation. In *Sustainability of Water Resources under Increasing Uncertainty* (ed. Rosjberg, D., Boutayeb, N., Gustard, A., Kundzewicz, Z., and Rasmussen, P.). IAHS Publication No. 240. Wallingford: IAHS Press, pp. 237–242.

Pomeroy, J. W. and Gray, D. M. (1990). Saltation of snow. *Water Resources Res.*, **26**(7), 1583–1594.

Pomeroy, J. W. and Gray, D. M. (1995). *Snowcover Accumulation, Relocation and Management*. NHRI Science Report No. 7. Saskatoon: National Hydrology Research Institute.

Pomeroy, J. W., Gray, D. M., and Landine, P. G. (1993). The prairie blowing snow model: characteristics, validation, operation. *J. Hydrol.*, **144**, 165–192.

Pomeroy, J. W., Gray, D. M., Shook, K. R., *et al.* (1998b). An evaluation of snow accumulation and ablation processes for land surface modelling. *Hydrol. Process.*, **12**(15), 2339–2367.

Pomeroy, J. W., Hedstrom, N., and Parviainen, J. (1999c). The snow mass balance of Wolf Creek. In *Wolf Creek Research Basin: Hydrology, Ecology, Environment* (ed. Pomeroy, J. and Granger, R.). Saskatoon: National Water Research Institute, Minister of Environment, pp. 15–30.

Pomeroy, J. W. and Li, L. (2000). Prairie and Arctic areal snow cover mass balance using a blowing snow model. *J. Geophys. Res.*, **105**(D21), 26 619–26 634.
Pomeroy, J. W. and Male, D. H. (1992). Steady-state suspension of snow. *J. Hydrol.*, **136**, 275–301.
Pomeroy, J. W., Marsh, P., and Gray, D. M. (1997). Application of a distributed blowing snow model to the Arctic. *Hydrol. Process.*, **11**, 1451–1464.
Pomeroy, J. W., Parviainen, J., Hedstrom, N., and Gray, D. M. (1998a). Coupled modelling of forest snow interception and sublimation. *Hydrol. Process.*, **12**, 2317–2337.
Pomeroy, J. W. and Schmidt, R. A. (1993). The use of fractal geometry in modelling intercepted snow accumulation and sublimation. *Proc. Eastern Snow Conf.*, **50**, 1–10.
Raderschall, N. (1999). *Statistische Uebertragung von Modelldaten eines Numerischen Wettervorhersagemodells auf Alpine Standorte*. Diplomarbeit des Meteorologischen Instituts der Rheinischen Friedrich-Wilhelms-Universitaet Bonn, unpublished.
Radionov, V. F., Bryazgin, N. N., and Aleksandrov, E. I. (1996). *The Snow Cover of the Arctic Basin*. St. Petersburg: Gidrometeoizdat. (In Russian; English translation available as: Radionov, V. F., Bryazgin, N. N., and Aleksandrov, E. I. (1997). *The Snow Cover of the Arctic Basin*. Technical Report APL-UW TR 9701, Applied Physics Laboratory, University of Washington, Seattle.)
Schmidt, R. A. (1972). *Sublimation of Wind-transported Snow – A Model*. USDA Forest Service Research Paper RM-90. Fort Collins, CO: Rocky Mountain Forest and Range Experiment Station.
Schmidt, R. A. (1991). Sublimation of snow intercepted by an artificial conifer. *Agric. Forest Meteorol.*, **54**, 1–27.
Schmidt, R. A. and Gluns, D. R. (1991). Snowfall interception on branches of three conifer species. *Can. J. Forest Res.*, **21**, 1262–1269.
Shine, K. P. (1984). Parametrization of the shortwave flux over high albedo surfaces as a function of cloud thickness and surface albedo. *Q. J. Roy. Meteorol. Soc.*, **110**, 747–764.
Shook, K. (1993). Fractal geometry of snowpacks during ablation. M.Sc. Thesis, University of Saskatchewan, Saskatoon.
Shook, K. (1995). Simulation of the ablation of prairie snowcovers. Ph.D. Thesis, University of Saskatchewan, Saskatoon.
Shook, K. and Gray, D. M. (1994). Determining the snow water equivalent of shallow prairie snowcovers. *Proc. Eastern Snow Conf.*, **51**, 89–95.
Shook, K. and Gray, D. M. (1996). Small scale spatial structure of shallow snowcovers. *Hydrol. Process.*, **10**, 1283–1292.
Shook, K. and Gray, D. M. (1997). Snowmelt resulting from advection. *Hydrol. Process.*, **11**, 1725–1736.
Smeets, C. J. P. P., Duynkerke, P. G., and Vugts, H. F. (1998). Observed wind profiles and turbulence fluxes over an ice surface with changing surface roughness. *Bound.-Lay. Meteorol.*, **92**, 101–123.
Steppuhn, H. (1981). Snow and agriculture. In *Handbook of Snow: Principles, Processes, Management and Use* (ed. Gray, D. M. and Male, D. H.). Toronto: Pergamon Press, pp. 60–125.
Steppuhn, H. and Dyck, G. E. (1974). Estimating true basin snowcover. In *Advanced Concepts in the Technical Study of Snow and Ice Resources. Interdisciplinary Symposium*. Washington, DC: US National Academy of Sciences, pp. 314–328.
Stull, R. B. (1988). *An Introduction to Boundary Layer Meteorology*. Dordrecht: Kluwer Academic Publishers.

Sturm, M. (1992). Snow distribution and heat flow in the taiga. *Arctic Alpine Res.*, **24**(2), 145–152.

Sturm, M., Holmgren, J., König, M., and Morris, K. (1997). The thermal conductivity of seasonal snow. *J. Glaciol.*, **43**, 26–41.

Sturm, M., Holmgren, J., and Liston, G. E. (1995). A seasonal snow cover classification system for local to global applications. *J. Climate*, **8**, 1261–1283.

Sturm, M., Perovich, D. K., and Holmgren, J. (2002). Thermal conductivity and heat transfer through the snow on the ice of the Beaufort Sea. *J. Geophys. Res.*, **107**(C10), 8045, doi:10.1029/2000JC000466.

Sverdrup, H. H. (1936). The eddy conductivity of the air over a smooth snowfield. *Geophys. Publ.*, **11**(7), 5–49.

Tabler, R. D. (1980). Self similarity of wind profiles in blowing snow allows outdoor modelling. *J. Glaciol.*, **26**(94), 421–434.

Tabler, R. D. and Schmidt, R. A. (1986). Snow erosion, transport and deposition. In *Proc. Symposium on Snow Management for Agriculture* (ed. Steppuhn, H. and Nicholaichuk, W.). Great Plains Agricultural Council Publication No. 120. Lincoln: University of Nebraska, pp. 12–58.

Thomas, G. and Rowntree, P. R. (1992). The boreal forest and climate. *Q. J. Roy. Meteorol. Soc.*, **118**, 469–497.

Thorpe, A. D. and Mason, B. J. (1966). The evaporation of ice spheres and ice crystals. *Br. J. Appl. Phys.*, **17**, 541–548.

Uttal, T., Curry, J. A., McPhee, M. G., *et al.* (2002). Surface heat budget of the Arctic Ocean. *Bull. Amer. Meteor. Soc.*, **83**(2), 255–275.

Van den Broeke, M. R. (1997). Structure and diurnal variation of the atmospheric boundary layer over a mid-latitude glacier in summer. *Bound.-Lay. Meteorol.*, **83**, 183–205.

Varley, M. J., Beven, K. J., and Oliver, H. R. 1996. Modelling solar radiation in steeply sloping terrain. *J. Climate*, **16**, 93–104.

Webb, E. K. (1970). Profile relationships: the log-linear range and extension to strong stability. *Q. J. Roy. Meteorol. Soc.*, **96**, 67–90.

Weisman, R. W. (1977). Snowmelt: a two-dimensional turbulent diffusion model. *Water Resources Res.*, **13**(2), 337–342.

Woo, M-K. and Steer, P. (1986). Monte Carlo simulation of snow depth in a forest. *Water Resources Res.*, **22**(6), 864–868.

Yamazaki, T., Fukabori, K., and Kondo, J. (1996). Albedo of forest with crown snow. *Seppyo, J. Jpn. Soc. Snow Ice*, **58**. 11–18 (in Japanese with English summary).

Yang, D., Goodison, B. E., Metcalfe, J. R., *et al.* (1995). Accuracy of Tretyakov precipitation gauge: result of WMO intercomparison. *Proc. Eastern Snow Conf.*, **52**, 95–106.

Zhao, L., Gray, D. M., and Male, D. H. (1997). Numerical analysis of simultaneous heat and mass transfer during infiltration into frozen ground. *J. Hydrol.*, **200**, 345–363.

# 4

# Snow-cover parameterization and modeling

Eric Brun, Zong-Liang Yang, Richard Essery, and Judah Cohen

## 4.1 History of numerical modeling of snow cover

### *4.1.1 Introduction*

In 1976, Eric Anderson compiled a detailed history of numerical modeling of snow cover which is paraphrased below with Anderson's permission (personal communication, 2006). Anderson began his history in the 1930s with the first application of modern energy transfer theory to a snow cover by Sverdrup (1936). In the 1940s, the U.S. Army Corps of Engineers and the U.S. Weather Bureau initiated the Cooperative Snow Investigations (U.S. Army Corps of Engineers, 1955). The purpose of these investigations was to promote a fundamental understanding of snow hydrology for project design and streamflow forecasting, particularly for the western United States. Extensive data collection and analysis were performed over a 10-year period at three research watersheds. The publication *Snow Hydrology* (U.S. Army Corps of Engineers, 1956) summarized these investigations. The results and methods included in this publication are widely referenced and form the basis of many of the snow-cover models in use today. During this same period, the 1940s and 1950s, extensive studies were being conducted in the Soviet Union addressing the physical properties, formation, and melting of a snow cover. Kuzmin's summary of these studies (Kuzmin, 1961) presented a thorough and complete discussion of snow-cover energy exchange over a melting snow cover, from both a theoretical and a practical viewpoint.

### *4.1.2 Snow-cover simulation models*

In the 1960s with the development of digital computers, researchers were able to construct conceptual simulation models of the snow accumulation and ablation

*Snow and Climate: Physical Processes, Surface Energy Exchange and Modeling*, ed. Richard L. Armstrong and Eric Brun. Published by Cambridge University Press. 

process. These models were developed to solve many kinds of practical hydrologic problems. In a conceptual model, each major physical process is represented by a mathematical relationship. This is in contrast to degree-day techniques, which were commonly used to estimate snow-cover outflow directly from air temperature data. Degree-day techniques do not include the snow accumulation process, nor do they explicitly account for heat deficits, liquid-water retention and transmission, and the areal extent of the snow cover.

Two of the earliest snow-cover simulation models were developed by Rockwood (1964) as part of the SSARR model U.S. Army Corps of Engineers (1972) and by Anderson and Crawford (1964) for use in conjunction with the Stanford Watershed model. Both models used air temperature as the sole index to energy exchange across the air–snow interface. This is also true of subsequent models developed by Eggleston *et al.* (1971) and Anderson (1973). Generalized snowmelt equations from the publication *Snow Hydrology* (1956), based on theoretical and empirical considerations, were used in several snow-cover simulation models (Amorocho and Espildora, 1966; Carlson *et al.*, 1974), and as an option in the SSARR model. The net radiation balance, estimated from incoming solar radiation and air temperature, was used to compute energy exchange in the snow-cover simulation model developed by Leaf and Brink (1973). This model was developed to determine the probable hydrologic effects of forest management.

All of the models mentioned above compute snowmelt by one set of equations and the change in the heat deficit of the snow cover with a separate equation (the SSARR model does not compute the heat deficit). Thus, these models are not energy balance models. Some of these models use empirical procedures to estimate changes in the heat deficit during non-melt periods. The Eggleston *et al.* and the Leaf and Brink models used the one-dimensional Fourier heat-conduction equation and the assumption that the snow surface temperature ($T_0$) equals the air temperature ($T_a$) to compute temperatures at one or two points within the snow cover. Quick (1967) also used the heat-conduction equation and the assumption that $T_0 = T_a$ to compute changes in the snow-cover temperature profile during periods when depth can be considered to be constant. In addition, Quick accounts for the effect of the density profile on the snow-cover temperature profile.

In the 1970s, several energy balance snow-cover models were developed (Obled, 1973; Humphrey and Skau, 1974; Outcalt *et al.*, 1975). The energy balance included net radiation transfer, latent and sensible heat transfer, heat transfer by rain water, and the change in heat storage of the snow cover. The change in heat storage was also determined from the computed temperature profiles at the beginning (time $t$) and the end (time $t + \Delta t$) of each computational time interval ($\Delta t$). Finite-difference approximations to the Fourier heat-conduction equation were solved to determine the temperature profile at time $t + \Delta t$, knowing the profile at time $t$. The thermal conductivity of the snow varied with snow density. The snow surface temperature was determined by various iterative schemes that sought to reduce to an acceptable level the difference between the value of the change in the heat storage term in the energy balance equation and the value of the same term as determined from changes in snow-cover temperatures. None of these models included the densification of the snow cover. Humphrey and Skau used periodic measurements of the density profile to account for changes in snow density. The other two models merely used the measured total snow-cover density. Only Obled made comparisons between computed and observed values of snow-cover outflow and water equivalent. Humphrey and Skau, as well as Obled, showed comparisons of computed and observed snow-cover temperature profiles. Outcalt *et al.* only made comparisons between the computed and observed dates on which melt begins and ablation is complete.

Anderson's 1976 work evolved from an earlier study of snow-cover energy exchange (Anderson, 1968). In this earlier study, Anderson computed the change in heat storage only when an isothermal snow cover was cooling immediately following a melt period. Thus the model could only be used during extended snowmelt periods or periods of daytime melt and night-time heat loss. The model was tested on data collected as part of a lysimeter study of snowmelt at the Central Sierra Snow Laboratory (U.S. Army Corps of Engineers, 1956). The data included all the necessary meteorological input variables and snow-cover variables such as water equivalent, depth, temperature, and snow-cover outflow. There was very good agreement between computed and observed values of daily snow-cover outflow and mean night-time snow surface temperature. However, there were several problems. The period of record was quite short (17 days), plus there was some uncertainty regarding the accuracy of a portion of the data. More importantly, there was almost no variability in meteorological conditions during the period (warm days and cool nights with mostly clear skies and moderate winds). Thus, it was not possible to determine if the model was valid over a wide range of meteorological conditions.

Attempts at finding another high-quality data set that included measurements of all the necessary input and verification variables were unsuccessful. Because the data were not available to adequately test snow-cover energy exchange models, the National Oceanic and Atmospheric Administration (NOAA) and the Agricultural Research Service (ARS) (Johnson and Anderson, 1968) established a snow research station to study the physical processes in snowmelt and snow metamorphism. The station was located within the ARS's Sleepers River Research Watershed near Danville, Vermont. While not the most ideal location in terms of the amount of snow (average maximum water-equivalent of about 300 mm), the station was nearly ideal for observing the wide variety of meteorological conditions that occurred during snowmelt periods. Data collection at the NOAA–ARS snow research station began in December 1968. Measurement methods had changed over time due to advances in instrumentation and additional types of data were collected to provide more and better information for testing snow-cover energy exchange models. In addition to providing better data for model testing, this research enhanced the understanding of the energy transfer process and the numerical techniques needed to solve the basic snow-cover energy exchange equations. Clearly, a new, more theoretically sound, and more complete snow-cover energy balance model needed to be developed. The result was Anderson's point energy and mass balance model (Anderson, 1976). This model was based on surface energy balance equations and equations for energy transfer within the snow cover. The snow cover was divided into finite layers and the model included the mathematical representations of densification of the snow layers and the retention and transmission of liquid water.

It must also be mentioned that Navarre developed a numerical snow model named "Perce-Neige" (Navarre, 1975) with characteristics similar to those of Anderson's model (1976). Navarre's work was published in 1975 but only in French, which meant that very few subsequent papers on snow modeling referred to it.

From 1978 to 1983, the World Meteorological Organization (WMO) compared the various snowmelt runoff models (WMO, 1986a). The corresponding WMO report examined 11 snow models and described their degree of success in providing hydrological models with runoff data. These models used a very simple representation of internal physical processes, compared with Anderson's model (1976). Probably the simple construction of these models was due to limited computing resources at that time and to the fact that the effects of internal processes on energy balance were not considered to be as important as they are today.

During the late 1980s, with the increased knowledge of snow processes and meteorology, the availability of snow and weather data sets, and the general use of faster computers, more sophisticated snow models were developed and used in a research or operational context in the fields of hydrology, avalanche forecasting, and climate analysis. A major conceptual breakthrough came from snow metamorphism studies, in which researchers simulated layering, a fundamental characteristic of the snowpack (Fig. 4.1), (Brun *et al.*, 1992).

The challenge remains to determine how sophisticated snow models need to be for their intended scope of use. For that purpose, ICSI (International Commission on Snow and Ice) initiated a project called SNOWMIP, which compared the results from recent snow models when used on different climate conditions (Essery and Yang, 2001). In Section 4.2, Zong-Liang Yang describes some of these models and summarizes the results of a recent worldwide investigation on existing snow models which provides a precise view of the state of the art in snow modeling. It shows a very large variability in snow model features and designs.

Section 4.3 discusses the relative importance of major energy exchange processes at the snow surface or within the snowpack in terms of their impact on snow

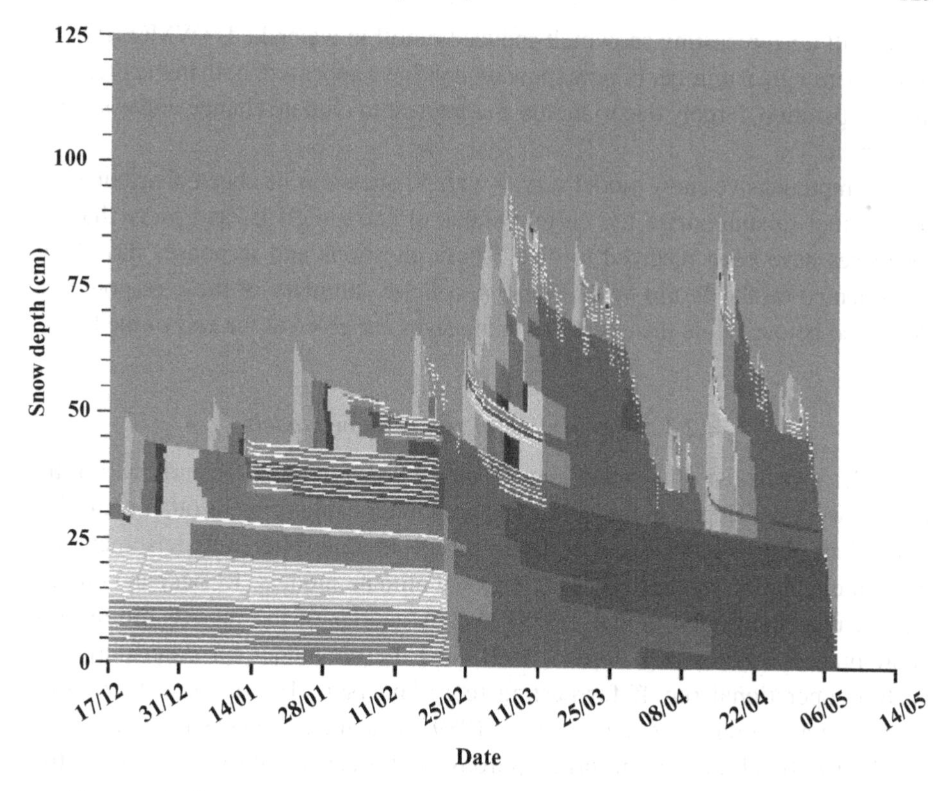


Figure 4.1. Simulation of temporal evolution of snowpack layering at Col de Porte during winter 1998/99. Each color represents a snow type (see Brun *et al.*, 1992). (Plate 4.1.)

simulations in a stand-alone model. Section 4.4 gives an overview of the representation of snow processes in different climate models.

## 4.2 Description of recent snow models

*Zong-Liang Yang*

### *4.2.1 Introduction*

The presence of snow on the earth's surface affects physical, chemical, and biological processes over a wide range of spatial scales and has important societal effects. In addition, the timing of snowmelt and the subsequent fate of melt water play an important role in the hydrological cycle and water resources (Gray and Prowse, 1993). To accommodate various applications, large numbers of snow models have been designed over the past few decades. In the mid 1980s, the World Meteorological Organization (WMO) conducted a survey of 11 snowmelt–runoff models that were built for forecasting snowmelt-induced runoff in watersheds (WMO, 1986b). Since then, a great number of new snow models have appeared, both in the literature and in operation, largely due to increasing interest in climate change and assessing its impact.

A comprehensive snow model survey with 50 questions has been distributed via the internet to summarize the current status of snow models, and more than 40 responses have been received to date. These questions and responses have been documented on the World Wide Web.[1] A concise summary of these responses is described below, while the questions themselves are given at the end of the book.

### 4.2.2 Description of recent snow models

Table 4.1 gives a list of the snow models and source references. Nearly half of the models were developed in the USA. The remainder came from 13 other countries. About half of the models were designed for use in atmospheric models, including general circulation models (GCMs), regional climate models (RCMs) and numerical weather prediction models (NWPMs). A quarter of the models are used in snow process studies (e.g. Jordan, 1991) and the remainder for other purposes, such as operational runoff forecasting (e.g. Anderson, 1973), snow–frozen-soil studies (e.g. Flerchinger and Saxton, 1989), avalanche forecasting (e.g. Brun *et al.*, 1989), climate monitoring (Grody and Basist, 1996), erosion control (Tarboton and Luce, 1996), downscaling GCM output (Hewitt 1997, personnal communication), and testing the optimum complexity of snow models for GCMs (Loth *et al.*, 1993). Some models have multiple purposes. For example, the model by Loth *et al.* (1993) is sufficiently sophisticated for snow process studies, but also computationally efficient enough for use in atmospheric models. Such a model may be an ideal tool for examining the optimum complexity of snow models for GCMs. Most of the models use precipitation, air temperature, wind speed, humidity, downward solar radiation, and downward longwave radiation for snow surface energy and water budget computations. Three snowmelt–runoff models simply use precipitation and air temperature as input (Anderson, 1973; Lindstrom *et al.*, 1997; Schreider *et al.*, 1997), while one model also uses wind direction to determine snow redistribution due to blowing snow events (Liston and Sturm, 1998).

The models display a wide range of complexity in coding, from tens of lines (Grody and Basist, 1996) to tens of thousands of lines (Jordan, 1991). In terms of model structure, almost all the models focus on the snow processes in the vertical dimension only and most of the models use a single layer to represent the whole snowpack. There is little consensus regarding the parameterization of snow albedo.

[1] http://www.geo.utexas.edu/climate/Research/SNOWMIP/snowmip.htm

Table 4.1 *List of snow models (in alphabetical order).*

| Model name | Application | Retention & percolation | Snow–vegetation interaction | Subgrid topography | Reference |
|---|---|---|---|---|---|
| Australian | Downscaling GCM output | No | No | No | Hewitt (1997, personal communication) |
| BAIM (biosphere–atmosphere interaction model) | GCM and RCM | Yes | Yes | No | Mabuchi *et al.* (1997) |
| BASE (best approximation of surface exchange) | GCM | No | Yes | No | Desborough and Pitman (1998) |
| BATS (biosphere–atmosphere transfer scheme) | GCM and RCM | No | Yes | No | Dickinson *et al.* (1993); Yang *et al.* (1997) |
| CLASS | GCM, RCM and NWPM | Yes | Yes | No | Verseghy (1991) |
| CROCUS | Understanding snow processes, operational avalanche forecasting in France, and for use in GCM | Yes | No | No | Brun *et al.* (1992) |
| DHSVM (distributed hydrology soil vegetation model) | GCM | Yes | Yes | Yes | Wigmosta *et al.* (1994) |
| DARSSM (division of atmospheric research snow and soil model) | Understanding snow processes, forecasting runoff | No | Yes | No | Kowalczyk (1999, personnal communication) |
| ECHAM snow energy and mass budget model | GCM and RCM | No | Yes | No | Roeckner *et al.* (1996) |
| ECHAM multi-layer snow model | Testing the optimal complexity of a snow cover model for GCM and RCM | Yes | Yes | No | Loth *et al.* (1993) |

(*cont.*)

Table 4.1 (*cont.*)

| Model name | Application | Retention & percolation | Snow–vegetation interaction | Subgrid topography | Reference |
|---|---|---|---|---|---|
| GFDL snow model | GCM | No | No | No | Manabe (1969) |
| Hadley centre/UKMO GCM land surface model | GCM and NWPM | No | Yes | No | Essery (1998a) |
| HBV | Forecasting runoff and hydropower operation | No | Yes | Yes | Lindstrom *et al.* (1997) |
| IAP94 | GCM | Yes | Yes | No | Dai and Zeng (1997) |
| IHACRES snow model | Forecasting runoff, flow (real time), and estimating possible climate change impacts | No | No | Yes | Schreider *et al.* (1997) |
| INM (snow model of the Instituto Nacional de Meteorologia) | Forecasting runoff | Yes | No | No | Fernandez (1999) |
| ISBA (interactions soil–biosphere–atmosphere) | GCM and NWPM | No | Yes | No | Douville *et al.* (1995) |
| Layered snow model for climate study | Understanding snow processes, forecasting runoff, and for use in GCM | Yes | Yes | No | Sun *et al.* (1999) Jin *et al.* (1999) |
| MAPS/RUC soil–vegetation–snow model | NWPM | No | No | No | Smirnova *et al.* (1997) |
| Mosaic | GCM | No | Yes | No | Koster and Suarez (1996) |
| MRI-CGCM ground hydrology model | GCM | No | No | No | Tokioka *et al.* (1995) |
| Melbourne University snow model (MU-SNW) | NWPM and GCM | No | Yes | Yes | Walland and Simmonds (1996) |
| NCEP/OH/OSU CAPS | NWPM | Yes | Yes | No | Koren *et al.* (1999) |

| | | | | | |
|---|---|---|---|---|---|
| NWSRFS SNOW-97 | Forecasting runoff | No | No | No | Anderson (1973) |
| RAMS snow model | Forecasting runoff | No | No | No | Lofgren (1997, personal communication) |
| RGM (regional geosystem model) | Understanding snow processes, forecasting runoff, and risk assessment of snow hazards | Yes | No | Yes | Scherer (1997, personal communication) |
| SEMS (snow evolution modeling system) | Understanding snow processes, NWPM, and RCM | Yes | Yes | Yes | Liston and Sturm (1998) |
| SHAW (simultaneous heat and water model) | Understanding snow/frozen soil/surface energy balance processes | Retention: yes; Percolation: no | Yes | No | Flerchinger and Saxton (1989) |
| SLURP | Forecasting runoff | No | Yes | Yes | Kite (1995) |
| SNAP (snowmelt numerical–analytical package) | Understanding snow processes and forecasting runoff | Yes | No | No | Albert and Krajeski (1998) |
| SNOWPACK | Understanding snow processes, forecasting runoff, and avalanche warning | Yes | No | No | Lehning *et al*. (2002a, b) Bartelt and Lehning (2002) |
| SNTHERM | Understanding snow processes and forecasting runoff | Yes | No | No | Jordan (1991) |
| SNTHERM (spatially distributed) | Understanding snow processes and forecasting runoff | Yes | Yes | Yes | Hardy *et al*. (1998) |

(*cont.*)

Table 4.1 (*cont.*)

| Model name | Application | Retention & percolation | Snow–vegetation interaction | Subgrid topography | Reference |
|---|---|---|---|---|---|
| SNTHERM.ver4 | Understanding snow processes | Yes | No | No | Stamnes *et al.* (1988) |
| SOIL | Understanding snow and frozen soil processes | Retention: yes; Percolation: no | Yes | No | Johnsson and Lundin (1991) |
| Special sensor microwave imager (satellite) derived snow cover model | Forecasting runoff, GCM, NWPM, and climate monitoring | Yes | Yes | No | Grody and Basist (1996) |
| SPONSOR | GCM, RCM, and NWPM | Yes | Yes | No | Shmakin (1998) |
| SPS (soil–plant–snow) | Understanding snow processes, RCM, and NWPM | No | No | No | Kim and Ek (1995) |
| SSiB (simplified simple biosphere model) | GCM and NWPM | No | Yes | No | Xue *et al.* (1991) |
| TSCM1 (Tohoku snow cover model with one-layer) | Understanding snow processes and forecasting runoff | Yes | No | Yes | Yamazaki (1995) |
| TSCMM (Tohoku snow cover model with multi-layer) | Understanding snow processes and forecasting runoff | Yes | No | No | Yamazaki (2001) |
| UEB (Utah energy balance snow accumulation and melt model) | Understanding snow processes and for runoff, erosion, and water balance forecasting, and modeling | Yes | No | Yes | Tarboton and Luce (1996) |

Some models assume that albedo is constant, or a function of snow age only, or a function of snow depth only, while some assume that it depends on several parameters, including grain size and impurity. There are many other approaches that are largely empirical. Most of the models neglect the spectral and directional differences in solar radiation transfer, while others make partial or full allowance for these differences. Half of the models use fixed values for thermal parameters (heat capacity and conductivity), while the remainder assume that these thermal parameters change with density. More than half of the models neglect the retention of snowmelt water and its percolation, and only 10% of the models treat vapor transfer processes within the snowpack. More than 50% of the models also treat the frozen soil process underneath snowpack, and half of the models take account of aspects of the snow–vegetation interaction, but few incorporate sophisticated radiative transfer and aerodynamic processes within the canopy and realistic simulations of snow cover under a forested floor. A quarter of the models incorporate the effects of subgrid scale topography on the distribution of precipitation, air temperature, and snow depth, and about 10% of the models use remote sensing data for input and validation.

Although detailed one-dimensional snow models do exist (e.g. Anderson, 1976; Jordan, 1991), for computational reasons atmospheric models use relatively simple snow models (Manabe, 1969; Verseghy, 1991; Dickinson *et al.*, 1993; Loth *et al.*, 1993; Lynch-Stieglitz, 1994; Marshall *et al.*, 1994; Dai and Zeng, 1997; Sun *et al.*, 1999). These GCM snow models, generally with one to five snow layers, are designed to resolve the diurnal and seasonal variations of surface snow processes such as surface temperature and heat fluxes, and simplify the treatment of the internal snow processes (e.g. the retention and transport of meltwater, melting and freezing, diffusion of temperature and water vapor, and the extinction of solar radiation). Because GCMs have been shown to be highly sensitive to snow processes (e.g. Yeh *et al.*, 1983; Cess *et al.*, 1991), it is critical to have a one-dimensional snow model with adequate realism that is efficient enough for long-term climate integrations.

Features of process-level snowmelt models could be helpful for capturing subgrid scale variability and improving snowmelt simulations. Consequently, an active line of research is being established that links physically based snowmelt models, geographical information system (GIS) analysis, remote sensing technology, and assimilated data sets from mesoscale meteorological models in studies of small catchments and watersheds (e.g. Hardy *et al.*, 1998; Liston and Sturm, 1998). However, relatively speaking, what is most lacking is the research and the application of snow water and energy budget analyses at continental and global scales. A question that remains unresolved is the level of complexity required for snow models at those scales. A closely related question is how to relate a one-dimensional, vertical snowpack model to heterogeneous surfaces in each of the GCM land grids. Specifically, what is the optimum methodology to scale up when snow cover is patchy? The heterogeneous distribution of vegetation and topography adds more complexity to this problem. Hopefully, the conclusions drawn from research at subresolution scales (e.g. local, catchment, and watershed scales) will provide an important guide towards developing and improving snow process models at GCM scales.

### 4.2.3 Summary and future directions

The preliminary analysis from a recent survey of snow models indicates that there are many sophisticated snow models currently available which are appropriate at the point or local scale, and that a wide range of models have been developed for application in small catchments and watersheds. These process-level snow models are useful in providing guidance towards improving GCM snow models. The future lines of research for snow modeling in climate models are:

- to develop, through testing against field data and detailed snow models, an optimum snowpack model, which not only simulates the surface snow processes but also captures the soil temperature variations under the snowpack;
- to link this model to heterogeneous vegetation and topography distribution; and
- to utilize remotely sensed data in deriving vegetation and snow parameters, and in validating model simulations.

## 4.3 Sensitivity of energy and mass fluxes at the snow–atmosphere interface to internal and interface parameters

*Eric Brun*

### 4.3.1 Introduction

Chapter 2 describes the complexity of snow physics. Most of the processes occurring inside the snowpack or at the snow–atmosphere interface can be represented in sophisticated snow models (see Sections 4.1 and 4.2) but not in snow parameterizations developed for General Circulation Models (GCMs) (see Section 4.4) because of the necessary limitation of computation time. Some snow processes have little effect on snow–atmosphere energy and mass exchanges while others are critical. The first group can be neglected in snow parameterizations for GCMs while particular attention should be paid to the second group.

This section describes the sensitivity of snow models to different interfaces or internal parameters. The effects of surface heterogeneity are not considered here; they are discussed in Section 4.4. Following Loth and Graf (1998b), this investigation used a reference run of a snow model and compared it with runs with the same model where input parameters or parameterizations were changed. In this investigation, we used the snow model CROCUS (Brun *et al.*, 1992) to compare simulated snow depth. Although snow water equivalent (SWE) is of greater interest and easier to interpret than snow depth in most applications, snow depth is better for assessing the performance of a snow model for the following two reasons: (1) it can be automatically measured with great accuracy, which means that daily and even hourly series of observed snow depth are available; (2) SWE is only sensitive to the energy and mass balance of the snow cover, while variations of snow depth also depend on the densification of the snow cover, which is a major process affecting the thermal properties of the snowpack. The CROCUS model runs cover the complete snow season 1993–1994 at Col de Porte (French Alps, 1320 m a.s.l.).

The sensitivity of the model's results to the following parameters has been investigated:

- snow albedo;
- liquid water retention;
- sensible and latent heat fluxes;
- rain–snow criterion;

other parameters also affect model results. That is particularly the case with snow emissivity but currently there is general agreement on its values, thus it is not necessary to investigate model sensitivity to it.

### *4.3.2 Sensitivity to albedo parameterization*

Albedo is certainly the first physical property of snow that climatologists need to consider in climate simulations because it is considered as a major source of positive feedback (Randall *et al.*, 1994). Since snow reflectance strongly depends on the size and shape of snow grains, a physically based calculation of albedo requires calculating the metamorphism of the different snow layers. Since it cannot be simply achieved, snow parameterizations designed for GCMs usually calculate snow albedo as a function of the age of the snow surface. This is meaningful because metamorphism generally induces grain growth when snow is ageing. Nevertheless, high temperature gradients prevailing in seasonal snowpacks may often create faceted grains (depth hoar), large grains but with small optical size because of their facets (see Sections 2.2.4 and 2.5.2). In such common cases, a calculation of snow albedo from the age of the surface snow can lead to a significant underestimation (Sergent *et al.*, 1998).

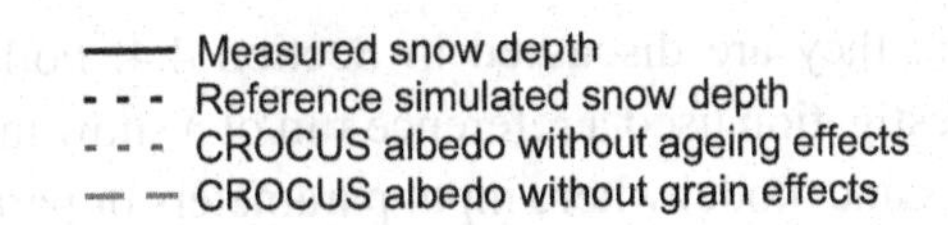


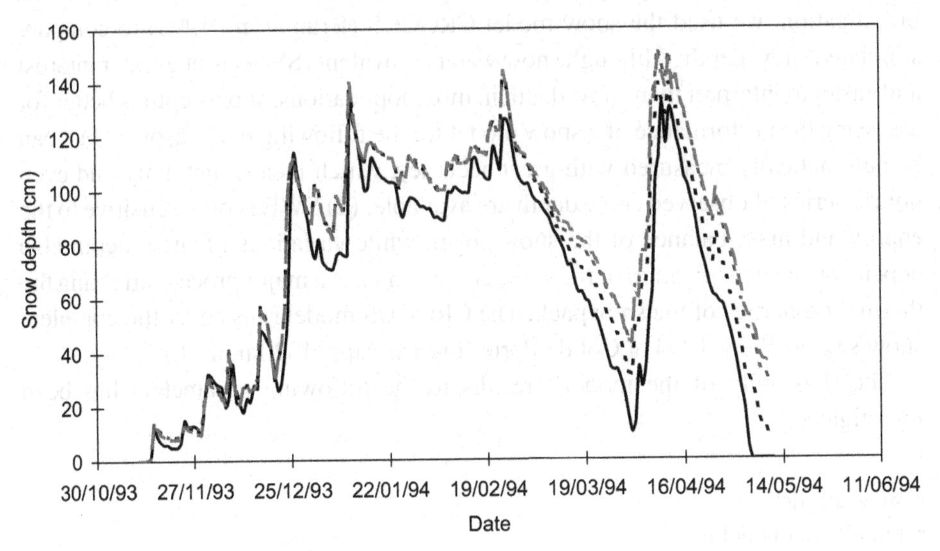


Figure 4.2. Sensibility to ageing and grain size in albedo calculation on snow depth simulations. (Plate 4.2.)

To emphasize the sensitivity of snow modeling to the parameterization of the albedo, six simulations have been performed with CROCUS where only the calculation of albedo differs.

- The first using the original version of CROCUS where albedo is calculated on three wavelength bands ($[0.3{-}0.8 \times 10^{-6\,\mathrm{m}}]$, $[0.8{-}1.5 \times 10^{-6\,\mathrm{m}}]$ and $[1.5{-}2.8 \times 10^{-6\,\mathrm{m}}]$) and depends on grain size and shape and on the snow surface age. Ageing acts only on the first band. It is supposed to represent the deposition of dust at the surface.
- The second with the CROCUS dependence of albedo on the grain's size and shape on the three bands but without taking ageing into account.
- The third with the CROCUS dependence of albedo on the age of the snow surface but not on the size and shape of the grains.
- The fourth to sixth simulations with a constant value for the albedo respectively equal to 0.85, 0.725, and 0.60 in all bands. Generally, snow albedo varies between 0.85 and 0.60.

In all simulations, forcing shortwave downward radiation was split into three wavelength bands from pyranometer measurements according to a parameterization of the shortwave spectral distribution as a function of cloudiness and of the ratio between direct and diffuse radiation.

Figure 4.2 represents simulations one to three and shows that the effects of grain size on snow albedo and the effects of ageing of the snow surface are in the same

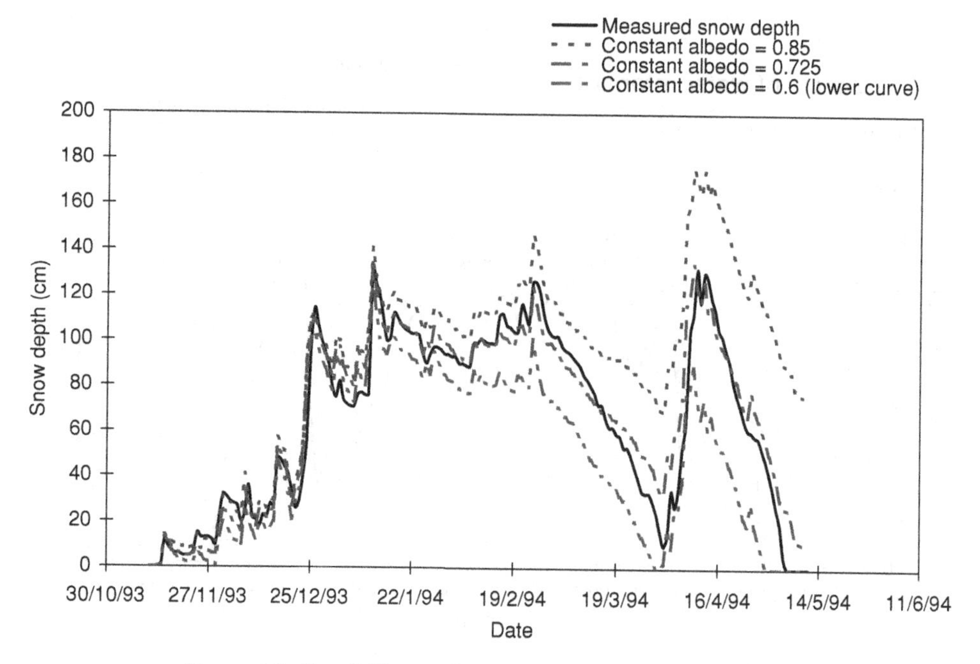


Figure 4.3. Sensibility to albedo on snow depth simulations. (Plate 4.3.)

order of magnitude. Figure 4.3 represents simulations four to six and illustrates the sensitivity of the snow model to the albedo. With the two extreme values of 0.85 and 0.60, complete melting of the snow cover differs by about two months. With the average value of 0.725, the model gives reasonable results but it is compensating between a too active melting during the accumulation period (end of November, February) and a less active melting during the melting periods in March, April, and May. This clearly shows that despite the extreme sensitivity of snow models to the albedo, some simple parameterizations can give reasonable results (Essery *et al.*, 1999).

### 4.3.3 Sensitivity to the parameterization of water retention

As seen in Section 2.4, wet snow retains a significant amount of immobile liquid water. Except in very high-latitude regions, the snowpack commonly undergoes diurnal melting–refreezing cycles during the melt season. A part of the liquid water produced during the day at the surface is retained by immobile saturation and may refreeze during the following night, typically under clear-sky conditions. If we consider a snow layer of 300 kg m$^{-3}$ with a common immobile saturation of 5% of the pore volume, a common refreezing depth of $15 \times 10^{-2}$ m corresponds to a

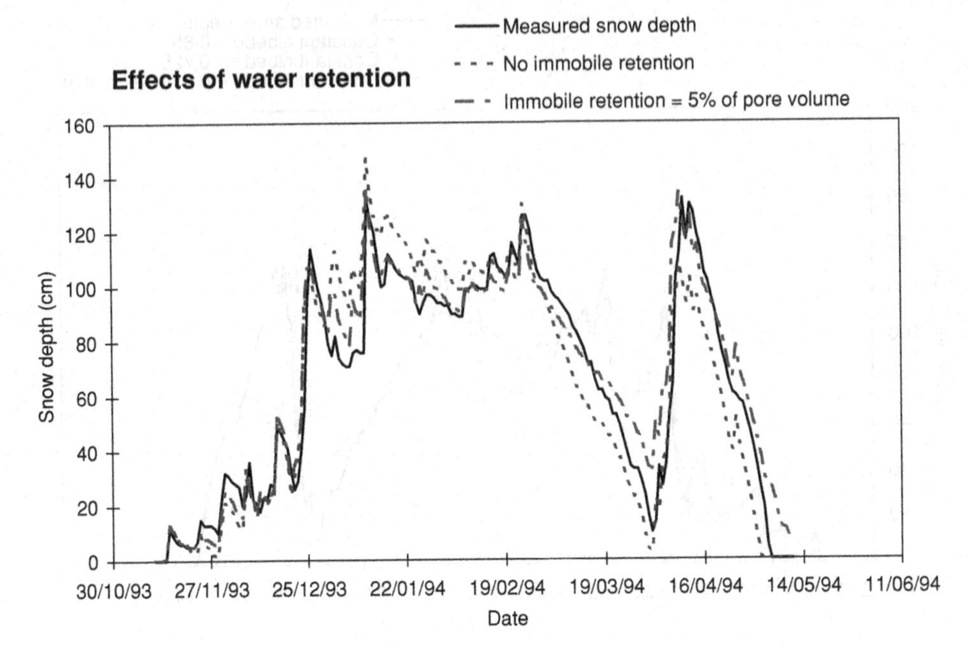


Figure 4.4. Sensibility to water retention on snow depth simulations. (Plate 4.4.)

refreezing of about $5 \times 10^{-3}$ m of liquid water, which corresponds to a release of latent heat equal to about $1.5 \times 10^{6}$ J m$^{2}$. In such a case, this amount of refrozen water may represent about 30% of the snow water equivalent produced by melting during the previous day. This refrozen water will be available for a second melting during the following day or later, increasing the energy necessary for the melting of the snowpack. Most snow parameterizations for GCMs cannot take this effect into account because its physically based simulation requires multi-layering with numerous thin layers close to the surface.

To illustrate the sensitivity of snow modeling to the parameterization of this process, we ran two versions of the CROCUS model which differ only by the simulation of immobile saturation. To avoid feedback effects, the albedo was fixed to a constant value of 0.725 in both versions because suppressing water retention in the model had a strong impact on metamorphism and on albedo. Figure 4.4 compares the results of both versions. Focusing on the very active melting period extending from mid-February to the end of March, it is obvious that the melting rate is significantly increased when retention is not taken into account in the snow model. An apparently contrary effect is observed at the beginning of January. Indeed, the absence of retention affects the compaction rate of wet snow, which means that the wetting event following the heavy snowfall recorded at the end of December induces much less densification without retention than with retention. Therefore, the simulated snow depth during this period when water retention is not taken into account is greater.

The increase of the melting rate during melting–freezing cycles simulated when water retention is not taken into account would be enhanced at locations where diurnal cycles show large amplitudes. This is typical at high elevation in mid- or low-latitude regions where meteorological situations with intense solar radiation during the day and strong refreezing during the night are common. A snow model that does not take into account retention and the refreezing of immobile liquid water would calculate a water outflow at the bottom of the snowpack as soon as surface melting occurs. Indeed, a shift of one month between the onset of surface melting and the runoff of melting water is common at high elevations in alpine regions. At least for hydrological purposes, this process must be considered in snow models.

### *4.3.4 Sensitivity to the parameterization of turbulent fluxes*

Sensible and latent heat exchanges significantly contribute to the energy and mass balance of the snowpack (see Section 3.2). A common feature of snow-covered regions is that the surface boundary layer is stable most of the time. Since the present knowledge of turbulent fluxes under stable conditions is still relatively poor, snow models can calculate these fluxes in different ways. In most parameterizations, stability and roughness length are key parameters that strongly affect the calculated fluxes. Large uncertainties still exist regarding the effect of high stability on the decrease of turbulent fluxes (Kondo *et al.*, 1978; Musson-Genon, 1995) as well as on the roughness length of the different types of snow surfaces (Martin and Lejeune, 1998). To show the sensitivity of a snow model to both stability and roughness length, we have performed different runs with various versions of CROCUS, which use a parameterization of turbulent fluxes deduced from Deardoff (1968).

The effects of stability are calculated by using the ratio between the transfer coefficient under stable or unstable conditions and the transfer coefficient under neutral conditions. This ratio depends on the bulk Richardson number $Ri_B$ (see Section 3.3.4). When $Ri_B$ tends to 0.2, the transfer coefficient rapidly tends to 0 and then inhibits any turbulent flux; 0.2 is quite a common value for $Ri_B$ above snow, and field observations show that turbulent fluxes cannot be neglected under such stable conditions (Martin and Lejeune, 1998). Figure 4.5 compares the results of three different versions of CROCUS. The first version uses Deardoff's parameterization (full effect of stability), which induces an obvious underestimation of the melting rate during spring. This is due to the underestimation of sensible heat fluxes during the relatively warm conditions prevailing in March. Indeed, since the snow surface temperature is limited to the melting point, $Ri_B$ is generally larger than 0.2 in such

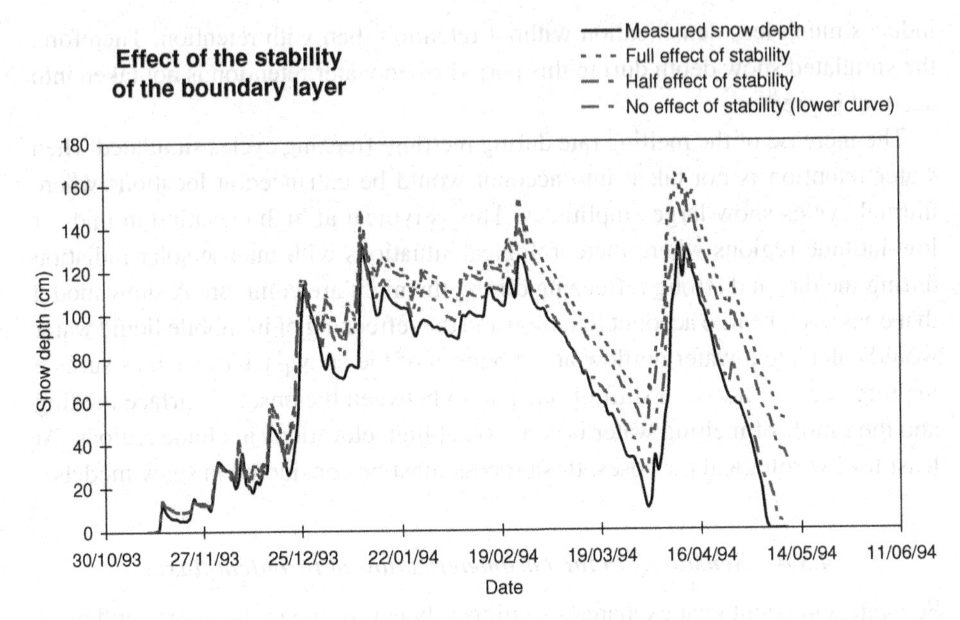


Figure 4.5. Sensibility to the stability of the boundary layer on snow depth simulations. (Plate 4.5.)

meteorological conditions and Deardoff's equations calculate negligible fluxes. The second version uses Deardoff's parameterization for unstable conditions but considers neutral conditions in the calculation when conditions are stable (no effect of stability). With this version, the melting rate is much more realistic. The third version is intermediate. It considers the effect of stability under stable conditions but it limits the decrease of turbulent fluxes to half of the turbulent fluxes calculated for neutral conditions (half of the effect of stability). The results are intermediate between both previous cases. This comparison shows that particular attention must be paid to the parameterization of the effects of the stability of the surface boundary layer on the calculation of sensible and latent heat fluxes. Because no completely suitable parameterization is presently available, numerous snow models neglect the effects of stability under stable conditions and consider that the conditions are neutral. However, GCMs take stability effects into account when calculating turbulent fluxes over snow-covered regions.

Under stable, neutral, or unstable conditions, the turbulent fluxes of sensible and latent heat depend on the roughness length $z_0$ of the surface. Even over homogeneous snow-covered surfaces, an accurate knowledge of the roughness length of snow surfaces is still lacking. The roughness length $z_0$ of seasonal snow commonly ranges around $2 \times 10^{-4}$ m but it varies over a large spectrum. Figure 4.6 compares the snow depth simulated with different versions of CROCUS where only $z_0$ changes. To

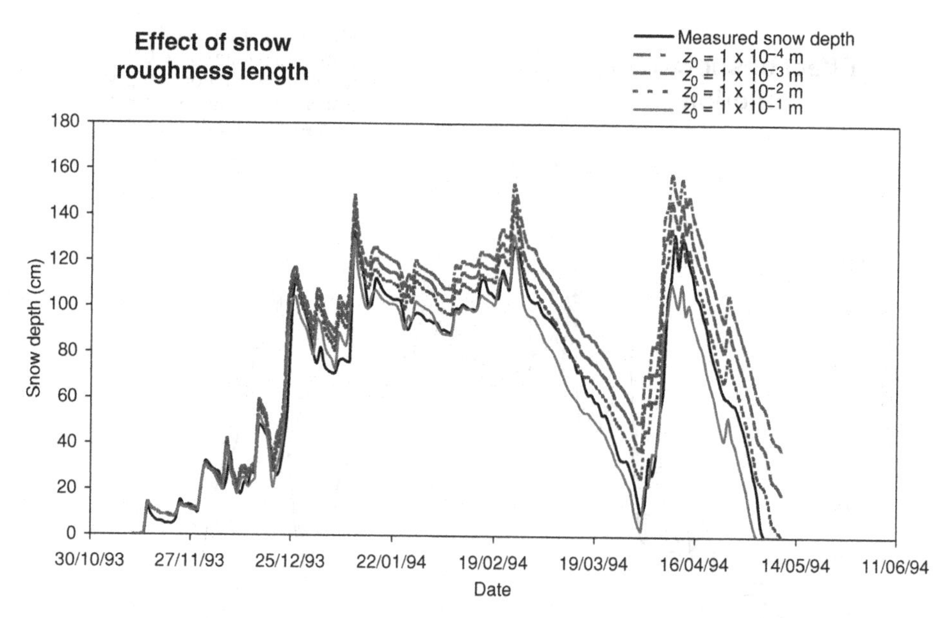


Figure 4.6. Sensibility to snow roughness length on snow depth simulations. (Plate 4.6.)

emphasize the comparison, turbulent fluxes under stable conditions are calculated as under neutral conditions. It shows that the value given to $z_0$ has a strong impact on the model results. The smaller the roughness length, the smaller the turbulent fluxes and consequently the smaller the melting rate because sensible fluxes are positive, i.e. downward most of the time. At Col de Porte, the best results are obtained when $z_0$ is equal to $3 \times 10^{-3}$ m. This rather high value for snow is partially due to the forest surrounding the test site.

### *4.3.5 Sensitivity to the snow–rain criterion*

In nature, snowfall and rainfall occur over a range of temperatures. The temperature at which snow flakes melt and turn to rain drops depends on the vertical profile of temperature and humidity in the atmosphere and on the precipitation rate. In snow models and parameterizations, separation between snow and rain is generally deduced from the air temperature when direct observations of the precipitation type are not available. Several studies on the snow–rain criterion have been conducted but no universal criterion exists. For example, the criterion is different in regions of flat terrain where the surface boundary layer is well developed compared with the top of a mountain where meteorological conditions are closer to the free atmosphere conditions. To illustrate the sensitivity of snow modeling to the snow–rain criterion,

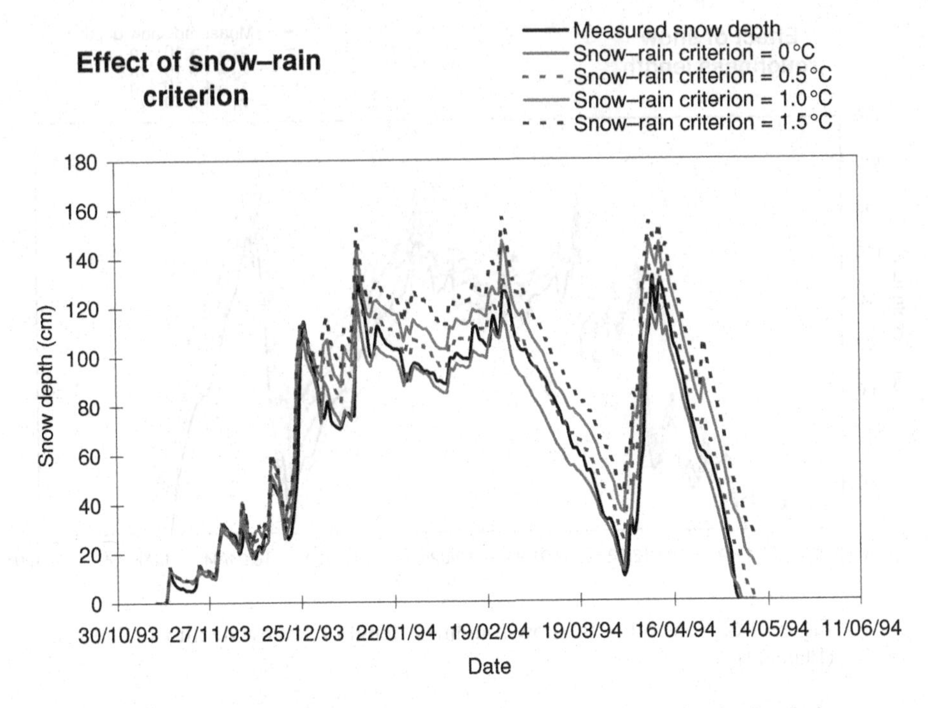


Figure 4.7. Sensibility to snow–rain criterion on snow depth simulations. (Plate 4.7.)

we ran four versions of CROCUS where the unique change was a criterion based on the air temperature varying from 0 to 1.5 °C. Figure 4.7 compares the snow depth simulated with the four versions. At Col de Porte, during the season 1993/1994, most of the differences came from a rainfall event occurring at the beginning of January at a temperature close to the melting point. It is obvious that the sensitivity of snow models to the snow–rain criterion strongly depends on the prevailing climate of the considered region. In temperate regions, winter snowfalls often occur at relatively warm temperatures and model results should be very sensitive to the criterion. In contrast, in cold and continental regions snowfalls occur at relatively cold temperature and the transition from winter to summer and from summer to winter is short enough to predict that snow modeling should not be very sensitive to this criterion.

### *4.3.6 Conclusions*

Snow models are very sensitive to the parameters that affect the energy balance of the snowpack. Therefore, particular attention should be paid to the choice of these coefficients or parameterizations. Fortunately, compensation for the effects of different processes often limits the consequences of a given parameterization, which means that different models may have similar results (Essery *et al.*, 1999).

In addition to the choice of parameter values, choosing the number of computational layers is of major importance to the model results and the model costs. Boone and Etchevers (2001) have shown that a three-layer snow model can perform almost as well as a more sophisticated model, at least in terms of energy balance, surface temperature, and snow–water equivalent.

In order to provide scientists with relevant guidelines for designing snow models appropriate to their region of interest, the International Commission on Snow and Ice initiated a project called SNOWMIP (snow model intercomparison project) (Essery and Yang, 2001). The objective of this project was to compare snow models of various complexities at four sites belonging to various climatic regions. A total of 24 models from 18 teams were involved. The models varied from simple models designed for hydrology to sophisticated ones designed for snow physics research. The main conclusions of the first phase of the project are given below (Etchevers, *et al.*, 2003).

- Some models showed a good ability to correctly simulate the snowpack features for all of the sites, whereas other models appeared to be more adapted to particular conditions. The high alpine site was the best simulated site, because the accumulation and melting periods are distinct.
- A detailed analysis showed that parameterization of the albedo was critical. A parameterization depending only on snow age gave less reliable results on the onset of snowmelt. Accounting for snow evolution in a better way (using a density evolution, or a grain size evolution as in detailed models) generally improves the models.
- The retention of water in the snow cover is also important. Bottom runoff usually began earlier in models without water retention. The water retention allows a part of the daily snow melt to refreeze during the night.
- This intercomparison based on data from open sites is now extended to forested sites. The second phase will be devoted to the representation of the interaction between snow and forest. At the end of this last phase, SNOWMIP will provide unique and comprehensive information on the ability of snow models to simulate various aspects of snow dynamics.

## 4.4 Snow parameterization in GCMs

*Richard Essery*

### *4.4.1 Introduction*

General circulation models (GCMs) are three-dimensional numerical models of the global climate system; an introductory review of their use in climate modeling is given by McGuffie and Henderson-Sellers (1997). The atmospheric component of a GCM may be coupled to an ocean model or run with prescribed sea-surface temperatures and sea-ice extents to provide surface boundary conditions over oceans. Land-surface models, used to supply boundary conditions over land, have to take account of the influences of snow cover on interactions between the surface and the atmosphere because the unique properties of snow, discussed in Chapter 2, can present marked and rapidly varying modifications in the characteristics of the land surface. The large contrast in albedo between snow-covered and snow-free land, in particular, is often cited as providing a possible positive feedback mechanism for climate change; reduced snow cover in a warmer climate will tend to increase the absorption of shortwave radiation at the surface, reinforcing the warming. This simple interpretation neglects other feedbacks involving changes in snow cover: the surface temperature of snow-covered land cannot exceed 0 °C, limiting the outgoing longwave radiation and giving a negative feedback; systematic changes in cloud cover resulting from changes in snow cover could lead to positive or negative feedbacks; and warming could lead to *increased* snow cover in cold regions where snowfall is currently limited by moisture supply rather than by temperature.

In comparison with satellite observations of snow cover on continental scales, Frei *et al.* (2003) found that 15 GCMs participating in the second phase of the atmospheric model intercomparison project gave better results than the 27 GCMs in the first phase (Frei and Robinson, 1998), although consistent model biases remained over Eurasia. Cess *et al.* (1991) investigated snow–climate feedbacks by comparing results from 17 GCMs and Randall *et al.* (1994) analysed results from 14 of the GCMs in more depth. Pairs of perpetual-April simulations were carried out with fixed sea-ice and homogeneous perturbations of ±2 °C in sea-surface temperatures. Dividing resulting differences in global-mean surface temperature between simulations by differences in global-mean net radiation at the top of the atmosphere gives a climate sensitivity parameter for each GCM. A second sensitivity parameter was calculated from simulations where the snow cover was fixed, rather than allowed to respond to the perturbations. The ratio of these two parameters is interpreted as a measure of snow feedback, with values of greater than one indicating positive feedbacks. The 17 GCMs produced results, shown in Fig. 4.8, ranging from weak negative to strong positive feedbacks. More recently, Hall and Qu (2006) investigated the strength of snow albedo feedbacks in simulations of climate change by 17 GCMs for the Intergovernmental Panel on Climate Change fourth assessment report and, again, found a wide range in results. Differences in GCM snow feedback results are not solely due to differences in their representations of snow processes, but improved representations will allow greater confidence in predictions of climate change.

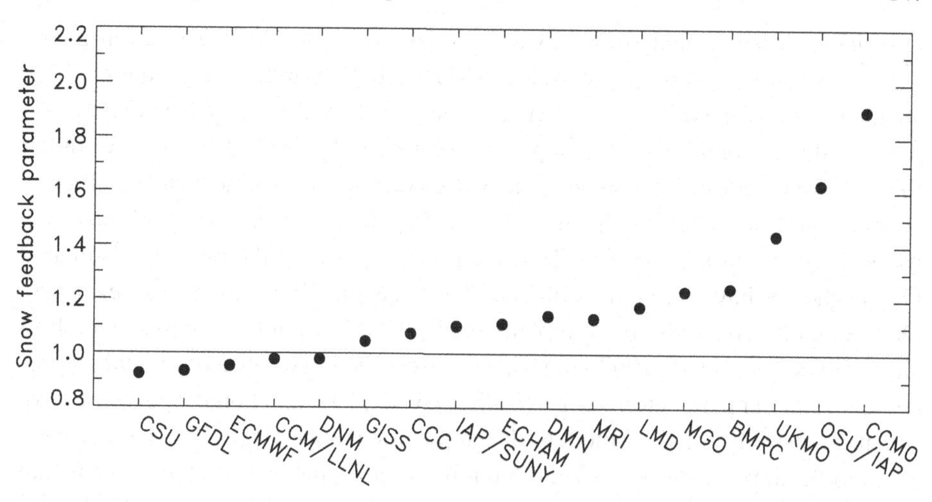


Figure 4.8. Snow feedback parameters from Cess *et al.* (1991) for 17 GCMs.

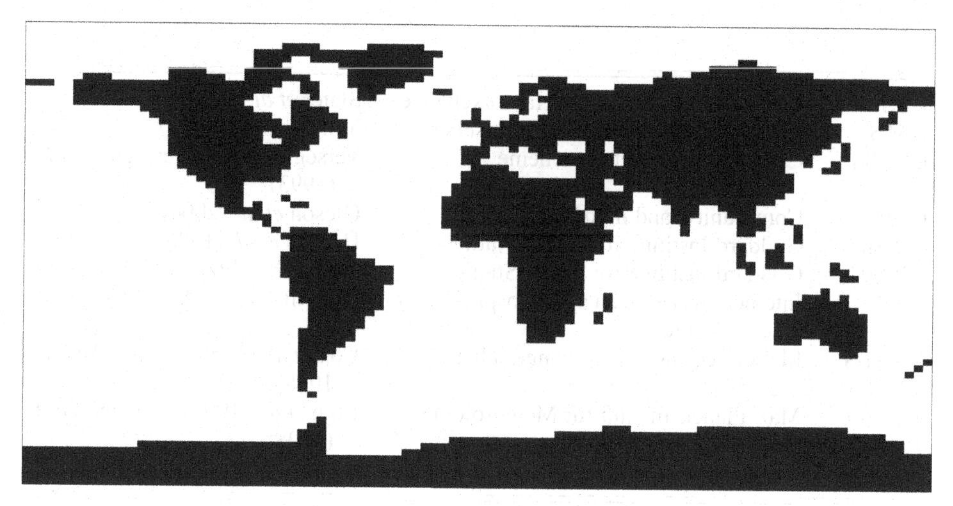

Figure 4.9. A global land mask with a typical GCM resolution of 2.5° latitude by 3.75° longitude.

The temporal and spatial resolutions of GCMs are strongly constrained by computational cost, particularly when investigating long-term changes in climate; the standard configuration of the HadAM3 climate model (Pope *et al.*, 2000), for example, has a horizontal resolution of 2.5° latitude by 3.75° longitude (Fig. 4.9), 19 vertical levels in the atmosphere and a 30 min timestep. The complexity with which physical processes can be represented is also limited, and processes on scales too small to be resolved by GCM grids have to be "parameterized" in terms of resolved quantities. Sophisticated snow-physics models, such as CROCUS (Brun *et al.*, 1989), SNTHERM (Jordan, 1991), and SNOWPACK (Bartelt and Lehning, 2002), are not suitable for use in GCMs, which ideally need snow models that are not computationally demanding, represent processes averaged over GCM grid scales rather than at a point, and are applicable in the widely varying environments in which snow cover occurs. Nevertheless, the insight gained from detailed snow models has been useful in the development of GCM snow models; many of the more sophisticated GCM schemes have adopted features of the Anderson (1976) model, for example. Detailed models have also been implemented in GCMs to study their impact on short simulations (Brun *et al.*, 1997) and used to assess the importance of individual processes in simpler parameterizations (Loth and Graf, 1998a,b; Pomeroy *et al.*, 1998).

An extensive list of snow models, developed for a range of applications, has been given in Section 4.2. Acronyms and references for a small selection of models that will be used to illustrate the discussion in this section are given in the table below, drawn from descriptions of GCMs, GCM land-surface schemes and snow models developed for use in GCMs.

| | | |
|---|---|---|
| BASE | Best approximation of surface exchange | Slater *et al.* (1998) |
| BATS | Biosphere–atmosphere transfer scheme | Yang *et al.* (1997) |
| CLASS | Canadian land-surface scheme | Verseghy (1991) Verseghy *et al.* (1993) |
| CLM | Community land model | Oleson *et al.* (2004) |
| GISS83 | Goddard Institute for Space Studies | Hansen *et al.* (1983) |
| GISS94 | Goddard Institute for Space Studies | Lynch-Stieglitz (1994) |
| ISBA | Interactions between soil, biosphere and atmosphere | Douville *et al.* (1995a, b) |
| MOSES | Met Office surface exchange scheme | Cox *et al.* (1999) Essery (1997a, 1998a) |
| MPI | Max-Planck-Institut für Meteorologie | Loth *et al.* (1993) Loth and Graf (1998) |
| SiB | Simple biosphere | Sellers *et al.* (1996) |

### 4.4.2 Thermal and hydraulic properties of snow

Changes in density and porosity due to compaction, crystal metamorphosis, and freezing of meltwater or rain cause the thermal and hydraulic properties of snow to vary with time (Sections 2.2–2.4). GCMs generally neglect snow hydrology and often simply adopt constant values for the density, heat capacity, and thermal conductivity of snow, but some more sophisticated parameterizations have been introduced.

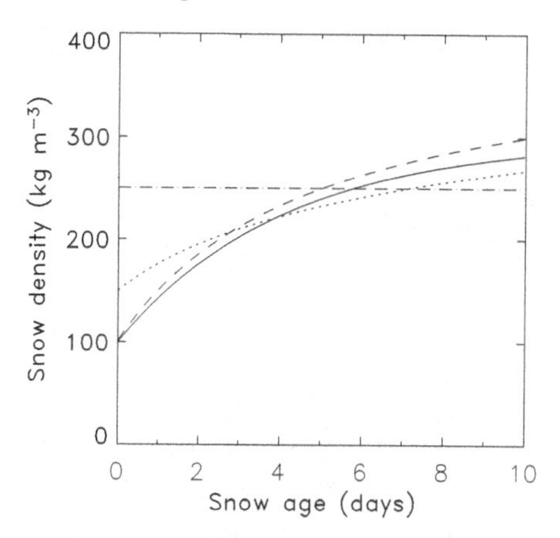


Figure 4.10. Snow density as a function of age from parameterizations used by CLASS (—), GISS94 (· · ·) and BATS (– – –), and the fixed value used by GISS83 and MOSES (–·–·–). Snow is assumed to be melting for GISS94 and BATS, and the snow mass is set to 100 kg m$^{-3}$ for GISS94.

CLASS and ISBA use compaction parameterizations in which the density of snow increases with time after snowfall. Given snow density $\rho_s$ at time $t$, the density at time $t + \Delta t$ is calculated as

$$\rho_s(t + \Delta t) = [\rho_s(t) - \rho_{max}] \exp\left(-\frac{\Delta t}{\tau_\rho}\right) + \rho_{max},$$

where $\rho_{max}$ is the maximum permitted snow density and $\tau_\rho$ is a time constant. BASE, BATS, GISS94, and MPI use similar parameterizations but with increasing compaction rates at higher temperatures. In both BASE and GISS94, the densification rate is given by

$$\frac{\partial \rho_s}{\partial t} = \frac{0.5\rho_s g M}{10^7 \exp(0.02\rho_s + 4000/T_s - 14.643)},$$

for snow mass $M$ (kg m$^{-2}$), snow temperature $T_s$ (K) and gravitational acceleration $g$ (m s$^{-2}$). CLM uses the more sophisticated scheme of Anderson (1976), which has separate rates for compaction due to overburden, metamorphosis, and melting. Results from these parameterizations are shown in Figure 4.10.

In all models with a variable snow density, the density is recalculated after snowfall as a weighted average of the new and old snow densities. Fresh snow density is usually set to a constant value (100 kg m$^{-3}$ in BASE, BATS, CLASS, and ISBA, 150 kg m$^{-3}$ in GISS94), but is parameterized as a function of air temperature or wet bulb temperature in some models; CLM, for example, follows Anderson (1976) in setting

$$\rho_{fresh} = 50 + 1.7(T + 15)^{3/2},$$

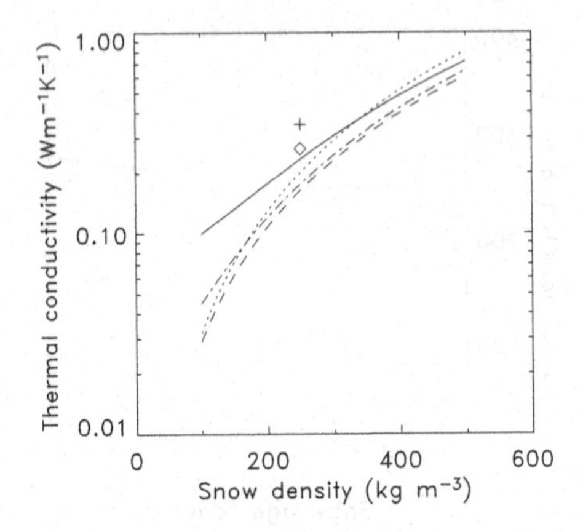


Figure 4.11. Thermal conductivity as a function of snow density from parameterizations used by CLASS (—), GISS94 (···), ISBA (– – –) and MPI (–·–·–), and fixed values used by GISS83 (+) and MOSES (◊).

for air temperature $T$ (°C), with the limits that $\rho_{\mathrm{fresh}}$ is between 50 and 169 kg m$^{-3}$.

The specific heat capacity of snow can be simply calculated as the sum of the heat capacities of the ice, water, and air mass fractions (Section 2.3), but transport of heat in snow is a complicated process involving conduction, advection, phase changes, and radiation (Section 2.3). Several empirical relationships have been proposed that give effective thermal conductivities for snow as functions of density (Anderson, 1976; Yen, 1981; Sturm *et al.*, 1997). Figure 4.11 compares parameterizations (cf. Fig. 2.10). GISS83 and MOSES, both of which assume a fixed snow density of 250 kg m$^{-3}$, and use fixed values of thermal conductivity. BASE, CLASS, GISS94, and MPI all use a quadratic function

$$k_{\mathrm{eff}} = a_1 + b_1 \rho_{\mathrm{s}}^2,$$

but choose different values for the parameters $a_1$ and $b_1$. CLM uses a different quadratic function

$$k_{\mathrm{eff}} = k_{\mathrm{a}} + \left(7.75 \times 10^{-5} \rho_{\mathrm{s}} + 1.105 \times 10^{-6} \rho_{\mathrm{s}}^2\right)(k_{\mathrm{i}} - k_{\mathrm{a}}),$$

from Jordan (1991), where $k_{\mathrm{air}}$ and $k_{\mathrm{ice}}$ are the conductivities of air and ice. ISBA uses a power-law relationship

$$k_{\mathrm{eff}} = k_{\mathrm{i}} \left(\frac{\rho_{\mathrm{s}}}{\rho_{\ell}}\right)^{1.88},$$

where $\rho_{\ell}$ is the density of water.

GCMs have mostly neglected the complexities of snow hydrology discussed in Section 2.4, instead draining meltwater instantaneously. As a simple improvement,

GISS94 allows a snow layer to retain up to 5.5% of its mass as liquid water, excess water draining through the bottom of the layer. MPI parameterizes this capacity as a function of density, decreasing from a maximum of 10% to 3% for densities of 200 kg $m^{-3}$ and greater. Liquid water may freeze within the snow in both of these models and CLASS.

### 4.4.3 Snow albedo

The albedo of a snow-covered surface is influenced by many factors, including the depth and grain structure of the snow, contaminants in the snow, the albedo of the underlying surface, heterogeneities in the snow cover, and masking by vegetation (Sections 2.5, 3.4). The simplest GCM snow models neglect these influences and assign a fixed albedo to any gridbox with snow cover. Snow albedos also vary greatly with the wavelength of incident radiation; although they split the solar spectrum into bands for calculating radiative transfers in the atmosphere, GCMs often use a single value of surface albedo for all wavelengths.

BASE, SiB, and the original version of MOSES represent snow ageing by making the albedo a function of temperature, decreasing as the temperature approaches the melting point. This approach gives an unrealistic increase in albedo when melting snow refreezes. A better representation is given by parameterizing the albedo as a function of the age of the snow surface. CLASS, ISBA, and MPI use exponential or linear relationships to increment snow albedos according to

$$\alpha_s(t + \Delta t) = [\alpha_s(t) - \alpha_{min}] \exp\left(-\frac{\Delta t}{\tau_\alpha}\right) + \alpha_{min}$$

or

$$\alpha_s(t + \Delta t) = \max\left[\alpha_s(t) - \frac{\Delta t}{\tau_\alpha}, \alpha_{min}\right],$$

where $\alpha_{min}$ is a lower limit on the snow albedo and the empirical time constant $\tau_\alpha$ is shorter for melting snow than cold snow. Values have to be assigned for the albedo of fresh snow and the depth of snowfall required to refresh the surface albedo. Maximum and minimum values of albedo are specified separately for visible and near-infrared bands in CLASS. In GISS83, the snow albedo is given by

$$\alpha_s = \alpha_{min} + (\alpha_{max} - \alpha_{min}) \exp(-a/5),$$

for snow surface age $a$ (in days) updated according to

$$a(t + \Delta t) = [1 - a(t)/50] \exp\left(-\frac{\Delta d_s}{d_c}\right),$$

where $\Delta d_s$ is the snowfall in a timestep and $d_c$ is the depth required to refresh the albedo.

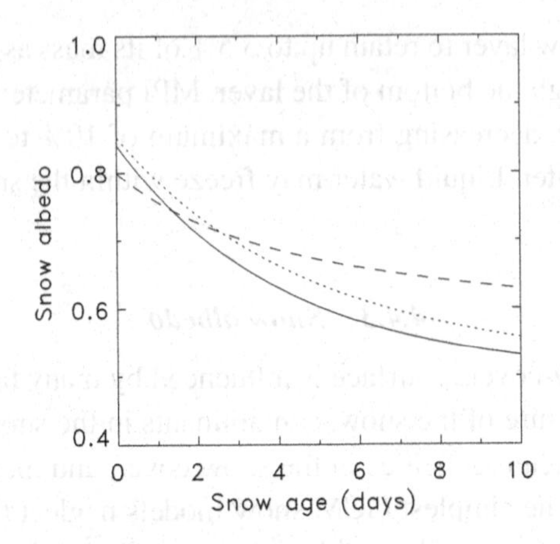


Figure 4.12. Albedo decay with age for melting snow from parameterizations used by CLASS (—), GISS83 (···) and BATS (– – –).

Marshall (1989) used results from a spectral snow albedo model (Warren and Wiscombe, 1980; Wiscombe and Warren, 1980) to develop a physically based parameterization for implementation in the NCAR GCM (Marshall and Oglesby, 1994). This parameterization has also been adopted as an option in MOSES (Essery *et al.*, 2003a). Given snow depth, grain size, soot content, and zenith angle, separate diffuse and direct beam albedos are calculated for visible and near-infrared wavelength bands. Grain growth rates are taken from measurements in cold and melting snow. BATS and CLM use a similar parameterization with a snow age parameter in place of the grain size. Figure 4.12 shows the albedo decay for melting snow in BATS, CLASS, and GISS83.

Many models parameterize gridbox-average albedos as functions of snow depth, since the influences of heterogeneity, vegetation masking, and absorption of radiation by the underlying surface are likely to be more significant for shallow snow. The albedo is typically taken to be a weighted average

$$\alpha = f_s\alpha_s + (1 - f_s)\alpha_0,$$

where $\alpha_s$ is the albedo of deep, homogeneous snow and $\alpha_0$ is the albedo of the snow-free surface. Functions used for the weighting have included

$$f_s = \frac{\text{HS}}{\text{HS} + a_2 z_0}$$

used by BATS, CLM, and ISBA,

$$f_s = 1 - e^{-a_2 \text{HS}}$$

used by GISS83 and MOSES, and

$$f_s = \min[a_2\text{HS}, 1]$$

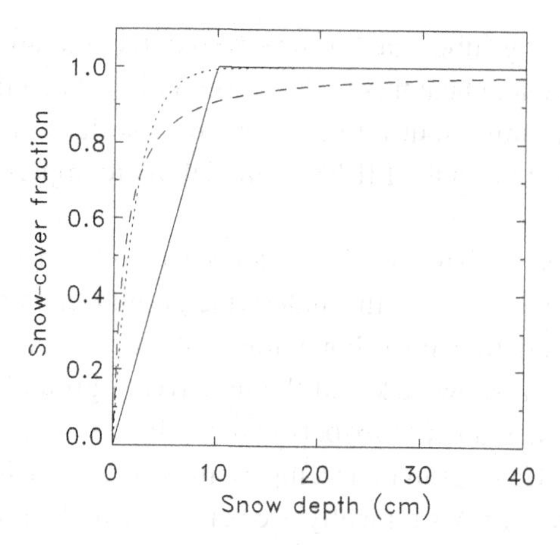


Figure 4.13. Snow-cover fraction as a function of average snow depth from parameterizations used by CLASS (—), MOSES (···) and ISBA with $z_0 = 10^{-3}$ m (– – –)

used by CLASS and SiB, where HS is the snow depth, $z_0$ is the surface roughness length, and $a_2$ is a parameter; the results are shown in Figure 4.13. Separate masking functions may be used for the vegetated and bare-ground fractions of a gridbox, and maximum snow-covered albedos may be assigned according to vegetation type. ISBA and a new scheme for the ECHAM4 GCM (Roesch *et al.*, 2001) additionally allow for subgrid distributions of snow with elevation by parameterizing the snow-cover fraction as a function of the standard deviation of surface heights within a gridbox.

Little theoretical or observational justification has been offered as yet for the parameterizations of snow-covered area and vegetation masking currently used in GCMs (Essery and Pomeroy, 2004), but attempts have been made to relate snow mass and fractional area to land-cover classes (e.g. Donald *et al.*, 1995; Liston, 2004). In an off-line test of BATS, Yang *et al.* (1997) found that an alternative function

$$f_s = \tanh\left(\frac{\mathrm{HS}}{a_2 z_0}\right)$$

gave a better match to observed snow-cover fractions for short vegetation.

### 4.4.4 Snow energy and mass balances

The components of snow surface energy and mass balances are discussed in Section 3.2. GCMs often neglect heat advected to the surface by precipitation and none, as yet, represents blowing snow. Surface roughness is reduced as snow depth increases or is calculated as an effective average for snow-covered and snow-free fractions of a gridbox, and snow-covered fractions are treated as saturated for the calculation of latent heat fluxes. The formulation of aerodynamic resistances was found to have an important influence on snow surface energy balances for a high-latitude environment in the PILPS 2e model intercomparison (Bowling *et al.*, 2003).

The energy balance of a snowpack is determined by the net surface energy flux, heat fluxes between the snow and the underlying ground, and phase changes within the snow (melting and freezing). For numerical calculations of heat fluxes and temperature changes, a snowpack and the underlying ground have to be divided into layers. The conducted heat flux between two layers is parameterized as being proportional to the difference in their temperatures and a weighted average of their thermal conductivities. GCMs typically use between 2 and 6 layers to represent the top few meters of soil but often represent a snowpack as a single layer or combine it with the surface soil layer. With increases in computational power, the use of multi-layer snow models such as GISS94 and CLM, which allow for 3 or 5 snow layers respectively, is likely to become more common.

When the calculated snow surface temperature reaches 0 °C, subsequent net energy input from radiation, turbulent heat fluxes, or ground heat fluxes (Section 3.3) is used to melt snow. In ISBA, an empirical approach is introduced to partition energy between melting snow and the warming of any overlying vegetation canopy, allowing the surface temperature to rise above 0 °C even while there is snow on the ground. Models differ in whether they calculate a single energy balance for the composite surface or separate energy balances for the snow-covered and snow-free parts of a partially snow-covered gridbox; composite energy balance calculations can melt snow too early (Liston, 2004; Essery *et al.*, 2005), whereas separate energy balances may lead to snow melting too late (Slater *et al.*, 2001).

Several off-line studies (using observed meteorological data to drive a surface model rather than coupling it to a GCM) have found snow models to be very sensitive to variations in the prescribed downward longwave radiation (Yang *et al.*, 1997; Slater *et al.*, 1998; Schlosser *et al.*, 2000). Considering uncertainties in GCM simulations of longwave radiation at high latitudes, Slater *et al.* (1998) argued that the use of complex snow models in GCMs may not yet be justified.

### 4.4.5 Heterogeneous snow cover

As can be seen from Figure 4.9, a typical GCM land gridbox spans a large area and may include regions with large differences in elevation, aspect, and vegetation cover. Many "land" gridboxes will actually contain significant fractions of inland or coastal water. GCMs, however, generally assume that land surface properties are homogeneous within each gridbox or can be characterized by effective parameters. Snow cover is frequently heterogeneous on length scales too small to be resolved by a GCM grid, introducing marked inhomogeneities in land surface properties such as albedo, roughness, and moisture availability. Most models diagnose a fractional snow cover, as discussed above, and use it to calculate effective gridbox parameters.

First-order turbulence closure schemes relate surface fluxes of momentum, heat, and moisture to windspeed, temperature, and humidity gradients between the surface and an atmospheric reference level, typically 10–30 m above the surface in GCMs, using transfer coefficients that depend on atmospheric stability and the roughness of the surface (Section 3.3). Due to the non-linearity of relationships between local fluxes and local gradients, gridbox-average fluxes are not simply related to gridbox-average gradients over heterogeneous surfaces, and the specification of effective parameters is not straightforward (Mahrt, 1996). The average sensible heat flux over a surface with heterogeneous snow cover can be dominated by the contribution from a small fraction of warm, dry snow-free land, giving an upward average flux counter to a downward average temperature gradient (Essery, 1997b). It has been suggested that the problem of calculating gridbox-average surface fluxes can be addressed by gathering distinct surface types within a gridbox into homogeneous "tiles" and calculating fluxes separately over each tile (Avissar and Pielke, 1989; Liston, 1995; Essery *et al.*, 2003a); this approach is adopted for snow-covered and snow-free fractions of gridboxes in CLASS. The TESSEL surface scheme (van den Hurk *et al.*, 2000) allows exposed snow, snow beneath tall vegetation, and snow-free exposed fractions to coexist within a gridbox.

Snow cover will often be heterogeneous for mid-latitude gridboxes covering areas with significant subgrid variations in surface elevation. To represent this in a GCM, Walland and Simmonds (1996) developed a model that divides each gridbox into cold snow, melting snow, and snow-free fractions above and below a diagnosed freezing level. Liston *et al.* (1996) found that nesting a snow model with a 5 km resolution within a regional atmospheric model with a 50 km resolution significantly improved the simulation of snow cover over the Rocky Mountains in Colorado. Fortunately, such high resolution may not be necessary; comparing simulations of snowmelt over a mountainous $1^\circ \times 1^\circ$ region in western Montana, Arola and Lettenmaier (1996) found that a distributed model gave very similar results whether used on a high-resolution grid or with just ten elevation bands.

### 4.4.6 Snow and vegetation

Snow can easily submerge short vegetation, greatly changing the albedo and roughness of the surface, but dense forest canopies retain low albedos even when snow covered. This can have a large influence on temperature forecasts (Viterbo and Betts, 1999). Betts (2000) has contrasted warming due to decreased albedo with cooling due to carbon sequestration if boreal forestation is used to offset anthropogenic carbon emissions.

Intercepted snow on a forest canopy (Section 3.5.5) has a large exposed surface area for sublimation, moistening and cooling the air, whereas snow on the ground below the canopy is sheltered from wind and solar radiation so sublimation is limited and melt is delayed. Few GCMs, however, distinguish between snow held on and below vegetation canopies. SiB and CLASS calculate separate moisture and heat fluxes for snow on the ground and snow held in vegetation canopies, but assume that the storage capacity of a canopy is the same for snow as it is for rain.

GCMs have often neglected the transmission of radiation through vegetation schemes, but CLASS uses a Beer Law parameterization to calculate canopy transmissivities as

$$\tau = \exp(-\kappa \Lambda),$$

where $\Lambda$ is the leaf area index and $\kappa$ is an extinction coefficient that depends on vegetation type and solar zenith angle; separate transmissivities are calculated for direct and diffuse radiation in visible and near-infrared wavebands. SiB and CLASS use a two-stream approximation (Sellers, 1985) that allows for multiple reflections but requires a large number of canopy parameters. Simpler schemes have been proposed by Nijssen and Lettenmaier (1999) and Yang *et al.* (2001).

Heat fluxes from snow-free forest canopies can raise air temperatures above 0 °C even when snow is on the ground (Yamazaki, 1995). Douville and Royer (1997) and Essery (1998b) observed that neglecting this heating degrades the ability of GCM land-surface schemes to match observations of heat fluxes and snowmelt for forested sites.

A number of GCM sensitivity studies have found boreal deforestation to lead to climate cooling as a result of reduced masking of snow-covered surface albedos (Thomas and Rowntree, 1992; Chalita and Le Treut, 1994; Bonan *et al.*, 1995; Douville and Royer, 1997). These investigations adopted extreme scenarios – typically instantaneous removal of all forests north of 45°N. The inclusion of vegetation dynamics (Cox *et al.*, 2000) and improved canopy models (Parviainen and Pomeroy, 2000; Essery *et al.*, 2003b) in GCMs, together with results from field campaigns such as BOREAS (Hall, 1999), MAGS (Stewart *et al.*, 1998), and NOPEX (Halldin and Gryning, 1999), will allow for more sophisticated investigations of the interactions between boreal forests, snow cover, and climate change.

## 4.5 The global snow coverage in climate change scenarios

*Judah Cohen*

One of the most important issues facing our society today is the prospect of rapid climate change forced by anthropogenic greenhouse gas emissions. Nowhere is the understanding of climate change more important than in the mid–high latitudes of the Northern Hemisphere, where most of the seasonally varying snow cover on the globe exists. As an example of the importance of snow cover as a regulator of the earth's climate, Hansen and Nazarenko (2004) estimate that soot on snow and ice in the Northern Hemisphere contributed to as much as a quarter of the observed warming from 1880–2000. In the case of soot on snow, the surface albedos are estimated to have decreased by only a few percent, demonstrating the potentially powerful role snow cover plays in modulating the climate system. Future climate scenarios predict large-scale decreases in snow-cover extent; thus the disappearance of snow cover in regions where snow cover is currently present could alter surface albedos by as much as 50% (Cohen, 1994). Such large decreases in surface albedo could force profound changes on the local and hemispheric climate due to the positive feedback of warming temperatures and ablating snow cover.

Of the many tools being utilized to analyze this problem, the most heavily relied upon for the study and prediction of global warming is the simulation of a future climate generated by general circulation models or global climate models (GCMs). The consensus from studies of most GCM simulations is that global temperatures will increase between 2 and 5 °C because of substantial $CO_2$ increase in the next century (Hansen *et al.* 1984; Washington and Meehl, 1984; Wilson and Mitchell, 1987; Wetherald and Manabe, 1988). Complicating the understanding of the degree and nature of global climate change are the many feedbacks and interactions with the ocean and clouds. It is also thought that the positive feedback between the high albedo of snow and ice with air temperatures will amplify mid–high latitude warming. Therefore the amount of warming at higher latitudes may be much greater than the global averages. Figure 4.14 from Bony *et al.* (2006) shows the surface air temperature response of 15 coupled atmosphere–ocean general circulation models to doubled $CO_2$ forcing.

The results are plotted as a function of latitude. All the models are consistent in showing "polar amplification" or the enhanced warming in the Arctic compared to the tropics. Observed surface air temperatures over the past century validate this hypothesis (Hansen *et al.*, 1996; Hurrell, 1996; Easterling *et al.*, 2000; Moritz *et al.*, 2002; Polyakov *et al.*, 2002, 2003). The probable influence of snow and ice cover on future global temperature trends highlights the importance of correctly parameterizing snow cover for natural and forced climate variability experiments involving GCMs.

We will discuss snow-cover variability in Section 4.5.1, snow–climate feedbacks in Section 4.5.2 and finally simulations of climate change scenarios in Section 4.5.3.

### 4.5.1 Snow-cover variability

Snow cover has been established as an important component of the land–atmosphere climate system. Snow cover experiences the largest fluctuations, both spatially and temporally, of all varying surface conditions. GCMs need to correctly simulate the

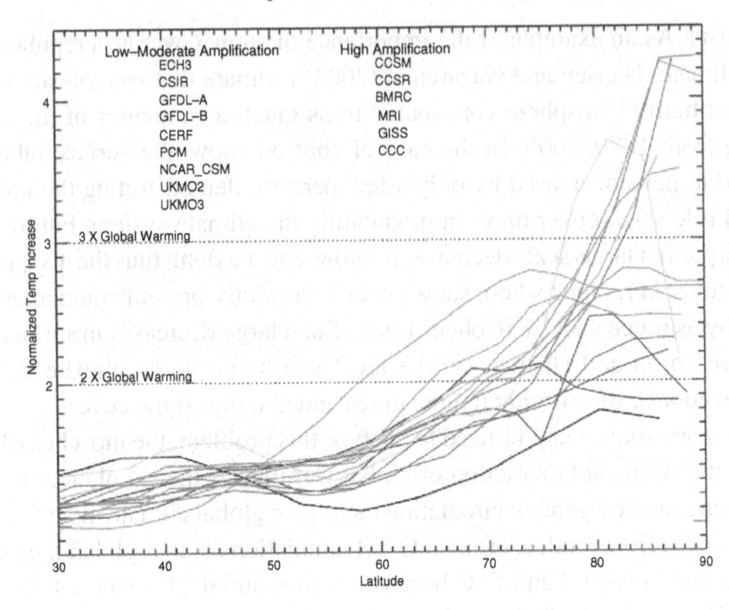


Figure 4.14. Surface air temperature change in 15 coupled GCMs that have doubled atmospheric $CO_2$ concentrations. All models show "polar amplification" and enhanced warming in the Arctic compared to tropical latitudes. Bony *et al.* (2006). (Plate 4.14.)

natural variability of snow cover from seasonal time scales out to interannual or even decadal time scales. GCMs also need to correctly parameterize the influence of snow cover on the climate. Several multi-GCM studies have evaluated how well models simulate snow-cover extent compared with remotely sensed data sets for the Northern Hemisphere. The first, Foster *et al.* (1996), (see Fig. 4.15), compared the snow cover and snow mass in seven GCMs with NOAA's visible remotely sensed snow-cover data set, the U.S. Air Force snow depth climatology (SDC), and microwave remotely sensed data sets for snow mass.

In the Northern Hemisphere, snow-cover extent is observed to be at a maximum in February (to about 40°N) and at a minimum in August, while Northern Hemisphere snow mass is observed to be at a maximum in March and at a minimum in August. It should be noted that although snow-cover extent historically peaked in February, as reported in Foster *et al.* (1996), January is currently the month with the observed peak in snow-cover extent, as reported in Frei *et al.* (2003). Also, snow cover is observed to retreat more quickly in the spring than it advances in the fall. Models are poorer at simulating snow cover during the transitional seasons of the spring and fall compared to the more stable winter and summer seasons; Frei *et. al.* (2003) found that model simulated snow cover tends to advance too slowly in the fall and to retreat too slowly in the spring, when compared with observations. Why observed

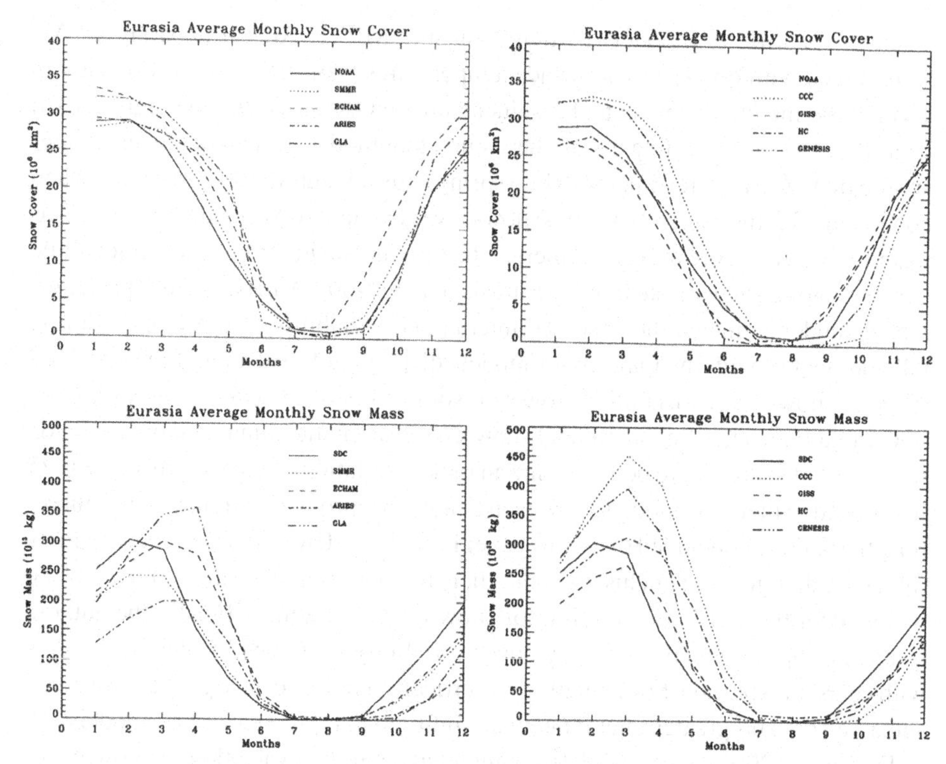


Figure 4.15. Snow cover and mass intercomparison for Eurasia from NOAA, SMMR and seven different GCMS. Taken from Foster *et al.* (1996).

snow cover melts and retreats faster in the spring than it advances in the fall and why models have greater difficulty simulating snow cover as observed during the transitional seasons is still not understood.

Comparisons between GCM simulated and observed snow cover indicate general differences at greater than the 95% significance level for every month of the year. The different GCMs show a wide range of values for snow cover and snow mass and even differ in respect to which months the maximum and minimum in snow cover and snow mass occur (see Fig. 4.15). Most of the errors in the GCM output probably result from inaccuracies in modeling surface temperatures, which stem from errors in the surface energy balance, and the parameterization schemes of precipitation, which often produce excessive or deficient precipitation. Properly simulating snow mass and snow cover involves improvements in model hydrodynamics, surface energy exchange, and other physical model parameterizations, all of which make the problem quite complex.

In the second snow-cover variability study, Frei and Robinson (1998) conducted an atmospheric model intercomparison project (AMIP)-type comparison of snow extent in 27 different models participating in AMIP-1. All participating GCMs were forced with observed sea surface temperatures from 1979–1988. The authors verified the models' snow extent results against NOAA's visible remotely sensed data set for snow cover. In general, they found that model-generated fall and winter snow extent was underestimated, while spring snow extent was overestimated when compared with the remotely sensed snow-cover extent. GCMs tended to underestimate snow cover more in North America than in Eurasia in the fall and winter while overestimating snow extent more in Eurasia than in North America in the spring. All models underestimated the observed interannual variability of snow cover during fall and winter, with the majority of models displaying less than half the observed range in snow-cover variability. However, some models do better in the spring.

Correlations between interannual snow-cover anomalies simulated in the GCMs with those observed were poor. Spearman rank correlations are generally $0.0 \pm 0.25$ for fall and winter. The models were even unable to reproduce extreme events; therefore, the authors concluded that snow cover is not forced by sea surface temperatures (SSTs), although this conclusion contradicts the observational study of Ye (2001), which identified SSTs as a forcing for snow-cover variability. Finally the authors examined the dependence of snow-cover variability on model resolution. Models with coarser resolution, horizontally and vertically, tended to have greater root mean square errors and greater interannual variability than higher resolution models.

Frei *et al.* (2003), (see Fig. 4.16) completed a similar evaluation of snow-cover variability in models participating in AMIP-2. In general, they found that model-simulated snow cover was greatly improved when compared to AMIP-1. Most notable was that the seasonal biases reported for AMIP-I had been corrected. Interannual variability was also notably improved in AMIP-2, although still less than that observed. Regional biases remained among the models, especially in Eurasia where biases were systematic.

Finally Frei *et al.* (2005) evaluated how well GCMs simulated snow water equivalent (SWE) over North America only among the suite of AMIP-2 GCMs. They conclude that the numerical models accurately simulate the seasonal timing of the relative spatial patterns of continental snow depth. However, the models do tend to ablate snow cover too quickly in the spring; in contrast to Foster *et al.* (1996) who found the models are too sluggish in melting spring snow cover. They also found that the multi-model mean was best at accurately simulating snow-cover climatology when compared with individual model results. Peak monthly SWE varied among the models by 50% of the observed peak value of $\sim$1500 km$^3$, which has important repercussions for the water balance of the North American continent; such large errors need to be improved. They found little correspondence in interannual variability between the models and observed snow mass and therefore conclude that at least in the models, SSTs are not forcing variations in continental-scale snow mass.

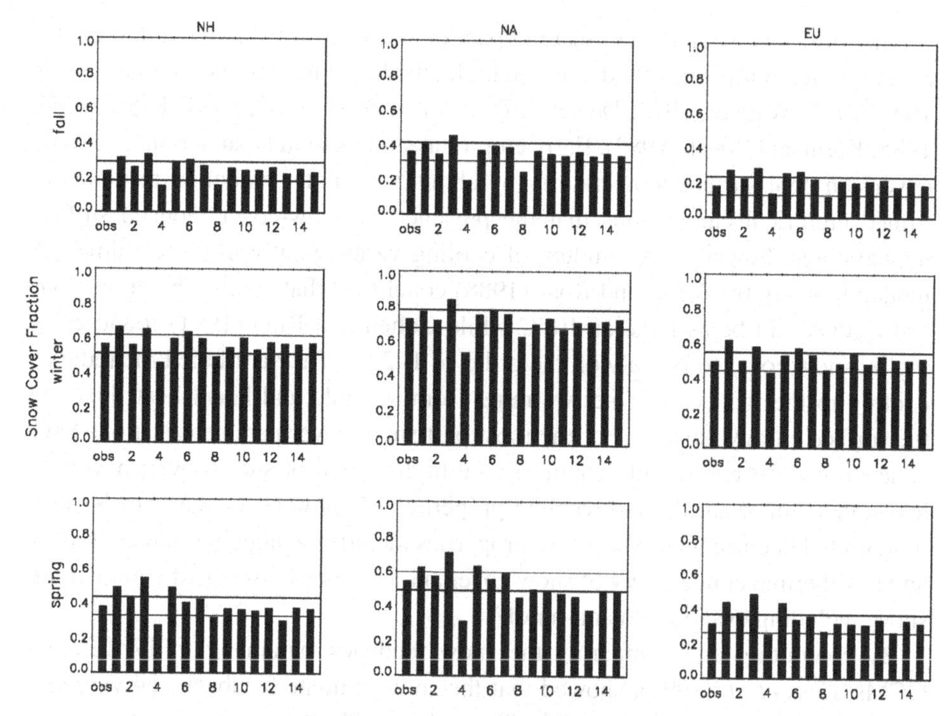


Figure 4.16. Observed and modeled mean seasonal snow-cover area for the Northern Hemisphere (NH), North America (NA) and Eurasia (EU). Observed values shown on left-hand side of each panel; model results are numbered. Horizontal lines indicate observed values ±0.05. Taken from Frei *et al.* (2003).

Not only do GCMs need to simulate the natural variability of snow cover from seasonal to interannual and even decadal time scales to represent the observed climate, but also to predict the role of snow cover in an anthropogenically changed climate accurately. Yet, as will be discussed below, not only do GCMs lack skill in simulating the natural variability of snow cover, the simulated influence of snow cover on the atmosphere also lacks consistency among different GCMs.

### 4.5.2 Snow–climate feedback

The impact of snow cover on surface air temperatures is well established. Snow cover is thought to influence surface temperatures in several ways: the high reflectivity of fresh snow cover can increase the surface albedo by 30–50%; fresh snow cover acts as a thermal insulator due to its low thermal conductivity; and melting snow is also a sink for latent heat since large amounts of energy are required for melt (Cohen, 1994).

The most widely studied climate impact of snow cover is the reduction of surface air temperatures mostly due to the high albedo of snow cover (Namias, 1960, 1962, 1985; Wagner, 1973; Dewey, 1977; Walsh *et al.*, 1982, 1985; Klein, 1983, 1985; Klein and Walsh, 1983). Early empirical studies found that anomalous snow cover can cool surface temperatures by about 5 °C. Later modeling results confirmed empirical calculations that the presence of snow-cover reduced surface temperatures; however, the amount of cooling varies greatly in those studies. A modeling study by Walsh and Ross (1988) concluded that cooling by anomalous snow cover can be as large as 10 °C while Cohen and Rind (1991) found cooling by snow cover to be only on the order of 1–2 °C. They attributed the reduced cooling to the inclusion of net warming from sensible and latent heat fluxes in the surface energy balance. For the most part snow-cover parameterizations have concentrated on realistically capturing the high albedo of snow cover; nevertheless, other important thermodynamic properties of snow cover may not be well represented in current GCMs. These properties include the high thermal emissivity and low thermal conductivity of snow cover and processes associated with melting snow, including evaporation and runoff.

By computing and comparing the snow feedback parameter in 17 different GCMs, Cess *et al.* (1991), showed that the snow–climate feedback can vary significantly among numerical models. The snow feedback parameter is a proxy to measure the impact of snow cover in a climate change scenario, where climate change was defined as a warming of SSTs by 4 °C uniformly across the globe. All GCMs in the experiment ran two climate change scenarios: one where snow cover remained fixed at the coldest SST run, and the second where snow cover was allowed to respond freely to the SST warming of 4 °C.

The response of the GCMs to snow–climate feedback varied considerably from strongly positive (where decreased snow cover increased global warming) to slightly negative (where decreased snow cover modified or damped global warming). However when only clear sky regions of the GCM were included in the snow–climate feedback, all GCMs had at least a small positive feedback (see Fig. 4.17). The authors concluded that the direct impact of snow cover on the incoming solar radiation is to reduce the amount absorbed by the earth and is therefore a net cooling forcing on the climate. However cloud feedbacks and changes in outgoing longwave radiation could enhance or moderate the cooling effect of the high albedo of snow cover. Randall *et al.* (1994) confirmed the results of Cess *et al.* (1991) by looking at the snow forced response of 14 atmospheric GCMs; here too a wide variation of responses occurred. Randall *et al.* (1994) found, however, that two GCMs had a negative feedback even under clear sky conditions. Both studies demonstrate that as more GCM groups examine the impact of a retreating snow cover on the future climate in a global warming scenario, the results will be highly varied (but the

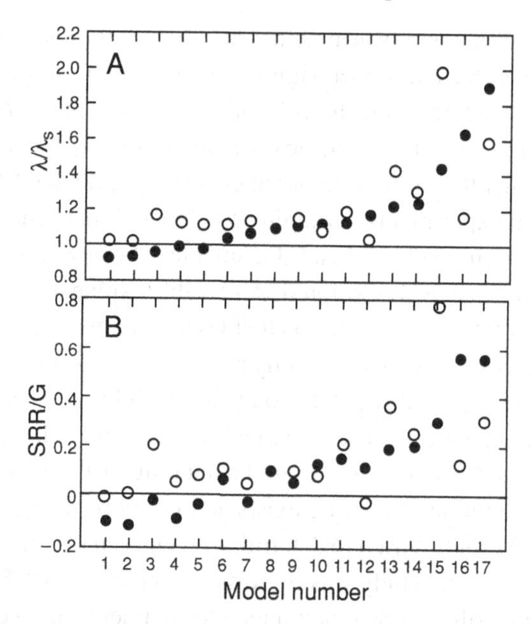


Figure 4.17. (a) The snow feedback parameter $\lambda/\lambda_s$ for 17 GCMs for both global (•) and clear (o) designations. (b) Values of SRR/G (a measure of the snow radiative response) from the 17 GCMs and for both global and clear designations. Taken from Cess *et al.* (1991).

predominant response should be an enhanced warming with high-latitude or polar amplification).

### 4.5.3 Climate change scenarios

The consensus among scientists is that global surface air temperatures have been increasing since the beginning of the industrial revolution (Folland *et al.* in IPCC 2001). The warming has not been uniform though, with the greatest warming occurring in the mid- to high-latitudes of the Northern Hemisphere. Also the warming has been largest in the winter and spring and least in the fall (Easterling *et al.*, 2000). What has been the role of snow cover on the observed warming and what may be its future role in a rapidly warming climate?

The easiest method of predicting future snow cover and its influence on a warming climate is to extrapolate from the most recent climate records. Accurate observations of snow cover from satellites have occurred over the past 35 years. Over that period the annual mean snow-cover extent has decreased. However, little trend or no trend is observed in the fall and winter with most of the decrease in observed snow cover occurring in the spring and summer. Groisman *et al.* (1994a,b) have attributed at least some of the spring warming in surface air temperatures to the decrease in spring snow cover. But winter warming has also been large and yet observed trends in winter snow cover have not been significant even at the snow margins and low latitudes. Therefore interpreting the role of snow cover on surface temperatures from the observations is not straightforward and ultimately GCMs will need to be relied upon to predict the future role of snow cover in a doubled $CO_2$ climate.

Many numerical experiments pertaining to global climate change due to greenhouse gases have been conducted and documented; however, the impact on snow cover has not been extensively explored. A possible explanation for why few studies exist that model future snow cover is that both air temperature and precipitation need to be accurately simulated to correctly reproduce snow cover at relatively small regional scales. This is probably still beyond the capabilities of current GCMs given all the inherent uncertainties. However, as GCMs are improved, more reliable experiments will be carried out and a better understanding of future snow cover should develop. Still, some limited literature exists and it will be discussed below.

Some early GCM studies tried to determine changes in soil moisture in a warmer climate. One soil moisture study was by Kellogg and Zhao (1988), who compared soil moisture over North America for winter and summer from five different GCMs. They found that during the winter, soil moisture increased over higher latitudes. It can be inferred from this that more precipitation fell in liquid rather than solid form and that more existing snow cover melted in a doubled $CO_2$ climate compared to the present climate. From their results the most probable conclusion is that snow cover would be less even at higher latitudes during winter in the warmer climate.

Another similar study that explored changes in soil moisture in an increased $CO_2$ climate was by Manabe and Wetherald (1987). In the paper, they compare the results from two GCM experiments; the GCM used is the Geophysical Fluid Dynamics Laboratory (GFDL) GCM. They conducted two types of experiments: a control run (atmospheric $CO_2$ concentration at 300 parts per million (ppm)) and a perturbation experiment with $CO_2$ increased to up to 600 and 1200 ppm. Snow cover is discussed in so far as it is relevant to soil moisture. As might be expected in a warmer climate, precipitation remains liquid longer during the fall and winter season and snow melt commences earlier in the winter and spring (resulting in an earlier date at which snow cover has completely disappeared) when compared to the control climate. Despite greater precipitation in the winter (due to a warmer and more moist atmosphere), total snowfall in the winter is less, so the duration and quantity of snow on the ground is less in the increased $CO_2$ modeling experiments when compared with the control climate.

In contrast, a study of a future doubled $CO_2$ climate by Ye and Mather (1997) found that global snow mass will increase in the polar regions (poleward of 60°N). They compared temperature and precipitation differences among three GCMs using current and doubled $CO_2$ values. The three GCMs compared are: the Goddard

Table 4.2 *Total changes in water equivalent in snow ($10^{12}$). Taken from Ye and Mather (1997).*

| | GFDL | GISS | UKMO |
|---|---|---|---|
| North polar region | −38.14 | −3.27 | 7.55 |
| South polar region | 500.08 | 925.20 | 1783.08 |
| Total north and south | 461.94 | 921.93 | 1790.63 |

Institute for Space Studies (GISS) GCM, the GFDL GCM and United Kingdom Meteorological Office (UKMO) GCM. In the UKMO GCM the liquid water equivalent (LWE) of snowfall increased in both the northern and southern polar regions. While in the GISS and GFDL GCMs, the LWE of snowfall only increased over Antarctica and decreased in the Northern Hemisphere's polar regions. The only regions in the Northern Hemisphere to show a net gain in snowfall were central and northern Greenland. However the total LWE of snowfall for both hemispheres increased in the doubled $CO_2$ simulations for all three GCMs (see Table 4.2). They estimate that the increase in total annual flux of water out of the oceans onto the polar ice caps for doubled $CO_2$ to be about one-tenth the magnitude of the current exchange of water from the oceans to the land from autumn (maximum sea level) to spring (minimum sea level). The increase of snow mass in the high latitudes agrees with an observational study by Miller and de Vernal (1992). They studied climate proxy data for the past 130 000 years before the present day and concluded that the conditions most favorable for glacial inception are climate conditions similar to the present climate. They conclude that the elevated winter temperatures predicted at high latitudes for doubled $CO_2$ will increase snowfall rates in the Canadian and Russian land areas adjacent to the Arctic Ocean. With increased snow mass and assuming negligible increases in summer temperatures, conditions are more favorable for ice sheet growth in a warmer climate.

A paper by Cohen (1994) briefly discusses different scenarios for snow cover in a doubled $CO_2$ climate. Using a control simulation and a doubled $CO_2$ simulation of the GISS GCM, Cohen found snow cover and snow mass to decrease uniformly across the Northern Hemisphere for all 12 months of the year. Snow cover decreases by nearly 30% and snow mass even more, by close to 40%.

More in-depth GCM studies include those by Boer *et al.* (1992) (see Fig. 4.18) and Essery (1997a) (see Fig. 4.19). Boer *et al.* (1992) compare simulations for the current climate and a doubled $CO_2$ climate using the Canadian Climate Center (CCC) GCM. They separately compare permanent and seasonal snow cover between the two climate simulations. Their findings for the CCC GCM is consistent with that of Ye and Mather (1997) for three different GCMs. Total snow mass

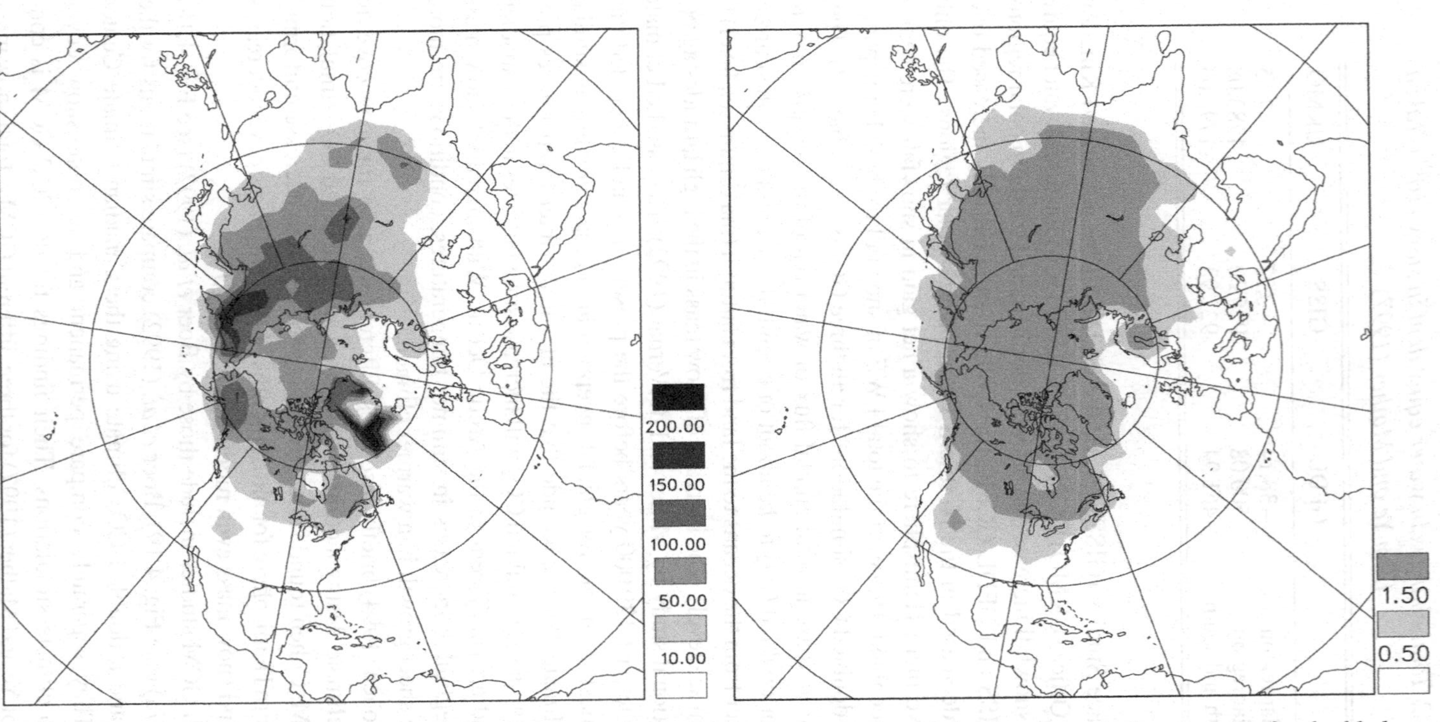


Figure 4.18. The left panel shows Northern Hemisphere winter (DJF) distribution of simulated seasonal snow mass for doubled $CO_2$. The right panel shows the change in the position of the simulated snow line between the control (fine shading) and the doubled $CO_2$ cases (coarse shading). The snow line is denoted by values of snow mass greater than 10 kg $m^{-2}$. Taken from Boer *et al.* (1992).

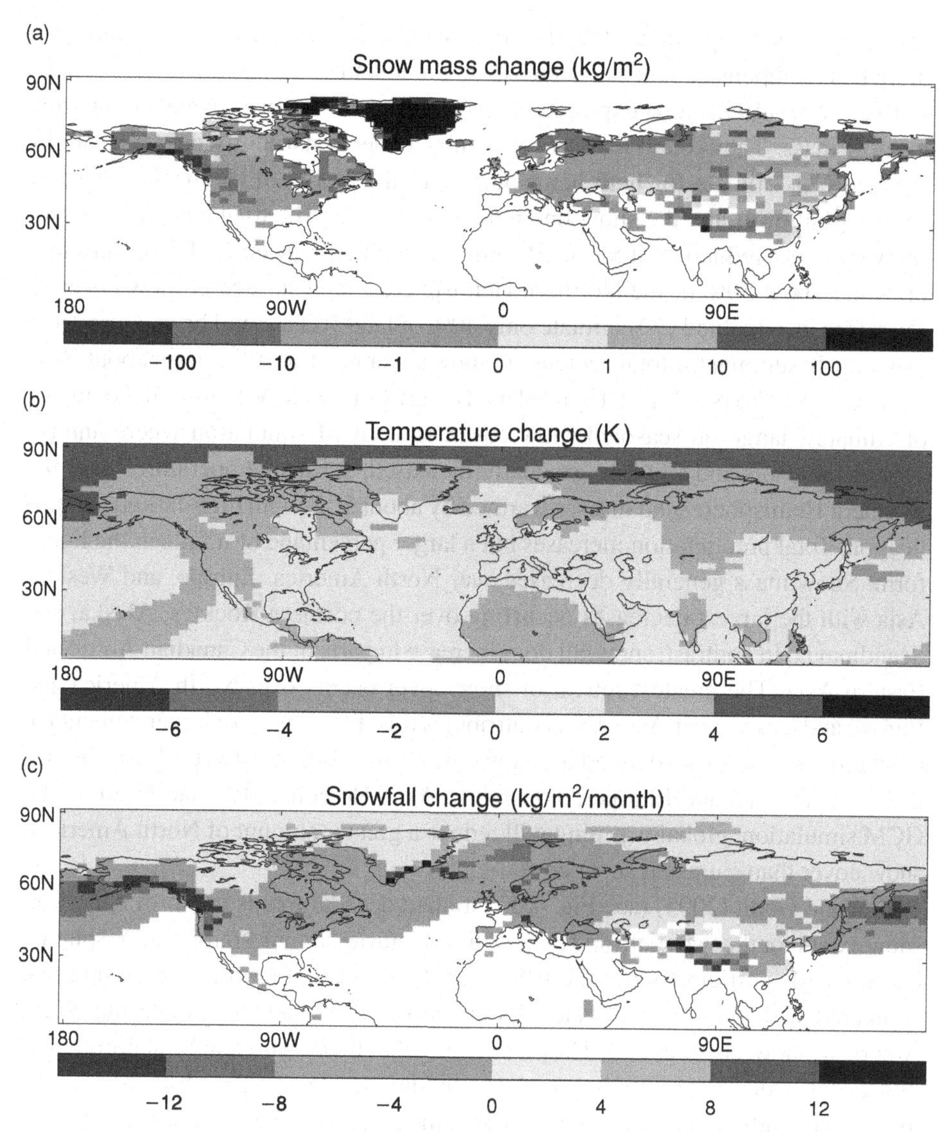


Figure 4.19. Differences between 2030–2050 averages from the climate-change simulation and 130-year averages from the control of the Hadley GCM for (a) snow mass (b) temperature and (c) snowfall. Taken from Essery (1997a). (Plate 4.19.)

for permanent snow cover for the globe increases for doubled $CO_2$. However the results differ when the two hemispheres are compared separately. In the Northern Hemisphere over Greenland there is a dramatic decrease in the snow accumulation rate, which would probably result in a depleted ice cap. In Antarctica, snow-cover accumulation increases and therefore would probably produce a mass increase in the southern ice cap. Taken together the increase in snow mass over Antarctica more than compensates for the decrease in Greenland.

Boer *et al.* (1992) also explored changes in the snow-cover extent in addition to snow mass. They found that seasonal snow cover consistently decreases in the doubled $CO_2$ climate for both hemispheres. In the Northern Hemisphere, winter snow is shallower and the snow line retreats poleward. The total change in winter snow cover is a reduction of $8.3 \times 10^{12}$ km$^2$ or about 20% (see Fig. 4.18). Currently in winter about 50% of the Northern Hemisphere's land surface is snow covered. Therefore in a doubled $CO_2$ climate only 40% of land surface will be snow covered in winter. In summer the total decrease in snow cover is $6.1 \times 10^{12}$ km$^2$ or about 50%.

Essery (1997a) (see Fig. 4.19) used the Hadley Centre GCM to look at the impact of climate change on seasonal snow cover. In a GCM simulation where anthropogenic aerosols and $CO_2$ are gradually increased, surface temperatures over the Northern Hemisphere land masses warmed by about 2 °C, with high-latitude amplification. Total precipitation increases but a larger percentage of it falls in the liquid form. Snow mass generally decreases over North America, Europe, and Western Asia with the largest decreases occurring over the northern Rockies, Alaska, and Scandinavia. In contrast, snowfall does increase in parts of the Canadian Arctic and Eastern Asia. The greatest retreat of snow cover occurs over North America and Europe and less so over Asia. Essery attributes this difference to a larger percentage of shallow snow cover (defined as snow cover with a depth between 1 and 10 cm) in North America and Europe compared to Asia. He concludes that based on the GCM simulation, global warming will reduce a greater amount of North American snow cover than Eurasian snow cover.

Frei and Gong (2005) (see Fig. 4.20) studied decadal trends in North American snow extent for the twentieth and twenty-first centuries among coupled atmosphere–ocean GCMs participating in the IPCC fourth assessment report. They found that compared with the observed snow extent most GCMs underestimate the North American snow-cover extent. They also found that there is no temporal correlation in decadal scale variability among the individual models or the observations and the models, individually or combined as a mean. In the twenty-first century they found a robust decreasing trend in the North American snow-cover extent that is statistically significant above the 99% confidence level. They conclude that the coupled models predict a significant decrease in snow extent, which can be used as an indicator of anthropogenically forced climate change. Table 4.3 summarizes the main findings for snow cover in a doubled $CO_2$ climate among the GCM studies discussed above.

Snow melt constitutes a large percentage of water used for consumption and irrigation. Runoff makes up 33% of the world's irrigation waters and can be as high as 100% (Rango, 1997). Given the importance of snow regionally, in addition

Table 4.3 *GCM studies evaluating changes in snow cover in a doubled $CO_2$ climate. The table lists the direction of change in snow-cover extent and/or mass for both hemispheres. If a region is not discussed, the column is left blank. Final column for the entire globe is for snow mass only.*

| Study | Snow extent NH | Snow mass NH | Snow extent SH | Snow mass SH | Globe |
|---|---|---|---|---|---|
| Kellogg and Zhao (1988) | Decrease | Decrease | | | |
| Manabe and Wetherald (1987) | Decrease | Decrease | | | |
| Boer *et al.* (1992) | Decrease | Decrease | Decrease | Increase | Increase |
| Cohen (1994) | Decrease | Decrease | | | |
| Ye and Mather (1997) | Decrease | Decrease[a] | Decrease | Increase | Increase |
| Essery (1997a) | Decrease | Decrease | | | |
| Frei and Gong (2005) | Decrease[b] | | | | |

[a] one GCM study shows an increase.
[b] North America only.

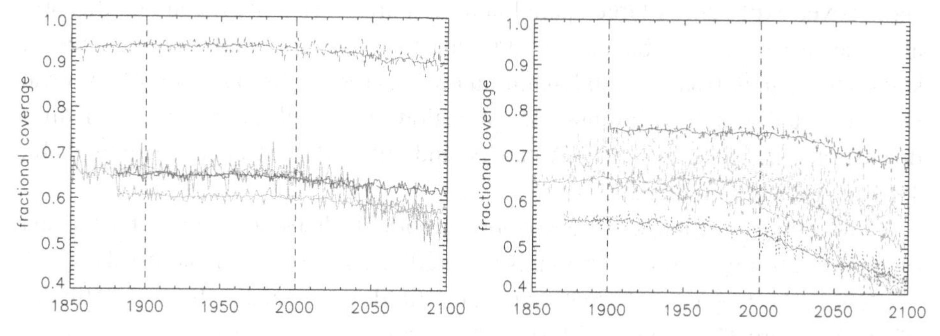


Figure 4.20. Annual time series (thin line), overlaid with nine-year running means (thick line), of ensemble-mean January North American snow-cover extent, including both twentieth and twenty-first century scenarios, for nine available coupled atmosphere–ocean GCMs. Taken from Frei and Gong (2005). (Plate 4.20.)

to the global-scale numerical simulations of the impacts of climate change on snow cover, more finer-scale experiments need to be done when the appropriate skill can be achieved. Some initial studies have been completed concentrating on local variations in snow cover in mountainous terrain. One study by Giorgi *et al.* (1997) looked at the elevation dependency of surface climate under doubled $CO_2$ conditions using a regional model of the Alps nested in a GCM. In the control simulation of their model, snow accumulates starting from an elevation of 800 meters in winter and 1100 meters in spring. While in the doubled $CO_2$ simulation, snow is completely depleted except at the highest elevations. This also results in the high-elevation amplification of warming in the doubled $CO_2$ simulation.

Similar results were found by Rango (1997) using a snowmelt runoff model for the Rio Grande basin. In general, it was found that for a doubled $CO_2$ climate there was an increase in rain events during winter, resulting in more immediate runoff and less snow accumulation. Also more of the snowpack melts during the winter with a net result of a reduced snowpack at the beginning of the traditional snow melt season (April 1). In particular, on April 1, the average LWE of snow cover was reduced from 58.1 to 20.4 cm. The spring runoff peaks would be two to four weeks earlier. Therefore there is increased runoff during winter and in the early part of the snow melt season and reduced runoff during later parts of the snow melt season (June–July).

### *4.5.4 Conclusions*

Snow cover is an integral component of the climate system, the importance of which is just starting to be realized. Snow cover cools the atmosphere, not only due to its high albedo, but also due to its high thermal insulation. Therefore, because snow cover experiences large spatial and temporal variations it can significantly alter the surface energy balance. In addition to the local climate forcings, recent studies have shown that thermodynamic anomalies forced locally by snow cover can be transported remotely through teleconnections and result in dynamic anomalies downstream in space and time (Watanabe and Nitta, 1998; Cohen and Entekhabi, 1999; Clark and Serreze, 2000).

Snow–climate feedbacks are even more important when considered in the context of global warming. Future warming is not predicted to be uniform across the globe but less in the tropics and greater at high latitudes, especially the high latitudes of the Northern Hemisphere. Most of the earth's snow cover exists in the same regions where global warming is predicted to be greatest. Therefore changes in future climate will significantly impact snow cover and in turn changes in future snow cover will have important feedbacks for predicted climate change. However given how difficult it is for models to simulate correctly the variability of observed snow cover and its feedback on the climate, it will continue to be a challenge for current GCMs to accurately predict the role of snow cover in a rapidly changing climate system.

As discussed earlier, GCMs do not correctly simulate snow-cover variability as observed and even lack consistency among themselves. Similarly, snow–climate interactions as parameterized by GCMs exhibit a wide dispersion, while the observed snow–climate interactions are not well defined. Therefore it is not surprising that the early diagnosis of changes in snow cover and snow mass in the coming decades is conflicting. Future snow-cover extent has been universally modeled to recede, both towards the poles and the tops of mountains, as greenhouse gases continue to increase in the twenty-first century. However the impact of a more limited snow extent on future lower troposphere air temperatures shows large variability. The range varies from significantly amplifying warming at mid–high latitudes to slightly dampening warming at these same latitudes. And even though snow cover is expected to decrease, models conflict as to whether total snow mass will also decrease or may actually increase (even resulting in a drop in sea level). How future snow cover and snow mass responds and its role in amplifying or moderating a changing climate is a challenge for climate and snow modelers, which will require increased collaborative effort by the hydrologic and atmospheric community to resolve. Our confidence in future climate projections depends on correctly simulating snow-cover variability, its interactions with the climate system, and its response to increased greenhouse gases.

## References

Albert, M. R. and Krajeski, G. (1998). A fast, physically-based point snow melt model for use in distributed applications. *Hydrol. Process.*, **12**(11), 1809–1824.

Amorocho, J. and Espildora, B. (1966). *Mathematical Simulation of the Snow Melting Processes*. Water Science and Engineering Papers No. 3001. Davis, CA: University of California.

Anderson, E. A. (1968). Development and testing of snow pack energy balance equations. *Water Resources Res.*, **4**(1), 19–37.

Anderson, E. A. (1973). *National Weather Service River Forecast System-Snow Accumulation and Ablation Model.* NOAA Technical Memorandum NWS HYDRO-17. Silver Spring, CO: US Dept. of Commerce.

Anderson, E. A. (1976). *A Point Energy Balance Model of a Snow Cover*. Office of Hydrology, National Weather Service, NOAA Technical Report NWS 19.

Anderson, E. A. and Crawford, N. H. (1964). *The Synthesis of Continuous Snowmelt Runoff Hydrographs on a Digital Computer*. Department of Civil Engineering Technical Report No. 36. Stanford, CA: Stanford University.

Arola, A. and Lettenmaier, D. P. (1996). Effects of subgrid spatial heterogeneity on GCM-scale land surface energy and moisture fluxes. *J. Climate*, **9**, 1339–1349.

Avissar, R. and Pielke, R. A. (1989). A parametrization of heterogeneous land surfaces for atmospheric numerical models and its impact on regional meteorology. *Mon. Wea. Rev.*, **117**, 2113–2136.

Bartelt, P. B. and Lehning, M. (2002). A physical SNOWPACK model for avalanche warning services. Part I: numerical model. *Cold Reg. Sci. Technol.*, **35**(3), 123–145.

Betts, R. A. (2000). Offset of the potential carbon sink from boreal forestation by decreases in surface albedo. *Nature*, **408**, 187–190.

Boer, G. J., McFarlane, N. A., and Lazare, M. (1992). Greenhouse gas-induced climate change simulated with the CCC second generation general circulation model. *J. Climate.*, **5**, 1045–1077.

Bonan, G. B., Chapin F. S., and Thompson, S. L. (1995). Boreal forest and tundra ecosystems as components of the climate system. *Climatic Change*, **29**, 145–167.

Bony, S., Colman, R., Kattsov, V. M., *et al.* (2006). How well do we understand climate change feedback processes?, *J. Climate*, **19**, 3445–3482.

Boone, A. and Etchevers, P. (2001). An intercomparison of three snow schemes of varying complexity coupled to the same land surface model: local-scale evaluation at an alpine site. *J. Hydrometeorol.*, **2**, 374–394.

Bowling, L. C., Lettenmaier, D. P., Nijssen, B., *et al.* (2003). Simulation of high latitude hydrological processes in the Torne-Kalix basin: PILPS Phase 2e. 1: Experiment design and summary intercomparisons. *Global Planetary Change*, **38**, 1–30.

Brun, E., David, P., Sudul, M., and Brunot, G. (1992). A numerical model to simulate snow-cover stratigraphy for operational avalanche forecasting. *J. Glaciol.*, **38**(128), 13–22.

Brun, E., Martin, E., Simon, V., Gendre, C., and Coleou, C. (1989). An energy and mass model of snow cover suitable for operational avalanche forecasting. *J. Glaciol.*, **35**(121), 333–342.

Brun, E., Martin, E., and Spiridonov, V. (1997). The coupling of a multi-layered snow model with a GCM. *Ann. Glaciol.*, **25**, 66–72.

Carlson, R. F., Norton, W., and McDougall, J. (1974). *Modeling Snowmelt Runoff in an Arctic Coastal Plain*. Institute of Water Resources. Fairbanks, AK: University of Alaska (available as PB-232–431).

Cess, R. D., Potter, G. L., Zhang, M.-H., *et al.* (1991). Interpretation of snow–climate feedback as produced by 17 general circulation models. *Science*, **253**, 888–892.

Chalita, S. and Le Treut, H. (1994). The albedo of temperate and boreal forest and the Northern Hemisphere climate: a sensitivity experiment using the LMD GCM. *Climate Dyn.*, **10**, 231–240.

Clark, M. P. and Serreze, M. C. (2000). Effects of variations in East Asian snow cover on modulating atmospheric circulation over the North Pacific ocean. *J. Climate*, **13**, 3700–3710.

Cohen, J. (1994). Snow cover and climate. *Weather*, **49**, 150–156.

Cohen, J. and Entekhabi, D. (1999). Eurasian snow cover variability and Northern Hemisphere climate predictability. *Geophys. Res. Lett.*, **26**, 345–348.

Cohen, J. and Rind, D. (1991). The effect of snow cover on the climate. *J. Climate*, **4**, 689–706.

Cox, P. M., Betts, R. A., Bunton, C. B., *et al.* (1999). The impact of new land surface physics on the GCM simulation of climate and climate sensitivity. *Climate Dyn.*, **15**, 183–203.

Cox, P. M., Betts, R. A., Jones, C. D., Spall, S. A., and Totterdell, I. J. (2000). Acceleration of global warming due to carbon-cycle feedbacks in a coupled climate model. *Nature*, **408**, 184–187.

Dai, Y.-J. and Zeng, Q.-C. (1997). A land surface model (IAP94) for climate studies, Part I: Formulation and validation in off-line experiments. *Adv. Atmos. Sci.*, **14**, 433–460.

Deardoff, J. F. (1968). Dependence of air–sea transfer coefficients on bulk stability. *J. Geophys. Res.*, **73**(8), 2549–2557.

Desborough, C. E. and Pitman, A. J. (1998). The BASE land surface model. *Global Planet. Change*, **19**(1–4), 3–18.

Dewey, K. F. (1977). Daily maximum and minimum temperature forecasts and the influence of snow cover. *Mon. Wea. Rev.*, **105**, 1594–1597.

Dickinson, R. E., Henderson-Sellers, A., and Kennedy, P. J. (1993). *Biosphere Atmosphere Transfer Scheme (BATS) Version 1e as Coupled to the NCAR Community Climate Model*. NCAR Technical Note, NCAR/TN-387+STR.

Donald, J. R., Soulis, E. D., Kouwen, N., and Pietroniro, A. (1995). A land cover-based snow cover representation for distributed hydrologic models. *Water Resources Res.*, **31**, 995–1009.

Douville, H., Royer, J.-F., and Mahfouf, J.-F. (1995a). A new snow parametrization for the Météo-France climate model. I. Validation in stand-alone experiments. *Climate Dyn.*, **12**, 21–35.

Douville, H., Royer, J.-F., and Mahfouf, J.-F. (1995b). A new snow parametrization for the Météo-France climate model. II. Validation in a 3-D GCM experiment. *Climate Dyn.*, **12**, 37–52.

Easterling, D. R., Karl, T. R., Gallo, K. P., *et al.* (2000). Observed climate variability and change of relevance to the biosphere. *J. Geophys. Res.*, **105**, 20 101–20 114.

Eggleston, K. O., Israelsen, E. K., and Riley, J. P. (1971). *Hybrid Computer Simulation of the Accumulation and Melt Processes in a Snowpack.* PRWG65–1, Utah Water Research Laboratory. Logan, UT: Utah State University.

Essery, R. (1997a). Seasonal snow cover and climate change in the Hadley Centre GCM. *Ann. Glaciol*, **25**, 362–366.

Essery, R. (1998a). Snow modelling in the Hadley Centre GCM. *Phys. Chem. Earth*, **23**(5–6), 655–660.

Essery R. and Yang Z. L. (2001). An overview of models participating in the snow model intercomparison project (SNOWMIP). *Proc. SnowMIP Workshop*, 11 July 2001, 8th Scientific Assembly of IAMAS, Innsbruck.

Essery, R. L. H. (1997b). Modelling fluxes of momentum, sensible heat and latent heat over heterogeneous snow cover. *Quart. J. Roy. Meteor. Soc.*, **123**, 1867–1883.

Essery, R. L. H. (1998b). Boreal forests and snow in climate models. *Hydrol. Process.*, **12**, 1561–1567.

Essery, R. L. H. and Pomeroy, J. W. (2004). Implications of spatial distributions of snow mass and melt rate on snow cover depletion: theoretical considerations. *Ann. Glaciol.*, **38**, 261–265.

Essery, R. L. H., Best, M. J., Betts, R. A., Cox, P. M., and Taylor, C. M. (2003a). Explicit representation of subgrid heterogeneity in a GCM land-surface scheme. *J. Hydrometeorol.*, **4**, 530–543.

Essery, R. L. H., Blyth, E. M., Harding, R. J., and Lloyd, C. M. (2005). Modelling albedo and distributed snowmelt across a low hill in Svalbard. *Nordic Hydrol.*, **36**, 207–218.

Essery, R. L. H., Pomeroy, J. W., Parvianen J., and Storck, P. (2003b). Sublimation of snow from boreal forests in a climate model. *J. Climate*, **16**, 1855–1864.

Essery, R., Martin, E., Douville, H., Fernandez, A., and Brun, E. (1999). A comparison of four snow models using observations from an alpine site. *Climate Dyn.*, **15**, 583–593.

Etchevers P., Martin E., Brown, R., *et al.* (2003). Intercomparison of the surface energy budget simulated by several snow models (SNOWMIP project). *Ann. Glaciol.*, **38**, 150–158.

Fernandez, A. (1998). An energy balance model of seasonal snow evolution. *Phys. Chem. Earth*, **23**(5–6), 661–666.

Flerchinger, G. N. and Saxton, K. E. (1989). Simultaneous heat and water model of a freezing snow–residue-soil system I. Theory and development. *Trans. ASAE*, **32**(2), 565–571.

Folland, C. K., Karl, T. R., Christy, J. R., *et al.* (2001). Observed climate variability and change. In *Climate Change 2001: The Scientific Basis.* Contribution of Working Group I to the Third Assessment Report of the Governmental Panel on Climate Change (ed. Houghton, J. T., Ding, Y., Griggs, D. J., *et al.*). Cambridge: Cambridge University Press, pp. 99–192.

Foster, J., Liston, G., Koster, R., *et al.* (1996). Snow cover and snow mass intercomparisons of general circulation models and remotely sensed datasets. *J. Climate*, **9**, 409–426.

Frei, A., Brown, R., Miller, J. A., and Robinson, D. A. (2005). Snow mass over North America: observations and results from the second phase of the Atmospheric Model Intercomparison Project (AMIP-2), *J. Hydrometeorol.*, **6**, 681–695.

Frei, A. and Gong, G. (2005). Decadal to century scale trends in North American snow extent in coupled atmosphere–ocean general circulation models. *Geophys. Res. Lett.*, **32**, L18502, doi: 10.1029/2005GL023394.

Frei, A., Miller, J. A., and Robinson, D. A. (2003). Improved simulations in the second phase of the atmospheric model intercomparison project (AMIP-2). *J. Geophys. Res.*, **108**, doi:10.1029/2002JD003030.

Frei, A. and Robinson, D. A. (1998). Evaluation of snow extent and its variability in the Atmospheric Model Intercomparison Project. *J. Geophys. Res.*, **103**, 8859–8871.

Giorgi, F., Hurrell, J. W., and Marinucci, M. R. (1997). Elevation dependency of the surface climate change signal: a model study. *J. Climate*, **10**, 288–296.

Gray, D. M. and Prowse, T. D. (1993). Snow and floating ice. In *Handbook of Hydrology* (ed. Maidment, D. R.). New York: McGraw-Hill, pp. 7.1–7.58.

Grody, N. C. and Basist, A. N. (1996). Global identification of snow cover using SSM/I instruments. *IEEE Trans. Geosci. Remore Sensing*, **34**, 237–249.

Groisman, P. Y., Karl, T. R., and Knight, R. W. (1994a). Changes of snow cover, temperature and radiative heat balance over the Northern Hemisphere. *J. Climate*, **7**, 1633–1656.

Groisman, P. Y., Karl, T. R., and Knight, R. W. (1994b). Observed impact of snow cover on the heat balance and the rise of continental spring temperatures. *Science*, **263**, 198–200.

Hall, A. and Qu, X. (2006). Using the current seasonal cycle to constrain snow albedo feedback in future climate change. *Geophys. Res. Lett.*, **33**, doi:10.1029/2005GL025127.

Hall, F. G. (1999). BOREAS in 1999: experiment and science overview. *J. Geophys. Res.*, **104**, 27 627–27 639.

Halldin, S. and Gryning, S.-E. (1999). Boreal forests and climate. *Agric. Forest Meteorol.*, **98–99**, 1–4.

Hansen, J., Lacis, A., Rind, D., *et al.* (1984). Analysis of feedback mechanisms. In *Climate Processes and Climate Sensitivity* (ed. Hansen, J. E. and Takahashi, T.). Washington DC: American Geophysical Union, pp. 130–163.

Hansen, J. and Nazarenko, L. (2004). Soot climate forcing via snow and ice albedos. *Proc. Natl. Acad. Sci.*, **101**, 423–428, doi:10.1073/pnas.2237157100.

Hansen, J., Reudy R., and Sato, M. (1996). Global surface air temperature in 1995: Return to pre-Pinatubo level. *Geophy. Res. Lett.*, **23**, 1665–1668.

Hansen, J., Russell, G., Rind, D., *et al.* (1983). Efficient three-dimensional global models for climate studies: models I and II. *Mon. Wea. Rev.*, **111**, 609–662.

Hardy, J. P., Davis, R. E., Jordan, R., Ni, W., and Woodcock, C. E. (1998). Snow ablation modelling in a mature aspen stand of the boreal forest. *Hydrol. Process.*, **12**, 1763–1778.

Humphrey, J. H. and Skau, C. M. (1974). *Variation of Snowpack Density and Structure with Environmental Conditions*. Center for Water Resources Research, Desert Research Institute (available as PB-238–000). Reno, NV: University of Nevada System.

Hurrell, J. W. (1996). Influence of variations in extratropical wintertime teleconnections on Northern Hemisphere temperatures. *Geophys. Res. Lett.*, **23**, 665–668.

Jin, J., Gao, X., Sorooshian, S., *et al.* (1999). One-dimensional snow water and energy balance model for vegetated surfaces. *Hydrol. Process.*, **13**, 2467–2482.

Johnson, M. L. and Anderson, E. (1968). The Cooperative Snow Hydrology Project-ESSA Weather Bureau and ARS Sleepers River Watershed. *Proc. Eastern Snow Conference*, pp. 13–23 (available from Atm. Sci. Library, NOAA, Silver Spring, MD).

Johnsson, H. and Lundin, L.-C. (1991). Surface runoff and soil water percolation as affected by snow and soil frost. *J. Hydrol.*, **122**, 141–159.

Jordan, R. (1991). *A One-dimensional Temperature Model for a Snow Cover*. U.S. Army Corps of Engineers, Cold Regions Research and Engineering Laboratory, Special Report 91–16.

Kellogg, W. W. and Zhao, Z.-C. (1988). Sensitivity of soil moisture to doubling of carbon dioxide in climate model experiments. Part I: North America. *J. Climate*, **1**, 348–366.

Kim, J. and Ek, M. (1995). *J. Geophys. Res.*, **100**(D10), 20 845–20 854.

Kite, G. W. (1995). The SLURP model. In *Computer Models of Watershed Hydrology* (ed. Singh, V. P.). Boulder, CO: Water Resources Publications, pp. 521–562.

Klein, W. H. (1983). Objective specification of monthly mean surface temperatures in the United States during the winter season. *Mon. Wea. Rev.*, **111**, 674–691.

Klein, W. H. (1985). Space and time variations in specifying monthly mean surface temperature from the 700 mb height field. *Mon. Wea. Rev.*, **113**, 277–290.

Klein, W. H. and Walsh, J. E. (1983). A comparison of pointwise screening and empirical orthogonal functions in specifying monthly mean surface temperature from 700 mb data. *Mon. Wea. Rev.*, **111**, 669–673.

Kondo, J., Kanechika, O., and Yasuda, N. (1978). Heat and momentum transfers under strong stability in the atmospheric surface layer. *J. Atmos. Sci.*, **35**(6), 1012–1021.

Koren, V., Duan, Q.-Y., Schaake, J., and Mitchell, K. (1999). *Validation of a Snow-Frozen Ground Parameterization of the Eta Model*. AMS Conference, paper J1.3.

Koster, R. and Suarez, M. (1996). *Energy and Water Balance Calculations in the Mosaic LSM*. NASA Technical Memorandum 104606, Vol. 9.

Kuzmin, P. P. (1961). *Protsess Tayaniya Shezhnogo Pokrova*. Leningrad: Gidrometeorologicheskoe Izdatel'stvo. (English translation: Melting of Snow Cover, Israel Program for Scientific Translations (TT 71–50095)).

Leaf, C. F. and Brink, G. E. (1973). *Computer Simulation of Snowmelt within a Colorado Subalpine Watershed*. USDA Forest Service Research Paper RM-99. Ft. Collins, CO: Rocky Mountain Forest and Range Experiment Station, U.S. Forest Service.

Lehning, M., Bartelt, P. B., Brown, R. L., Fierz, C., and Satyawali, P. (2002a). A physical SNOWPACK model for the Swiss Avalanche Warning Services. Part II: snow microstructure. *Cold Reg. Sci. Technol.*, **35**(3), 147–167.

Lehning, M., Bartelt, P. B., Brown, R. L., Fierz, C., and Satyawali, P. (2002b). A physical SNOWPACK model for the Swiss Avalanche Warning Services. Part III: meteorological boundary conditions, thin layer formation and evaluation. *Cold Reg. Sci. Technol.*, **35**(3), 169–184.

Lindstrom, G., Johansson, B., Persson, M., Gardelin, M., and Bergstrom S. (1997). Development and test of the distributed HBV-96 hydrological model. *J. Hydrol.*, **201**, 272–288.

Liston, G. E. (1995). Local advection of momentum, heat and moisture during the melt of patchy snow covers. *J. Appl. Meteorol.* **34**, 1705–1715.

Liston, G. E. (2004). Representing subgrid snow cover heterogeneities in regional and global models. *J. Climate*, **17**, 1381–1397.

Liston, G. E. and Sturm, M. (1998). A snow transport model for complex terrain. *J. Glaciol.*, **44**, 498–516.

Liston, G. E., Pielke, R. A., and Greene, M. E. (1996). Improving first-order snow-related deficiencies in a regional climate model. *J. Geophys. Res.*, **104**, 19 559–19 567.

Loth, B. and Graf, H.-F., (1998a). Modeling the snow cover in climate studies. Part I: Long-term integrations under different climatic conditions using a multilayered snow-cover model. *J. Geophys. Res.*, **103**, 11 313–11 327.

Loth, B. and Graf, H. (1998b). Modeling the snow cover in climate studies. Part II: The sensitivity to internal snow parameters and interface processes. *J. Geophys. Res.*, **103**(D10), 11 329–11 340.

Loth, B., Graf H.-F., and Oberhuber, J. M. (1993). Snow cover model for global climate simulations. *J. Geophys. Res.*, **98**, 10 451–10 464.

Lynch-Stieglitz, M. (1994). The development and validation of a simple snow model for the GISS GCM. *J. Clim.*, **7**, 1842–1855.

Mabuchi, K., Sato, Y., Kida, H., Saigusa, N., and Oikawa, T. (1997). A biosphere–atmosphere interaction model (BAIM) and its primary verifications using grassland data. *Papers Meteorol. Geophys.*, **47**(314), 115–140.

Mahrt, L. (1996). The bulk aerodynamic formulation over heterogeneous surfaces. *Bound.-Lay. Meteorol.*, **78**, 87–119.

Manabe, S. (1969). Climate and the ocean circulation I. The atmospheric circulation and the hydrology of the Earth's surface. *Mon. Wea. Rev.*, **97**, 739–774.

Manabe, S. and Wetherald, R. T. (1987). Large scale changes of soil wetness induced by an increase in atmospheric carbon dioxide. *J. Atmos. Sci.*, **44**, 1211–1235.

Marshall, S. (1989). *A Physical Parametrization of Snow Albedo for Use in Climate Models.* NCAR cooperative thesis 123.

Marshall, S. and Oglesby, R. J. (1994). An improved snow hydrology for GCMs. *Climate Dyn.*, **10**, 21–37.

Marshall, S., Roads, J. O., and Glatzmaier, G. (1994). Snow hydrology in a general circulation model. *J. Clim.*, **7**, 1251–1269.

Martin, E. and Lejeune, Y. (1998). Turbulent fluxes above the snow surface. *Ann. Glaciol.*, **26**, 179–183.

McGuffie, K. and Henderson-Sellers, A. (1997). *A Climate Modelling Primer*. New York: John Wiley & Sons.

Miller, G. H. and de Vernal, A. (1992). Will greenhouse warming lead to Northern Hemisphere ice-sheet growth? *Nature*, **355**, 244–246.

Moritz, R. E., Bitz, C. M., and Steig, E. J. (2002). Dynamics of recent climate change in the Arctic. *Science*, **297**, 1497–1502.

Musson-Genon, L. (1995). Comparison of different turbulence closures with a one-dimensional boundary layer model. *Mon. Wea. Rev.*, **123**(1), 163–180.

Namias, J. (1960). Snowfall over the eastern United States: factors leading to its monthly and seasonal variations. *Weatherwise*, **13**, 238–247.

Namias, J. (1962). Influences of abnormal heat sources and sinks on atmospheric behavior. In *Proc. International Symposium on Numerical Weather Prediction*, 1960. Tokyo: Meteorological Society of Japan, pp. 615–627.

Namias, J. (1985). Some empirical evidence for the influence of snow cover on temperature and precipitation. *Mon. Wea. Rev.*, **113**, 1542–1553.

Navarre, J. P. (1975). Modèle unidimensionnel d'évolution de la neige déposée: modèle perce-neige. *La Météorologie*, **4**(3), 103–120.

Nijssen, B. and Lettenmaier, D. (1999). A simplified approach for predicting sortwave radiation transfer through boreal forest canopies. *J. Geophys. Res.* **104**, 27 859–27 868.

Obled, C. (1973). *Modèles Mathématiques de la Fusion Nivale Etude des Risques d'Avalanches*. Laboratoire de Mécanique des Fluides, Institut National Polytechnique de Grenoble, Grenoble, France. (English translation: *Mathematical Models of Snowmelt Study of Avalanche Risks*, Translated for NOAA, available on loan from Language Services Division, F43, National Marine Fisheries Service, NOAA, Washington, DC.)

Oleson, K. W., Dari, Y., Bonan, G. B., *et al.* (2004). *Technical Description of the Community Land Model (CLM)*. NCAR Technical Note NCAR/TN-461+STR. Boulder, CO: National Center for Atmospheric Research.

Outcalt, S. I., Weller, G. D., and Brown, J. (1975). *A Digital Computer Simulation of the Annual Snow and Soil Thermal Regimes at Barrow, Alaska*. Cold Regions Research and Engineering Laboratory Research Report 331, Hanover, NH.

Parviainen, J. and Pomeroy, J. W. (2000). Multiple-scale modelling of forest snow sublimation: initial findings. *Hydrol. Process.*, **14**, 2669–2681.

Polyakov, I., Alekseev, G., Bekryaev, R., *et al.* (2002). Observationally based assessment of polar amplification of global warming. *Geophys. Res. Lett.* **29**(25), 1–4.

Polyakov, I., Walsh, D., Dmitrenko, I., Colony, R. L., and Timokhov, L. A. (2003). Arctic Ocean variability derived from historical observations. *Geophys. Res. Lett.*, **30**(6), 1298, doi:10.1029/2002GL016441.

Pomeroy, J. W., Gray, D. M., Shook, K. R., *et al.* (1998). An evaluation of snow accumulation and ablation processes for land surface modelling. *Hydrol. Process.*, **12**, 2339–2367.

Pope, V. D., Gallani, M. L., Rowntree, P. R., and Stratton, R. A. (2000). The impact of new physical parametrizations in the Hadley Centre climate model: HadAM3. *Climate Dyn.*, **16**, 123–146.

Quick, M. C. (1967). A comparison of measured and theoretical snowpack temperatures. *J. Hydrol.*, **5**, 1–20.

Randall, D. A., Cess, R. D., Blanchet, J.-P., *et al.* (1994). Analysis of snow feedbacks in 14 general circulation models. *J. Geophys. Res.*, **99**(D10), 20 757–20 771.

Rango, A. (1997). The response of areal snow cover to climate change in a snowmelt runoff model. *Ann. Glaciol.*, **25**, 232–236.

Rockwood, D. M. (1964). *Streamflow Synthesis and Reservoir Regulation*. Technical Bulletin No. 22. Portland, OR: U.S. Army Engineer Division, North Pacific.

Roeckner, E., Arpe, K., Bengtsson, L., *et al.* (1996). *The Atmospheric Circulation Model ECHAM-4: Model Description and Simulation of Present-day Climate*. MPI-Rep. 218, MPI für Meteorologie, Hamburg.

Roesch, A., Wild, M., Gilgen, H., and Ohmura, A. (2001). A new snow cover fraction parametrization for the ECHAM4 GCM. *Climate Dyn.*, **17**, 933–946.

Schlosser, C. A., Slater, A. G., Robock, A., *et al.* (2000). Simulations of a boreal grassland hydrology at Valdai, Russia: PILPS Phase 2(d). *Mon. Wea. Rev.*, **128**, 301–321.

Schreider, S. Yu., Whetton, P. H., Jakeman, A. J., Pittock, A. B., and Li, J. (1997). Runoff modelling for snow-affected catchments in the Australian Alpine Region, Eastern Victoria. *J. Hydrol.*, **2**(1), 35–47.

Sellers, P. J. (1985). Canopy reflectance, photosynthesis and transpiration. *Int. J. Remote Sens.*, **6**, 1335–1372.

Sellers, P. J., Randall, D. A., Collatz, G. J., *et al.* (1996). A revised land surface parametrization (SiB2) for atmospheric GCMs. Part 1: Model formulation. *J. Climate*, **9**, 676–705.

Sergent, C., Leroux, C., Pougatch, E., and Guirado, F. (1998). Hemispherical-directional reflectance measurements of natural snow in the 0.9–1.45 μm spectral range: comparison with adding-doubling modelling. *Ann. Glaciol.*, **26**, 59–68.

Shmakin, A. B. (1998). The updated version of SPONSOR land surface scheme: PILPS-influenced improvements. *Global Planet. Change*, **19** (1-4), 49–62.

Slater, A. G., Pitman, A. J., and Desborough, C. E. (1998). The validation of a snow parametrization designed for use in general circulation models. *Int. J. Climatol.*, **18**, 595–617.

Slater, A. G., Schlosser, C. A., Desborough, C. E., *et al.* (2001). The representation of snow in land surface schemes: results from PILPS 2(d). *J. Hydrometeorol.*, **2**, 7–24.

Smirnova, T. G., Brown, J. M., and Benjamin, S. G. (1997). Performance of different soil model configurations in simulating ground surface temperature and surface fluxes. *Mon. Wea. Rev.*, **125**, 216–261.

Stamnes, K., Tsay, S.-C., Wiscombe, W., and Jayaweera, K. (1988). Numerically stable algorithm for discrete-ordinate-method radiative transfer in multiple scattering and emitting layered media. *Appl. Opt.*, **27**(12), 2502–2509.

Stewart, R. E., Leighton, H. G., Marsh, P. *et al.* (1998). The Mackenzie GEWEX Study: the water and energy cycles of a major North American river basin. *Bull. Am. Meteorol. Soc.*, **79**, 2665–2683.

Sturm, M., Holmgren, J., Konig M., and Morris, K. (1997). The thermal conductivity of seasonal snow. *J. Glaciol.*, **43**, 26–41.

Sun, S. F., Jin, J. M., and Xue, Y. (1999). A simple snow–atmosphere–soil transfer model. *J. Geophys. Res.*, **104**, 19 587–19 597.

Sverdrup H. U. (1936). The eddy conductivity of the air over a smooth snow field. *Geofys. Publ.* **11**, 5–69.

Tarboton, D. G. and Luce, C. H. (1996). *Utah Energy Balance Snow Accumulation and Melt Model (UEB).* Computer model technical description and users guide. Utah Water Research Laboratory and USDA Forest Service Intermountain Research Station.

Thomas, G. and Rowntree, P. R. (1992). The boreal forests and climate. *Q. J. Roy. Meteor. Soc*, **118**, 469–497.

Tokioka, T., Noda, A., Kitoh, A., *et al.* (1995). A transient $CO_2$ experiment with the MRI CGCM -Quick Report. *J. Meteor. Soc. Japan*, **73**, 817–826.

U.S. Army Corps of Engineers. (1955). *Snow Investigations. Lysimeter Studies of Snow Melt.* Snow Investigations Research Note 25, Portland, OR: U.S. Army Corps of Engineers, North Pacific Division.

U.S. Army Corps of Engineers. (1956). *Snow Hydrology. Summary Report of the Snow Investigations.* Portland, OR: U.S. Army Corps of Engineers, North Pacific Division (available as PB-151660).

U.S. Army Corps of Engineers. (1972). *SSARR Model. Program Description and User's Manual.* Program 724-K5-G0010. Portland, OR: U.S. Army Corps of Engineers, North Pacific Division.

Van den Hurk, B. J. J. M., Viterbo, P., Beljaars, A. C. M., and Betts, A. K. (2000). *Offline Validation of the ERA40 Surface Scheme.* ECMWF Technical Memorandum 295, European Centre for Medium-Range Weather Forecasts.

Verseghy, D. L. (1991). CLASS – A Canadian land surface scheme for GCMs. Part I: Soil model. *Int. J. Climatol.*, **11**, 111–133.

Verseghy, D. L., Lazare, M., and McFarlane, N. A. (1993). CLASS – A Canadian land-surface scheme for GCMS, II. Vegetation model and coupled runs. *Int. J. Climatol.*, **13**, 347–370.

Viterbo, P. and Betts, A. K. (1999). Impact on ECMWF forecasts of changes to the albedo of the boreal forests in the presence of snow. *J. Geophys. Res.*, **104**, 27 803–27 810.

Wagner, A. J. (1973). The influence of average snow depth on monthly mean temperature anomaly. *Mon. Wea. Rev.*, **101**, 624–626.

Walland, D. J. and Simmonds, I. (1996). Sub-grid-scale topography and the simulation of Northern Hemisphere snow cover. *Int. J. Climatol.*, **16**, 961–982.

Walsh, J. E., Jasperson, W. H., and Ross, B. (1985). Influences of snow cover and soil moisture on monthly air temperature. *Mon. Wea. Rev.*, **113**, 756–769.

Walsh, J. E. and Ross, B. (1988). Sensitivity of 30-day dynamical forecasts to Continental snow cover. *J. Climate*, **1**, 739–754.

Walsh, J. E., Tucek, D. R., and Peterson, M. R. (1982). Seasonal snow cover and short term climatic fluctuations over the United States. *Mon. Wea. Rev.*, **110**, 1474–1485.

Warren, S. G. and Wiscombe, W. J. (1980). A model for the spectral albedo of snow. II. Snow containing atmospheric aerosols. *J. Atmos. Sci.*, **37**, 2734–2745.

Washington, W. M. and Meehl, G. A. (1984). Seasonal cycle experiments on the climate sensitivity due to a doubling of $CO_2$ with an atmospheric general circulation model coupled to a simple mixed layer ocean, *J. Geophys. Res.*, **89**, 9475–9503.

Watanabe, M. and Nitta, T. (1998). Relative impacts of snow and sea surface temperature anomalies on an extreme phase in the winter atmospheric circulation. *J. Climate*, **11**, 2837–2857.

Wetherald, R. T. and Manabe, S. (1988). Cloud feedback processes in a general circulation model, *J. Atmos. Sci.*, **45**, 1397–1415.

Wigmosta, M. S., Lettenmaier, D. P., and Vail, L. W. (1994). A distributed hydrology–vegetation model for complex terrain.*Water Resources Res*, **30**(6), 1665–1679.

Wilson, M. F. and Mitchell, J. F. B. (1987). A doubled $CO_2$ climate sensitivity experiment with a GCM including a simple ocean. *J. Geophys. Res.*, **92**, 13 315–13 343.

Wiscombe, W. J. and Warren, S. G. (1980). A model for the spectral albedo of snow. I. Pure snow. *J. Atmos. Sci.*, **37**, 2712–2733.

World Meteorological Organization (1986a). *Intercomparison of Models of Snowmelt Runoff.* Operational Hydrology Report 23 WMO 646.

World Meteorological Organization (1986b). *Methods of Correction for Systematic Error in Point Precipitation Measurement for Operational Use*. Operational Hydrology Report No. 21, Geneva.

Xue, Y., Sellers, P. J., Kinter, J. L., III, and Shukla, J. (1991). A simplified biosphere model for global climate studies. *J. Climate*, **4**, 345–364.

Yamazaki, T. (1995). The influence of forests on atmospheric heating during the snowmelt season. *J. Appl. Meteorol.* **34**, 511–519.

Yamazaki, T. (2001). A one-dimensional land surface model adaptable to intensely cold regions and its applications in eastern Siberia. *J. Meteorol. Soc. Japan*, **79**, 1107–1118.

Yang, R., Friedl, M., and Ni, W. (2001). Parameterization of shortwave radiation fluxes for nonuniform vegetation canopies in land surface models. *J. Geophys. Res.*, **106**, 14 275–14 286.

Yang, Z.-L., Dickinson, R. E., Robock, A., and Vinnikov, K. Y. (1997). On validation of the snow sub-model of the biosphere–atmosphere transfer scheme with Russian snow cover and meteorological observational data. *J. Climate*, **10**, 353–373.

Ye, H. (2001). Quasi-biennial and quasi-decadal variations in snow accumulation over Northern Eurasia and their connections to the Atlantic and Pacific oceans. *J. Climate*, **14**, 4573–4584.
Ye, H. and Mather, J. R. (1997). Polar snow cover changes and global warming. *Int. J. Climatol.*, **17**, 155–162.
Yeh, T.-C., Wetherald, R. T., and Manabe, S. (1983). A model study of the short-term climate and hydrologic effects of sudden snow-cover removal. *Mon. Wea. Rev.*, **111**, 1013–1024.
Yen, Y. (1981). *Review of Thermal Properties of Snow, Ice and Sea-ice*. U.S. Army Cold Regions Research and Engineering Laboratory Report 81–10, Hanover, NH.

# 5

# Snow-cover data: measurement, products, and sources

Ross Brown and Richard L. Armstrong

## 5.1 Introduction

In the past several decades, the growing importance of the climate change issue has prompted new needs for snow-cover information over a wide range of spatial and temporal scales for climate change detection, for the development of snow process models, and for validation activities (e.g. satellite snow-cover products, physical snow models, output from global and regional climate models). Validation of multi-layer physical snowpack models requires detailed information on snowpack structure, surface albedo (reflectivity), temperature profiles, snowmelt, and surface energy fluxes. The inclusion of small-scale processes such as canopy interception and sublimation in physical snow-cover models (e.g. Pomeroy *et al.*, 1998) has prompted a new need for information on the mass of snow intercepted and stored in the vegetation canopy. At a larger scale, land surface process and hydrological models (e.g. Verseghy, 1991; Cline *et al.*, 1998) require information on the spatial distribution of snow-cover properties to develop and validate methods and approaches to take account of subgrid scale variations in snow with terrain and vegetation cover (e.g. Essery, 1997). At even larger scales, global climate models (GCMs) have generated new needs for information on the global distribution of snow cover and water equivalent at monthly and climatologically averaged time-scales for validating snow-cover simulations. Barry (1997) provided an overview of the requirements and the status of cryospheric data for model validation activities.

These new needs for snow-cover information and snow properties over a wide range of spatial and temporal scales pose a number of challenges to the snow research community for observing and analyzing snow-cover information. The systematic collection of information on snow accumulation (snowfall, precipitation, precipitation type) and redistribution requires the development and application of

*Snow and Climate: Physical Processes, Surface Energy Exchange and Modeling*, ed. Richard L. Armstrong and Eric Brun. Published by Cambridge University Press. 

new sensor technology and methods for sensor interpretation. For example, ultrasonic snow depth sensors with high temporal resolution require the application of data processing and quality control logic to remove spurious data caused by blowing snow. One of the greatest challenges is to develop automated systems for the reliable and accurate measurement of winter precipitation (Goodison *et al.*, 1998b).

Satellites have provided the means for global snow-cover monitoring since the late 1960s (Robinson *et al.*, 1993) in support of climate change detection (Hall, 1988), and for the validation of GCM snow-cover simulations (Foster *et al.*, 1996; Frei and Robinson, 1998). While satellites have been able to monitor snow-cover extent with sufficient accuracy for climate monitoring purposes, there is an ongoing research effort to develop reliable methods for observing snow water equivalent (SWE) from space for a wide range of typical land covers and terrain types as described in Section 5.3 below. There is thus a continuing need for surface-based observations of snow depth and SWE to support satellite algorithm development and validation. As noted by Barry *et al.* (1995), no single sensor or methodology can provide the snow-cover information required to satisfy current needs. The merging of information from ground observations, models and satellite observations is consequently an integral feature of modern snow science, which is why "scaling issues" (Blöschl, 1999; Erxleben *et al.*, 2002; Moltoch *et al.*, 2004) have become so important e.g. how does one relate a single snow course observation to a 25 km by 25 km satellite-derived estimate of SWE?

## 5.2 *In situ* snow data

Because of the profound influence snow has on natural ecosystems (Jones *et al.*, 2001) and the human environment, snow measurement science has a long and rich history extending from Chinese snow cages in the first millennium (Biswas, 1970) to ultrasonic snow sensors employed on automatic weather stations (Gubler, 1981; Goodison *et al.*, 1988). A recent review of snow-cover observations and data sources was provided by Groisman and Davies (2001). Properties of a snow cover that have been routinely monitored are: snow depth, snow water equivalent, and snowfall (amount and water equivalent). Strictly speaking snowfall is not a property of a snow cover, but it is included here because of its essential role in the initiation and accumulation of a snowpack. In addition, numerous paleo and proxy sources of information related to snow cover are available. A brief overview of data sources and measurement issues are provided below.

### 5.2.1 *Pre-instrumental data*

Prior to the late 1800s, there were few systematic measurements of snow depth. However, routine snow-cover-related observations were made of variables such as the number of days with snowfall, number of days with snow on the ground, or first/last dates of snow on the ground. The presence of snow on the ground is one of the earliest regularly recorded snow-cover observations. In Great Britain, for example, regular observations of "snow lying" (when snow covers more than half the ground at 0900 GMT) began in 1912 (Jackson, 1978b) but it was only in the late 1940s that snow depth was routinely recorded (Jackson, 1978a). It has been shown (Foster, 1986; Robinson, 1991; Brown and Goodison, 1996) that regionally averaged *in situ* observations of snow-cover duration agree closely with satellite observations for open, non-mountainous areas. This offers the potential for using historical *in situ* observations to extend the more recent satellite-based observations of snow-cover extent back to the early part of the twentieth century (Frei *et al.*, 1999; Brown, 2000). Observations of related parameters such as the first and last dates of snow on the ground, or dates of the first thaw have been made for extended periods of time at a number of locations around the world. For example, Ball (1992) was able to investigate spring thaw variability from 1715 to 1840 at York Factory and Churchill, Canada, using records of the first date of snow thaw made by Hudson Bay Company personnel. Information on snow-cover conditions can also be inferred from observations of the number of days when snow or solid precipitation fell. For example, Manley (1969) used data on the "frequency of occurrence of days with snow or sleet observed" in the UK to construct a normalized "snowiness index" back to the 1660s. Manley found good agreement between the index and snow on the ground at elevations above 300 m, and on the basis of this relationship was able to conclude that in the severe winter of 1695, some upland Aberdeenshire farms probably had snow cover for five months.

Inferences concerning past snow-cover conditions can also be made from various biophysical markers that are sensitive to snow cover or to snowmelt runoff. For example, Lavoie and Payette (1992) used changes in basal abrasion levels and changes in growth forms of black spruce to document an increase in twentieth-century snow cover along the treeline in subarctic Québec. Paleo reconstruction of lake water levels in semi-arid areas with snowmelt-fed reservoirs (e.g. Vance *et al.*, 1992) can also provide insights into spring snowpack conditions over time periods spanning several thousand years.

### 5.2.2 *Snow depth*

Snow depth is the most obvious property of a snowpack, but it is one of the less useful values from a water and energy budget perspective since snow depth can change *independently* from snow mass due to processes such as settling, metamorphism, and melt/refreeze events (Fitzharris *et al.*, 1992). Nonetheless, daily snow depth data are used in a multitude of applications such as estimating building snow loads, snow clearing contracts, winter survival of crops, ground frost penetration, and biological studies. Daily snow depth is considered a high priority variable for global climate monitoring, validation of climate models, and climate change impact assessment (Cihlar *et al.*, 1997).

Manual snow depth measurements are made using a ruler or with one or more fixed snow stakes (Fig. 5.1c). Where daily observations of the total depth of snow on the ground are made by ruler, some judgement is required to obtain a "representative" value. This is evident in the instructions to Canadian observers (AES, 1977):

*The total depth of snow on the ground at the time of the observation shall be determined, (in whole centimetres) by making a series of measurements and taking the average. The area selected for the measurement shall be chosen with a view to avoiding drifts. Care shall be taken to ensure that the total depth is measured including the depth of any layers of ice which are present.*

Determining a representative snow depth for an incomplete, patchy snow cover is particularly subjective and the Canadian manual for weather observations offers no instructions or guidance on how to do this. According to Doesken and Judson (1997), a visual average should be determined for the area surrounding the weather station, and when the snow cover is less than 50%, the depth recorded as a trace. This problem is avoided when a series of fixed stakes are used for measuring snow depth. However, care is required when reading fixed stakes as snow can preferentially accumulate or melt around a fixed object. Even where snow depth observing practices are carried out consistently, the measurements are only truly representative of the snow conditions at the measuring site. Weather observing sites are usually located in open areas, often at airports, thus the data are mainly representative of exposed sites.

Usually manual snow depth measurements are taken once per day in the morning or afternoon, although twice daily readings are made in some countries. The timing of the depth observation is not critical, but it can create inhomogeneities in derived snow-cover data series such as the number of days with snow on the ground if not followed consistently over time e.g. a shift from PM to AM readings occurred in the United States in the 1960s (D. Robinson, personal communication, 1998). Since a snowpack is more likely to experience melting and settling during the daytime, especially in situations of small amounts of new snow on the ground from overnight snowfall events, a shift to morning readings could introduce a spurious increase in snow-cover statistics such as the number of days with snow on the ground.

Manual observations of daily snow depth have been carried out in association with regular meteorological observation programs at synoptic and climate stations in most countries with a seasonal snow cover. Typically, these networks evolved

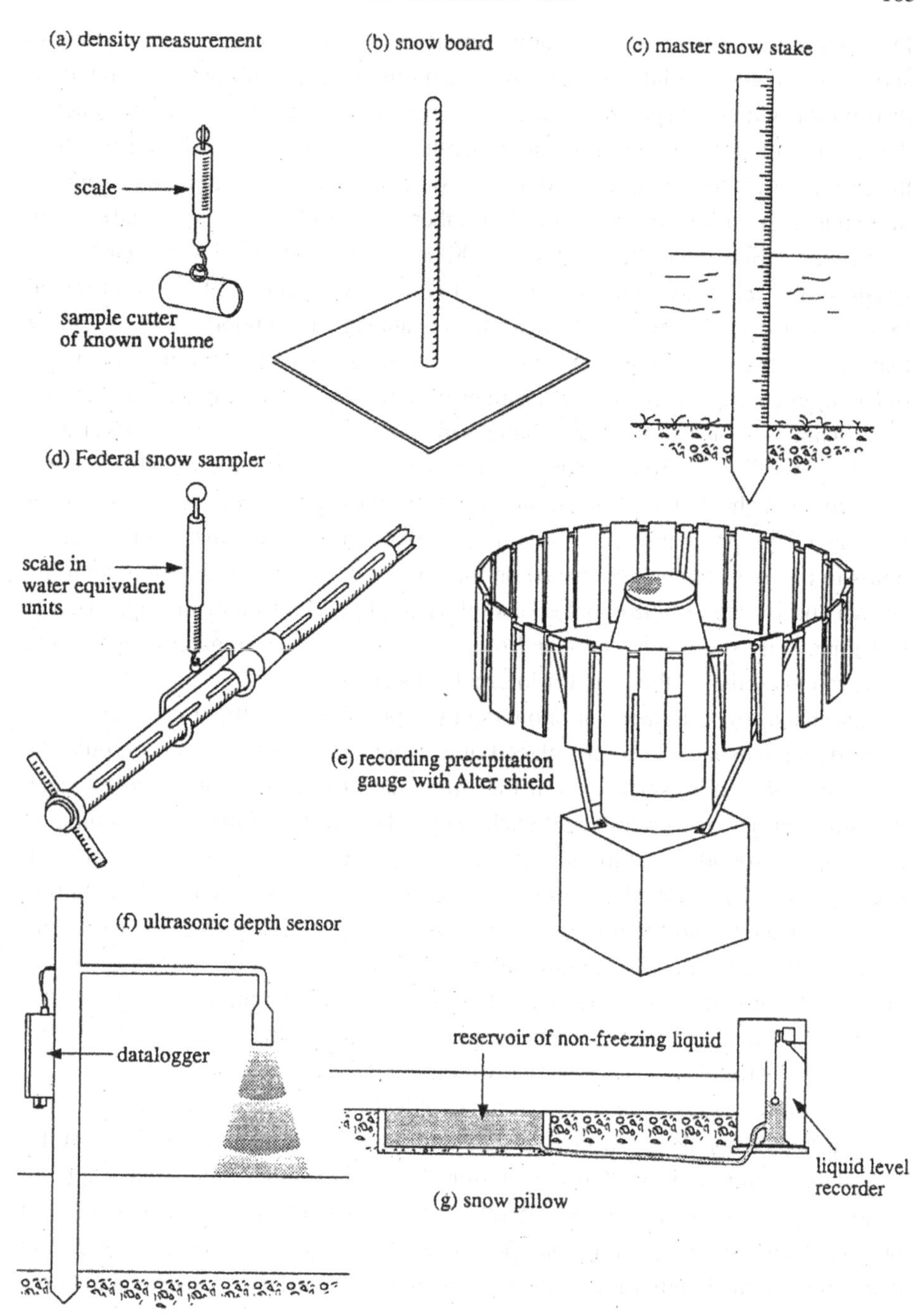


Figure 5.1. Instruments for measuring snow accumulation and melt (Fitzharris *et al.*, 1992). Reproduced with permission from the New Zealand Hydrological Society.

in response to needs for weather and climate information in support of economic activities and the spatial distribution of stations tends to follow the population distribution, with poor spatial coverage in mountainous and remote areas. In general, the number of stations reporting snow depth increases over time during much of the twentieth century, with a reduction in the 1980s and 1990s in many countries in response to budgetary cutbacks. A number of countries have systematic snow depth observations going back to the late 1800s, e.g. Switzerland (Föhn, 1990), USA (Easterling *et al.*, 1999), the former Soviet Union (Armstrong, 2001), and Finland (Kuusisto, 1984). When working with these data sets it is important to remember that snow depth observations are subject to the same sources of inhomogeneity as other climatological elements e.g. changes in station location, changes in observing time, changes in measuring units, changes in observers, and urban effects (warming and dirtying of snow from pollution). Robinson (1989) developed quality control procedures to check the internal consistency of daily snow depth observations for U.S. data. The procedure compares the observed change in snow depth over 24 hours to the expected change based on the observed snowfall and air temperature. Brown and Braaten (1998) applied a slightly modified version of this approach to Canadian daily snow depth data and found that over 98% of non-zero snow depth observations satisfied the tests for internal consistency.

Automated measurement of snow depth is possible with the use of ultrasonic snow depth sensors. These are placed above the snow surface, and compute the distance to the snow surface from the time taken for a pulse of sound to reach the snow surface and be reflected back (Fig. 5.1f). On-board electronics compute the speed of sound as a function of the ambient air temperature. As outlined in Pomeroy and Gray (1995), anomalous measurements may occur during falling or blowing snow, and the ultrasound cannot distinguish newly fallen, low density snow. At other times snow depth can be measured with an accuracy of $\sim$1 cm, and the sensor can be interrogated at frequent regular time intervals (e.g. hourly) to obtain a detailed history of snow depth changes from settling, wind erosion and melt that is particularly useful for the validation of physical snow process models. The other major advantage of the automated depth sensor is that the snow layer is undisturbed. When the snow layer is gone, ultrasonic sensors continue to faithfully record the height of the growing vegetation. This requires the application of quality control logic to remove spurious "depths." A criticism of the auto sensor is that it only measures snow depth at "one point," a circular area with diameter 0.2–2 m depending on the height of the sensor above the snow surface (Pomeroy and Gray, 1995). This makes careful site selection essential.

Daily snow depth is a versatile measurement in that a variety of snow-cover statistics can be derived from daily depth observations for monitoring snow cover. These include: the duration of snow on the ground for various depth thresholds;

the dates of the start and end of continuous snow cover; the maximum snow depth; and the date of maximum snow depth. Monitoring these kinds of statistics provides more detailed information on changes in snowpack amount and timing (Goodison and Walker, 1993) than traditional monthly statistics such as the average or median snow depth. A summary of snow data sources for climate studies was provided by Groisman and Davies (2001). Major compilations of historical daily snow depth data have been published for the USA (Easterling *et al.*, 1999), the former Soviet Union (Armstrong, 2001), and Canada (MSC, 2000). There is relatively little published snow depth data for Europe and Asia, although there are published records of monthly snow-cover duration at selected stations for the People's Republic of China (Shiyan *et al.*, 1997). The U.S. National Snow and Ice Data Center (NSIDC) is continually working to rescue important snow data sets and make them freely available to the research community. The reader should visit their website to check the latest data sets as well as the U.S. Carbon Dioxide Data Information and Analysis Center (CDIAC), which has made a number of important snow data sets available for downloading, e.g. Easterling *et al.* (1999) and Shiyan *et al.* (1997). U.S. daily snow depth data are archived at the National Climatic Data Center (NCDC) and were compiled and analyzed by Heim (1998) to produce a monthly snowfall and snow-cover climatology for 5525 stations in the contiguous U.S.A. and Alaska.

The World Meteorological Organization (WMO) Global Telecommunication System (GTS), which transmits synoptic and aviation weather data internationally, includes daily snow depth observations in its messages. These observations are used by a number of operational meteorological agencies to generate daily snow depth fields for input to numerical weather prediction models. However, on any given day only a small fraction of the stations observing snow depth may actually report the data over the GTS. Snow depth and snow-cover information are included in the global climatological data summaries and standard climate normals, which are compiled on a regular basis by the WMO.

### 5.2.3 Snow water equivalent

The water equivalent of a snowpack, or SWE, is the vertical depth of the water layer which would be obtained by melting the snow cover over a given area (WMO, 1981). SWE is related to the depth and density of the snowpack as shown in Equation (5.1) from Pomeroy and Gray (1995):

$$\mathrm{SWE\,(mm)} = HS\rho_{\mathrm{s}} \tag{5.1}$$

where $HS$ is the depth of snow (m) and $\rho_{\mathrm{s}}$ is the density of snow (kg m$^{-3}$). The conversion from a mass of snow (kg m$^{-2}$) to a depth of water (mm) is based on the fact that 1 mm of water spread over an area of 1 m$^2$ weighs 1 kg.

SWE information is essential for water resource management (e.g. flood forecasting, reservoir management, irrigation scheduling) and much of the effort spent in collecting SWE information around the world is firmly grounded in the economic and safety benefits of water management. SWE information is also required for validating snow models and GCM snow-cover simulations. Correct simulation of seasonal and annual variation in the mass of snow on the ground is needed to capture important direct (e.g. snow albedo) and indirect (e.g. runoff, soil moisture recharge, surface evaporation, clouds) feedbacks between snow and the climate system. For example, snow-related direct and indirect feedbacks are known to be important factors in the strength of the Indian summer monsoon circulation (Barnett *et al.*, 1989; Bamzai and Shukla, 1999). Monitoring of SWE is also required for climate change detection. Global climate warming is projected to be associated with increased precipitation, and the net effect of these trends requires monitoring both snow-cover extent and SWE. For example Brown (2000) presented evidence of increases in early winter snow depth and SWE over North American mid-latitudes during the twentieth century in response to increasing precipitation, while spring snow cover and SWE were observed to decrease in response to warmer temperatures. Groisman *et al.* (1994) demonstrated that reductions in spring snow cover contribute to spring warming through an enhanced positive feedback at this time of the year.

The most commonly used approach for determining SWE is the gravimetric method which involves taking a vertical core through the snowpack, and weighing or melting the core to obtain the SWE (Fig. 5.1a, d). A variety of coring and weighing systems have been used around the world with varying lengths and diameters depending on measurement units and local snow conditions (see Sevruk, 1992). One of the earliest national SWE observing networks was established in Finland in 1909 (Kuusisto, 1984). However, systematic observation of SWE was not widespread until the middle of the twentieth century. In the U.S.A. for example, the National Weather Service began regular point measurements of SWE at first-order stations during the winter of 1952/53 (Schmidlin, 1990). In order to obtain representative values of SWE, measurements are often carried out at regular marked intervals along a permanently marked transect or "snow course." Many factors are involved in the design of a snow course (e.g. purpose, accessibility, terrain) and the reader is referred to Goodison *et al.* (1981) for a detailed discussion. The length of a snow course and the number of sampling points depends on the desired level of accuracy and the spatial variability of the snow cover, usually represented by the coefficient of variability (COV or CV) which is the ratio of the standard deviation to the mean. There is extensive literature on this subject e.g. Goodison *et al.* (1981), WMO (1981), Sevruk (1992), and Pomeroy and Gray (1995). The spatial variability of snow depth is typically higher than snow density, which means that more depth measurements are required along a snow course than snow density. For example, in hilly terrain, a snow course is generally 120–270 m long with depth and density measurements at about 30 m intervals. In open environments with a shallow snow cover, a snow course may need to be as long as 1–2 km, with density measurements taken 100–500 m apart and depth measurements at about five equally spaced points between the density locations (Pomeroy and Gray, 1995).

The accuracy of manual SWE measurements from snow samplers is discussed in detail in Sevruk (1992). The main systematic error (due to instrumentation) is related to a tendency for additional snow to be forced up into a tube as it is pushed through a snowpack. Random errors associated with observers and snow conditions include the difficulty in keeping loose granular snow in the corer, drainage of water from very wet snow, ice crusts, and sampling very shallow, patchy snow cover. Snow course measurements are usually carried out on a weekly or bi-weekly basis. However, not all courses have regular measurements throughout the snow-cover season. In many cases, measurements made by operational agencies or utilities for runoff management are confined to the late winter and early spring period as the main interest is in determining the peak SWE prior to melt. This can limit the usefulness of some snow course data for climate-related studies. Other limitations for using SWE surface observations in climate-related studies are: the uneven spatial distribution of data, the relatively short periods of continuous observations, the data quality (see Schmidlin, 1990), and a general lack of data availability. The National Snow and Ice Data Center is attempting to address the last problem through data rescue initiatives such as hydrological snow surveys from the former Soviet Union (Krenke, 1998). The uneven spatial and temporal distribution of snow course observations poses a major challenge for the development of global-scale SWE climatologies for validating climate model simulations. The blending of *in situ*, satellite- and model-derived information is required to provide consistent spatial SWE information over a range of land cover surfaces and terrain types (Hartman *et al.*, 1995; Carroll *et al.*, 1999, 2001).

Automated surface-based observations of SWE are possible from devices such as snow pillows (Fig. 5.1g), which measure the mass of snow over a small area from displaced fluid or a pressure transducer. Snow pillows are usually octagonal or circular in shape, with an area of $\sim$5–10 m$^2$. Snow pillows are most effective for monitoring relatively deep snowpacks in sheltered environments. Interpretation of data can be complicated by "bridging" (from ice or hard snow layers in the snowpack) or by the draining of wet snow (Pomeroy and Gray, 1995). Snow pillows are ideal for remote locations because they are low maintenance and are usually linked to a land-line or satellite data transmission systems to provide real-time information such as the U.S. Department of Agriculture (USDA) Snow Telemetry (SNOTEL) network over the western United States (Rallison, 1981). An advantage of these automated systems is that they can be interrogated on a daily basis to provide more detailed information during the melt season than regular weekly or two-weekly snow course observations. Daily values of SWE from over 600 snow pillow sites in the western U.S.A. from 1979 are available from the USDA National Water and Climate Center. A comprehensive analysis of these data for the 1980–1998 period was provided by Serreze *et al.* (1999).

### 5.2.4 *Snowpack stratigraphy*

Detailed information on snowpack stratigraphy (e.g. layer density, hardness, grain size, grain type, chemistry, temperature) is required for a wide range of needs such as monitoring avalanche potential, snow trafficability studies, and atmospheric transport and deposition of pollutants. Most of these observations must be made manually by digging a "snow pit" and observing/measuring snow properties. This is a time- and labour-intensive process and snow pits are usually only dug when needed to support critical activities such as avalanche risk assessment, and for research projects. Snow pit data are required to validate detailed snowpack layer models such as CROCUS (Brun *et al.*, 1992), SNOWPACK (Bartelt and Lehning, 2002; Lehning *et al.*, 2002a, b), or SNTHERM (Jordan, 1991) that simulate layer development and snow grain evolution. The terminology and symbols employed in classifying the layers in a snowpack are given in the *International Classification for Seasonal Snow on the Ground* (Colbeck *et al.*, 1990). Extensive collections of snow pit data from Greenland and the Antarctic are archived at the U.S. National Snow and Ice Data Center. Procedures and methods for the observing and recording of data from snow pit profiles can be found in McClung and Schaerer (2006).

### 5.2.5 *Snowfall and solid precipitation*

There are compelling reasons for collecting accurate information on snowfall. First, accurate precipitation data (adjusted for systematic errors) are essential to balance the energy and water cycle in the climate system, for climate monitoring, for determining the global and regional hydrological balance, and for understanding key components of the cryosphere such as the snow-covered area, snow water equivalent, and glacier mass balance. Precipitation is expected to increase in response to global warming (IPCC, 1996) and there is evidence (Bradley *et al.*, 1987; Vinnikov *et al.*, 1990; Groisman and Easterling, 1994; Mekis and Hogg, 1999) that precipitation has exhibited an upward trend during the twentieth century. Warming is also likely to result in changes in the solid–liquid fraction of precipitation in shoulder seasons. Accurate information on precipitation intensity, timing, and the solid/liquid fraction are also needed to correctly simulate snowpack development and snowmelt.

Snowfall, however, is notoriously difficult to observe accurately. Before elaborating on some of these difficulties, it is important to make clear the distinction between snowfall and snowfall precipitation (or solid precipitation). *Snowfall* is the depth of freshly fallen snow that accumulates on a snow board (Fig. 5.1b) during the observing period and has been traditionally measured with a ruler. *Snowfall precipitation* is the amount of liquid water in the snowfall intercepted by a precipitation gauge. At climate stations the depth of new snowfall is measured once or twice per day using a ruler and snow board (Fig. 5.1b). In many countries, snowfall precipitation is estimated from daily total snowfall assuming a fresh snowfall density of 100 $kg\,m^{-3}$. However, the density of freshly fallen snow varies widely as snowfall density is a function of air temperature and crystal type (Judson and Doesken, 2000). Values can range from 10–30 $kg\,m^{-3}$ for dry, cold "wild snow" (Seligman, 1980) to more than 150 $kg\,m^{-3}$ for warm, wet snow. Most fresh snowfall densities fall within a range of 30–150 $kg\,m^{-3}$ (Pomeroy and Gray, 1995; Judson and Doesken, 2000). Manual ruler observations of snowfall are subject to numerous sources of error; the most important are the blowing and drifting of snow, and the melting or rapid settling of snow when it reaches the ground. Therefore, new snow density also depends strongly on the length of the observation period (1 h, 24 h etc.). For example, Goodison *et al.* (1981) monitored average density increases between 8 and 13 $kg\,m^{-3}\,h^{-1}$ during snow storms with a duration of less than 12 h. Doesken and Judson (1997) provide a good overview of some of the practical problems associated with observing snowfall.

Snowfall precipitation is measured using a snowfall gauge; gauges can range from a standard rain gauge to a specially shielded snowfall gauge. A fundamental problem of gauge measurement of snowfall is that most precipitation gauges catch less snowfall than the "true" amount because accelerated wind flow over the top of the gauge reduces the number of snowflakes able to enter the orifice. In alpine regions, catch efficiency is at most 80% with shielded devices placed 1.5 m above the snow surface (Föhn, 1985). The under-catch effect increases with wind strength, and for a relatively modest wind speed of 15 $km\,hr^{-1}$, an unshielded US standard 8 inch precipitation gauge is estimated to catch only ~50% of the "true" snowfall (Yang *et al.*, 1998). This effect can be reduced by using shielding devices such as the Alter shield shown in Figure 5.1e, which will increase the catch by 20–70% (Yang *et al.*, 1999). However, this obviously creates a discontinuity in precipitation time series that must be corrected if the data are to be used in climate change studies (Yang *et al.*, 1999). Precipitation gauges, shields, and observing practices vary considerably from country to country, and over time, e.g. USSR (Groisman *et al.*, 1991), U.S.A. (Groisman and Easterling, 1994), and Canada

(Metcalfe *et al.*, 1997). Other important systematic sources of error include instrument siting, trace precipitation amounts, and wetting loss (the amount of water sticking to the side of a gauge each time it is emptied). Metcalfe and Goodison (1993) showed that when trace precipitation amounts were adjusted for wetting loss, they could account for a significant increase in corrected precipitation totals (~30% for a prairie site). This correction is particularly important in the Arctic where some stations report over 80% of precipitation observations as trace amounts (Metcalfe and Goodison, 1993).

Correction of these systematic sources of error is required to obtain unbiased, homogeneous estimates of snowfall precipitation. The World Meteorological Organization recognized this problem in 1985 when it initiated a gauge intercomparison project to document sources of systematic error in the various national systems for precipitation measurement (Goodison *et al.*, 1998b). The recent trend toward increased automation of climate observations also has important consequences for the homogeneity of precipitation measurement series. Goodison *et al.* (1998b) concluded that heated tipping bucket rain gauges were not recommended for winter use due to excessive evaporation loss. Weighing gauges were found to be the most practical but these can introduce a "timing" error due to snow or freezing precipitation sticking to the inside of the gauge and melting at some later time. Automated gauges can also catch blowing snow and provide no information on the liquid/solid fraction of precipitation. These problems complicate the real-time interpretation of the data as well as the application of procedures to adjust for systematic errors.

## 5.3 Remote sensing data

As noted earlier, the extent and variability of seasonal snow cover are important parameters in large-scale climate and hydrologic systems. Satellite remote sensing offers the opportunity to monitor and evaluate various snow parameters and processes at regional to global scales (Hall and Martinec, 1985; Hall *et al.*, 2005). This section describes the types of satellite remote sensing currently being applied to snow-cover studies. Similar sensor configurations are typically available for aircraft use but this section concentrates on data from polar orbiting satellites because of the potential for monitoring snow cover at the hemispheric to global scale with daily or near-daily spatial coverage. More importantly, a few of the data sets described below provide a time series of sufficient length to evaluate climate patterns and climate change.

### *5.3.1 Visible-band satellite data*

Snow cover is often easily identifiable in visible-band satellite images because it typically possesses an albedo that exceeds almost all other land surface types. Snow may be identified manually by noting the magnitude of the reflectance or it may be identified automatically by the application of an algorithm that recognizes the specific spectral signature of snow. The spectral reflectivity of snow depends on a number of parameters including grain size and shape, impurity content, near-surface liquid water content, surface roughness, and solar elevation. The visible albedo is highly sensitive to the impurity content (Warren and Wiscombe, 1981). In the near-infrared region snow reflectance decreases strongly with wavelength and becomes primarily dependent on grain size (Grenfell and Perovich, 1981; Wiscombe and Warren, 1981). The presence of liquid water in a melt state has very little effect on the albedo. However, the larger grains which result from melt metamorphism result in a reduced near-infrared reflectance. At wavelengths above 1.4 μm the reflectance of snow amounts to only a few percent enabling good discrimination between snow and clouds since the reflectance of clouds remains high at that wavelength.

Some of the earliest applications of satellite remote sensing involved efforts to map and monitor the areal extent of snow cover. In fact, snow-cover extent is the longest available environmental product provided by satellite remote sensing. In 1966, the National Oceanographic and Atmospheric Administration (NOAA) began an operational program to map the Northern Hemisphere snow extent using available visible-band satellite data (Matson and Wiesnet, 1981; Matson *et al.*, 1986; Robinson *et al.*, 1993). Within the following ten years, researchers began to present results demonstrating the operational capabilities of satellite remote sensing in snow hydrology (Rango, 1975; Schneider *et al.*, 1976).

During the past four decades much important information on continental to hemispheric scale snow extent has been provided by satellite remote sensing in the visible wavelengths. From 1966 to 1999 NOAA-NESDIS produced weekly snow extent charts for Northern Hemisphere land surfaces using visible-band satellite imagery (Robinson *et al.*, 1993; Frei and Robinson, 1999). These NOAA charts were derived from the manual interpretation of Advanced Very High-Resolution Radiometer (AVHRR), Geostationary Operational Environmental Satellite (GOES), the European satellite (METEOSAT), Japan's geostationary meteorological satellites, and other visible satellite data by trained meteorologists. The charts were then digitized on a weekly basis using an 89 by 89 Northern Hemisphere polar stereographic grid with a nominal resolution of 190.5 km. The data values are binary and grid cells are classified as snow covered or snow free if the cell has more or less than 50% snow cover (Dewey and Heim, 1982).

In 1997, NOAA-NESDIS began the process of migrating to a more automated procedure for generating a daily, higher resolution (1024 by 1024 polar stereographic grid with a resolution of approximately 25 km) snow-cover analysis as the coarse resolution of the weekly charts had been shown to cause errors in the National Meteorological Center's Numerical Weather Prediction (NWP) models. The result was the Interactive Multisensor Snow and Ice Mapping System (IMS) which incorporates a wide variety of satellite imagery (AVHRR, GOES, SSM/I) as well as derived mapped products (USAF Snow/Ice Analysis) and surface observations, and allows a trained meteorologist to produce a hemispheric analysis in one hour as opposed to 10 hours with the old weekly product (Ramsay, 1998). The new daily analysis and the old weekly product were overlapped for two winters to determine if the switch introduced any inhomogeneity into the existing weekly product (Robinson *et al.*, 1999). The weekly product was phased out on June 1, 1999, but a "pseudo-weekly" map is generated by taking the Sunday IMS map and interpolating this back to the coarse resolution of the earlier weekly product.

The NOAA-NESDIS data set has been used extensively in analysis of snow-cover variability (e.g. Robinson and Dewey, 1990; Gutzler and Rosen, 1992; Groisman *et al.*, 1994; Frei and Robinson, 1999), for investigating snow-cover linkages to atmospheric circulation and climate (e.g. Leathers and Robinson, 1993; Gutzler and Preston, 1997; Clark *et al.*, 1999; Watanabe and Nitta, 1999), and for evaluating climate models (e.g. Foster *et al.*, 1996: Frei *et al.*, 2005). However, it is important to keep in mind that this product has well-documented limitations for regional-scale analysis of snow cover (Scialdone and Robock, 1987; Wiesnet *et al.*, 1987; Robinson and Kukla, 1988; Wang *et al*, 2005). Sources of error include difficulties in the discrimination of clouds from snow, the use of previous estimates of snow cover in regions with persistent cloud cover, masking of snow on the ground by forests, and accurate charting in areas of patchy snow. To facilitate these types of analyses the National Snow and Ice Data Center (NSIDC), University of Colorado, has developed the Northern Hemisphere EASE-Grid Weekly Snow Cover and Sea Ice Extent Version 3 (Armstrong and Brodzik, 2005), a Northern Hemisphere cryospheric product that combines snow cover and sea-ice extent at weekly intervals (Fig. 5.2). The snow data set is based on the weekly NOAA charts, revised by Robinson *et al.* (1993), for the period 1966–2005, while the sea-ice data set, based on passive microwave remote sensing, covers the period 1978–2005. This data set also includes monthly climatologies describing snow and sea-ice extent in terms of average conditions, probability of occurrence, and variance. The data set is produced in an azimuthal equal area projection (NSIDC Equal Area Scalable Earth Grid or EASE-Grid).

### 5.3.2 High-resolution/multi-spectral data

Several polar orbiting satellite sensors have multiple spectral bands located in the visible (0.4–0.7 μm), near-infrared (0.7–1.1 μm), and shortwave infrared (1.1–3.0 μm) as well as the thermal infrared (3.0–100 μm) regions of the electromagnetic spectrum. These sensors typically have very high spatial resolution compared to the actual snow products derived from sensors such as AVHRR described

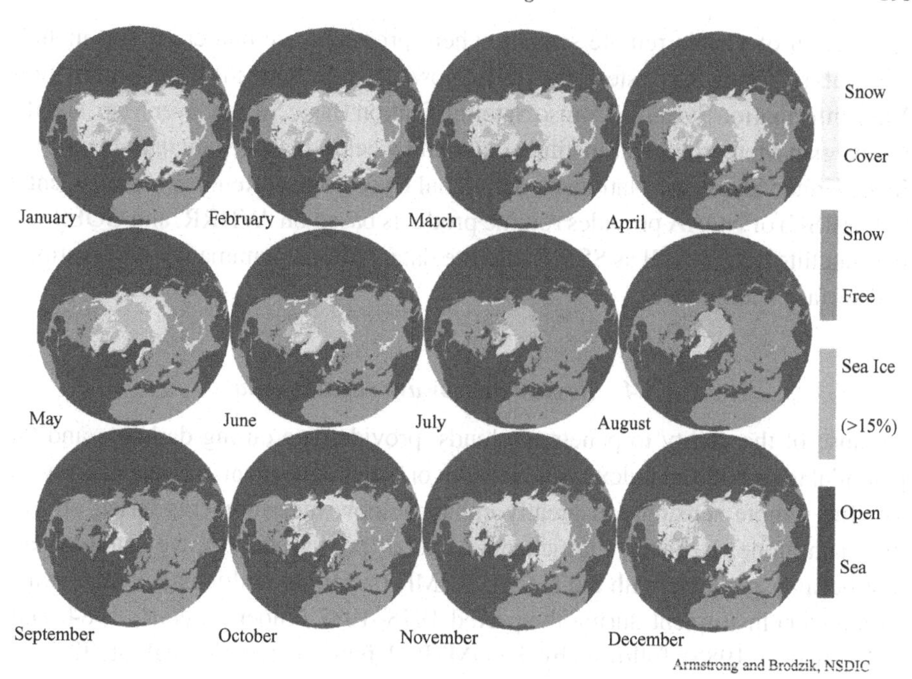


Figure 5.2. Monthly Northern Hemisphere snow cover (1966–2005) and sea ice extent (1978–2005) climatologies (Source: NSIDC *Northern Hemisphere EASE-Grid Weekly Snow Cover and Sea Ice Extent Version 3, 2005*). (Plate 5.20.)

above, for example, Landsat Thematic Mapper at 80 m and the Systeme Probatoire pour l'Observation de la Terre (SPOT) at 1.0 m. The multi-spectral sensors provide the capability to map snow based on the specific spectral signature of snow. For a subpixel analysis of snow-covered areas a high number of spectral bands is of great value for use in spectral unmixing algorithms, i.e. for decomposing the effective spectrum of a single pixel into its primary spectra related to the different snow-free classes present (spectral end members) (e.g. Nolin *et al.*, 1993; Rosenthal and Dozier, 1996: Dozier and Painter, 2004). However, the disadvantages include a limited opportunity to view the snow surface when the low temporal resolution, up to 18 day repeat times, is combined with the problems associated with cloud cover. These limitations, plus the typically greater cost of these high spatial resolution data sets, have limited their use to local and short-term experimental applications.

### 5.3.3 *Other regional-scale snow products based on visible data*

In locations where accurate snowmelt runoff forecasts are required for hydropower management the regional-scale application of satellite data often plays a very important role. For example, Norway is one of the most active countries in the application of visible remote sensing where programs are numerous and include e.g. Statkraft SnowSat System, the NVE Snow Cover System, the Cap Gemini Snow View Information System, Tromso Satellite Station Basic Snow Cover Map, and the Norwegian Meteorological Institute daily and weekly maps (Solberg *et al.*, 1997). In the United States, the National Operational Hydrological Remote Sensing Center (NOHRSC) of NOAA provides routine products based on AVHRR, and GOES visible satellite data as well as SSM/I satellite data, airborne gamma ray observations, and surface station data.

### 5.3.4 Passive microwave satellite data

Because of the ability to penetrate clouds, provide data during darkness and the potential to provide an index of snow depth or water equivalent, passive microwave satellite remote sensing can greatly enhance snow measurements based on visible data alone. Reliable, multichannel, global passive microwave satellite data first became available with the NASA SMMR (scanning multichannel microwave radiometer) instrument during the period 1978–1987 (Gloersen *et al.*, 1984; Hall and Martinec, 1985), followed by the DMSP (defense meteorological satellite program) SSM/I (special sensor microwave/imager) from 1987 to 2008 and beyond (Hollinger *et al.*, 1990). The SMMR and SSM/I instruments provided a combined range of microwave frequencies from 6 to 89 GHz in both horizontal and vertical polarizations. Techniques used to derive snow parameters from passive microwave data have relied most heavily on channels at 18–19, 37, and 85 GHz. When snow covers the ground, the microwave energy emitted by the underlying soil is scattered by the snow grains. Therefore, when moving from snow-free to snow-covered land surfaces, a sharp decrease in emissivity and associated brightness temperatures provides an indication of the presence of dry snow (Mätzler, 1994). In addition, snow exhibits a negative spectral gradient which means that as the microwave frequency increases, for example from 19 to 37 GHz, the emissivity and associated brightness temperatures decrease. Brightness temperature is the product of emissivity at the given microwave frequency and the physical temperature of the target (Staelin *et al.*, 1977). Nearly all other land surface types exhibit a positive spectral gradient. Theoretical and empirical studies have demonstrated that the amount of scattering, or decrease in brightness temperature, can be correlated with the thickness and density of the snow cover and specific wavelength. Based on these relationships, algorithms have been developed that indicate the presence of snow (Grody and Basist, 1996; Hiltbrunner, 1996) and compute either snow water equivalent or depth, given an assumed density (examples include, Chang *et al.*, 1987; Goodison, 1989; Nagler, 1991; Tait, 1998; Pulliainen and Hallikainen, 2001). Nearly all of these algorithms have been developed and tested for dry snow conditions only.

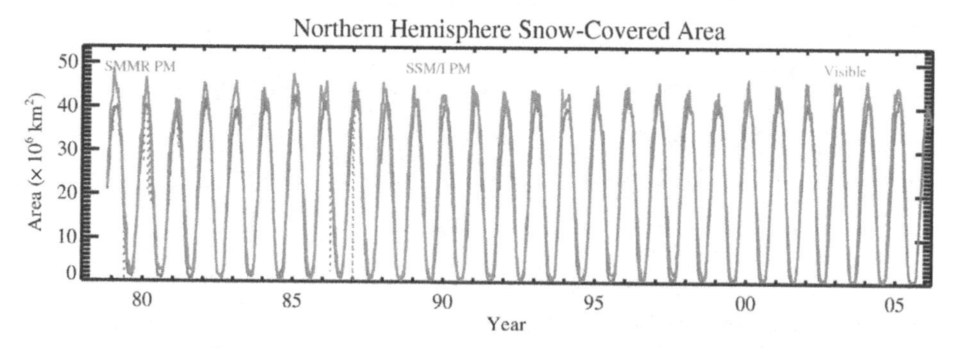


Figure 5.3. Northern Hemisphere monthly SCA, 1978–2005, from NOAA snow charts (orange) and microwave satellite (purple/green) data sets. (Plate 5.3.)

Snow water equivalent cannot be determined when the snow is wet (i.e. liquid water is present on the snow grain surface) because wet snow is primarily an emitter at microwave frequencies and thus the information derived from the scattered portion of the signal is lost. However, it is still possible to detect the presence of wet snow due to its high polarization difference (Walker and Goodison, 1993; Mätzler, 1994) and emission signature (Basist *et al.*, 1998).

Validation of this microwave data set has been accomplished by comparisons with the weekly NOAA snow extent data set described above (Armstrong and Brodzik, 2001a, b). This study compared data derived from two satellite products which involve completely different sensor systems and analysis techniques. The visible data are manually interpreted while a single numerical algorithm is applied to the microwave brightness temperature record throughout the time series. In the comparisons shown here (Fig. 5.3) the Chang *et al.* (1987) snow depth algorithm is used to determine snow extent for the SMMR period and a modified version of the same algorithm is used for the SSM/I period (Armstrong and Brodzik, 2001a). Since it is the *mass* of snow that essentially controls the scattering of microwave energy in a snowpack, derivation of snow depth from passive microwave data requires that certain assumptions be made about snow density and grain size (Mätzler, 2002). For example, the global SSMR-derived monthly snow depth climatology used by Foster *et al.* (1996) assumed a mean snow density of 300 kg m$^{-3}$ and an average snow grain size of 0.3 mm (Chang *et al.*, 1987). Thus passive microwave-derived snow depths have additional sources of uncertainty included in them.

Figure 5.3 shows the similar inter-annual variability of the two data sets where both consistently indicate Northern Hemisphere maximum extents exceeding 40 million km$^2$. Figure 5.4 compares mean monthly Northern Hemisphere snow extent for the period 1978–2005 and shows the difference between the two data sets. The microwave data indicate less snow-covered area than the visible data

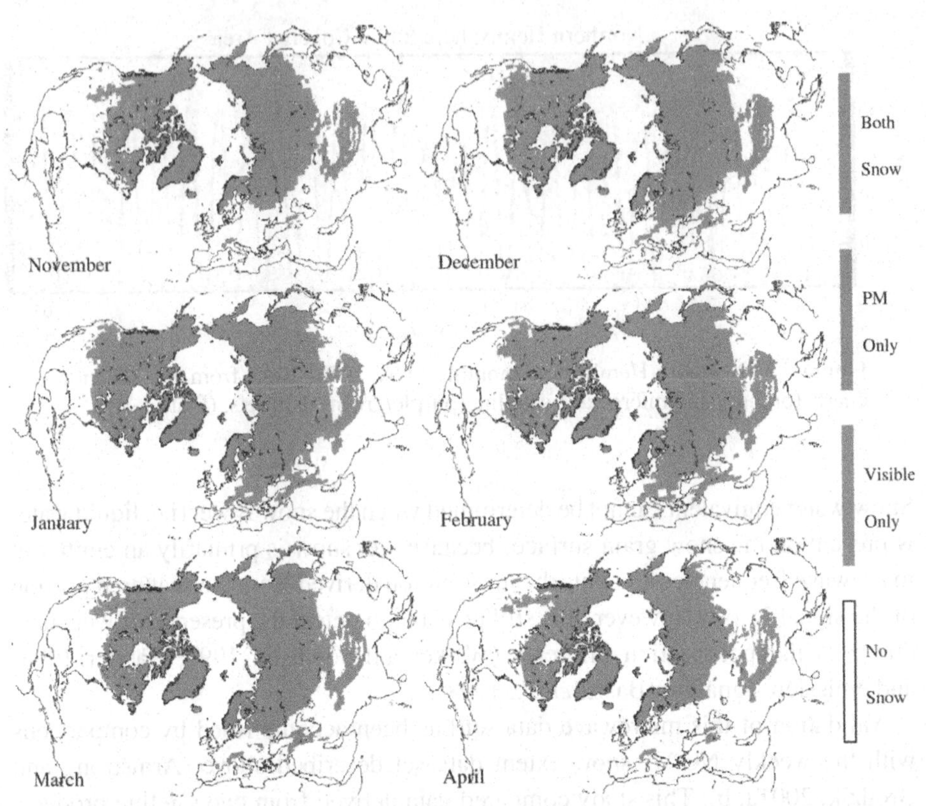


Figure 5.4. Comparison of mean monthly Northern Hemisphere snow extent derived from visible and passive microwave satellite data, 1978–2005 (50% or more of the weeks in the particular month over the total time period classified as snow covered). (Plate 5.4.)

throughout the year with a mean difference during the winter months (November–April) of about 4 million square kilometers, decreasing from about 8 million square kilometers (twenty-five percent) in November to about 0.3 million square kilometers (one percent) in April. During November and, to a lesser degree, December, the passive microwave data underestimate the southern-most snow extent in most geographic regions. The difference is greatest in November at the lower elevations across North America, Europe, and western Asia where the snow cover is more likely to be spatially intermittent and shallow (less than about 5.0 cm) and may exist at the melting temperature. Although the algorithm used here is not able to consistently detect wet snow, only night-time or early morning ("cold") orbits were used in this comparison, reducing the chance that wet snow is present. Robinson *et al.* (1993) and Brown (2000) showed that the NOAA data set was capturing the area of snow with depths as shallow as 1–2 cm. Because visible data are better

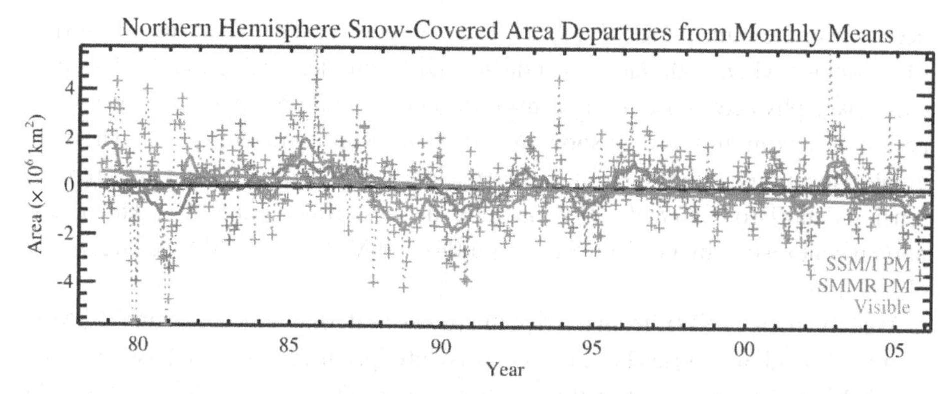


Figure 5.5. Northern Hemisphere SCA departures from monthly means, 1978–2005, from NOAA snow charts (orange) and microwave satellite (purple/green) data sets. The NOAA time series for this period exhibits a significant decreasing trend of −2.0% per decade (solid orange line); the microwave snow-cover time series exhibits a decreasing trend of −0.7% per decade that is not significant at a 90% level (dashed green line). (Plate 5.5.)

suited to detecting shallow snow, they also monitor the fluctuating snow extent edge more precisely, resulting in greater spatial variability. This is the probable reason for the NOAA data indicating greater departures from the monthly means (Fig. 5.5) as the region of most rapid fluctuation in snow cover is typically the snow-edge boundaries where the snow is thin and intermittent. Therefore it is likely that the large differences in snow extent during fall and early winter are due primarily to the inability of the current passive microwave algorithms to detect thin and intermittent snow cover. A similar pattern during fall and early winter was also observed by Basist *et al.* (1996). There is no reason to expect that the NOAA data would overestimate snow extent during fall and early winter.

As the snow cover continues to build during the months of January–March, as well as on into the melt season, agreement between the two data types continually improves. This is logical because as the winter season progresses the amount of area covered by deeper snow increases, thus enabling the microwave algorithm to detect a greater percentage of the total snow cover. As the Northern Hemisphere seasonal snow cover retreats towards the Arctic Basin during April and May the microwave data show very close agreement with the visible data. The increased accuracy during the period of spring melt is fortuitous as this is the most important period of the snow-cover season in terms of snow hydrology, thus allowing confident application of these snow water equivalent data to hydrologic forecasting and modeling.

Over the 1978–2005 period the NOAA time series indicates a significant decreasing trend of −2.0% per decade in Northern Hemisphere snow-cover extent while the passive microwave gives a non-statistically significant decrease of −0.7% per decade over the same period (Fig. 5.5). The strongest seasonal signal occurs during May–August when both data sets indicate significant decreasing trends. This pattern makes physical sense in the context of increasing air temperatures during the period of maximum seasonal snow melt over the Northern Hemisphere. The geographic locations with the strongest decreasing trends within the satellite remote sensing data (Brodzik *et al.*, 2006) include the western United States, which supports recent results by Groisman *et al.* (2004) and Mote *et al.* (2005) using *in situ* observations.

Weekly maps of SWE from passive microwave satellite remote sensing (SSM/I) over the Canadian prairies have been consistently produced over the past 15 years. The SWE algorithms used in this program were developed for open areas with shallow snow covers, such as prairies and steppes (Goodison, 1989; Walker and Goodison, 1993) and they provide an accuracy on the order of ±10 mm, which is sufficient for operational purposes and climate-related studies. The current ability for reliable, all-weather monitoring of SWE and snow cover over the steppe and prairie regions of the world is important because these regions produce much of the world's wheat and corn, and because snow-cover fluctuations over these regions are closely linked to larger continental-scale variations in snow cover and air temperature.

The development of SWE climatological information from polar orbiting passive microwave satellite sensors is not a trivial task as this requires a methodology to handle problems such as incomplete orbital data coverage and temporary wet snow events (Piwowar *et al.*, 1999). However, since 1978 it has been possible to monitor the global fluctuation of snow cover by taking advantage of passive microwave data from both SMMR and SSM/I. This requires the adjustment of the SMMR brightness temperature data to remove a bias in computed SWE values (Derksen and Walker, 2003; Derksen *et al.*, 2003; Armstrong and Brodzik, 2005). NSIDC has produced a global monthly SWE climatology for the period 1978–2005 by combining SMMR and SSM/I data in a common format (EASE-Grid) (Armstrong and Brodzik, 2005). The algorithms used to develop this data set include adjustments for known underestimates of snow water equivalent in forested areas based on earlier work by Chang *et al.* (1996). In addition, the algorithm incorporates snow extent from the NOAA data set described above in order to correct for the consistent underestimate of snow extent by the microwave algorithms during the shallow and intermittent snow-cover conditions of early winter. This snow product, available from NSIDC, includes individual files of global monthly average snow extent and water equivalent at a 25 km spatial resolution, as well as monthly climatologies describing the average snow extent, snow water equivalent, probability of occurrence, and variance.

### 5.3.5 Active microwave systems

Active microwave systems (radar) provide significantly better spatial resolution (a few tens of meters) than passive systems, but radar possesses its own set of limitations as described below. Active microwave instruments operate at wavelengths of millimeters up to several tens of centimeters. Sensors measuring in the wavelength range of about 7.5–1.5 cm (frequencies of 4–18 GHz) are most commonly used in snow research. Global scale monitoring of snow cover with active microwave or radar systems has become feasible in recent decades through the advent of synthetic aperture radar (SAR) satellites. The first successful SAR data for snow studies were collected during the European SAR-850 Campaign in 1981 (Mätzler and Schanda, 1984; Rott and Mätzler, 1987). Recent generations of active microwave satellite sensors provide high temporal and spatial resolution, unhampered by cloud cover or darkness during the winter season.

Interactions between the emitted radar pulse and the snow cover are influenced by two sets of parameters: (a) sensor parameters, including frequency, polarization, and viewing geometry and (b) snow parameters including density, liquid water content, grain size and shape, stratification, and surface roughness. These snow properties determine the dielectric properties, and therefore the backscattering characteristics. The backscatter coefficient is determined by reflection at the snow surface, scattering within the snow cover, and reflection at the snow–soil boundary; thus, the measured backscatter is a combination of surface and volume scattering. The penetration depth for a given frequency depends mainly on the liquid water content and snow density. For dry snow, maximum penetration depths vary with frequency but can be of the order of tens of meters for the SAR frequencies typically used for snow studies. Consequently, separation of dry snow and bare ground is not possible, at least with single polarization, single frequency SAR and therefore their capabilities are primarily limited to the mapping of wet snow (Rott and Kunzi, 1985). The fact that passive microwave techniques are conversely greatly limited in the presence of wet snow results in the obvious conclusion that much of the success of future work will come from a multi-sensor approach.

Because of the imaging geometry of SAR sensors, data are subject to geometric distortions. These terrain-induced distortions include foreshortening, layover, and shadow effects. Therefore the dependence of backscattering on local incidence angles has to be analyzed and modeled explicitly based on DEM data of suitable spatial resolution and accuracy (Nagler and Rott, 2000). Standardized and generally acceptable correction procedures are still lacking. In this respect, polarimetric SAR has been determined to be more effective than single-polarized imagery for studying wet snow or ice surfaces, since topographic information is not required (Shi *et al.*, 1994). An additional possibility for overcoming terrain-induced problems is to use multi-temporal or repeat pass SAR data as described by Rott and Nagler (1994), Strozzi *et al.* (1999), and Shi *et al.* (2001).

As an additional example of the applications of radar frequencies, recent work has described the specific capabilities of a spaceborne $K_u$ – band scatterometer (12–18 GHz) in monitoring global snow cover (Nghiem and Tsai, 2001). This study was based on the analysis of data from the National Aeronautics and Space Administration (NASA) scatterometer NSCAT operated on the Advanced Earth Observing Satellite (ADEOS) from September 1996 to July 1997. The data continued to be available with the launch of the SeaWinds $K_u$ – band scatterometer on the QuickSCAT satellite in 1999. QuickSCAT/SeaWinds provides global backscatter data with a fixed incidence angle over a swath width of 1800 km with daily coverage at latitudes greater than 37 degrees.

Some examples of the application of active microwave systems to the remote sensing of snow are provided by Ulaby and Stiles (1980), Mätzler and Schanda (1984), Shi *et al.* (1994), Nagler and Rott (2000), Rau *et al.* (2000), and Shi *et al.* (2001). For additional references and more detailed coverage of this topic, see Duguay and Pietroniro (2005) and Massom and Lubin (2006).

### 5.3.6 New sensor systems

The launch of the NASA Earth Observing System (EOS) platforms of Terra, in December of 1999, and Aqua, in May of 2002, provided additional and enhanced opportunities for mapping of snow at the global scale. Both Terra and Aqua carry a MODIS (moderate resolution imaging spectroradiometer), which provides global snow maps at a 500 m resolution as both daily and eight-day products in both a sinusoidal (500 m) and a "modelers grid" projection (0.05°) (Hall *et al.*, 2001). Additional research products are also being developed, which include subpixel snow-covered area, albedo, and grain size (Painter *et al.*, 2003).

The Aqua platform also carries the AMSR-E (advanced microwave scanning radiometer – EOS), which has been providing passive microwave-derived snow water equivalent at the global scale since 2002 (Kelly *et al.*, 2003). AMSR-E incorporates several enhancements to SMMR and SSM/I in that it provides twice the spatial resolution of the previous sensors as well as additional channels (AMSR channels are 6.9, 10.6, 18.7, 23.8, 36.5, and 98.0 GHz, all V and H pole). AMSR-E snow-cover products include snow extent and snow water equivalent at a 25 km spatial resolution. Data are available for both the Northern and Southern Hemispheres at three time intervals: daily, weekly maximum, and monthly average. These standard MODIS and AMSR-E snow products are available from NSIDC.

As noted above, there are clear advantages and corresponding disadvantages in applying only visible or passive microwave methods to snow mapping; thus, newer products represent a blend of both MODIS and AMSR-E (Armstrong *et al.*, 2005).

## 5.4 Operational snow-cover products

Operational products commonly involve several different agencies in numerous countries throughout the world. They may or may not be available to users outside of the agencies that produce them. These data products are often derived from research level methods and algorithms that may be modified and upgraded at any time throughout their distribution. This is in contrast with the data sets that are considered appropriate for climate change studies, which have been produced as a result of consistent and systematic processes throughout the full time series of the product. The individual user of a given data set must obtain the necessary information to determine the internal consistency of the data set and if it is appropriate for time series or trend studies. These products may be local, regional, or national and a selected sample of data sources are provided below. Access is typically through the agency web sites, which are provided, but which may of course change. Therefore the reader is encouraged to visit the snow data web site of the National Snow and Ice Data Center (NSIDC) http://nsidc.org/snow/data.html.

Because of increasing costs to maintain manually operated surface networks, many existing measurement networks are severely contracting, particularly in Russia and Canada (Barry, 1995), and networks are increasingly being automated with major implications for data quality and homogeneity (e.g. Milewska and Hogg, 2002). In many locations, manual measurements are planned to be replaced by satellite remote sensing, yet the satellite programs rely on at least limited surface measurements for calibration and validation. National efforts will be required to provide the necessary resources to improve and make available existing archives of snow data. These can be reinforced by strong international endorsement of the importance of snow-cover information for scientific reasons as well as for practical socio-economic reasons in areas with seasonal snow cover, especially in mountains and adjacent lowlands. It is hoped that these goals will be greatly implemented through such international programs as the WCRP Climate and Cryosphere (CliC) and the Global Climate Observing System (GCOS).

### *5.4.1 Snow depth*

Examples of readily available data sets, often in near-real time, are listed below along with access information. Examples of archived snow data sets representing varying historical periods are presented in Section 5.2. At the full global scale snow depth is measured once daily at weather stations and is reported over the global telecommunications system (GTS). Although there are about 7000 land stations in the GTS, stations report snow depth with varying frequency and consistency. These global snow depth data are available from the WMO-GTS synoptic reports for stations that do report that specific code group in real time (see ftp://ftp.ncdc.noaa.gov/pub/data/globalsod). Snowfall is not differentiated in six-hourly/daily precipitation measurements at weather stations although the type of precipitation is reported in the synoptic weather code.

Operational global daily snow depth analysis by the Canadian Meteorological Centre can be obtained at http://weatheroffice.ec.gc.ca/analysis/index_e.html.

### 5.4.2 Snow extent

The Northern Hemisphere EASE-Grid Weekly Snow Cover and Sea Ice Extent, Version 3, (Armstrong and Brodzik, 2005) contains consistently processed Northern Hemisphere snow extent for the period 1966–2005 (http://nsidc.org/data/nsidc-0046.html).

The National Operational Hydrologic Remote Sensing Center (NOHRSC) is the National Weather Service (NWS) center of expertise in satellite and airborne remote sensing and geographic information systems (GIS) used to support the NWS operational hydrology program. The center provides remotely sensed and GIS hydrology products, GIS applications, and spatial data sets in the format and time frame required to support various research, developmental, and operational hydrology programs conducted at local, regional, and national scales. Products are both operational and experimental and include numerous variables such as snow extent, SWE, snow temperature, and snow melt (http://www.nohrsc.nws.gov).

NOAA-NESDIS provides daily operational snow products for the Northern Hemisphere as gridded data (approximately 25 km) with monthly statistics. The Interactive Multisensor Snow and Ice Mapping System (IMS) Daily Northern Hemisphere Snow & Ice Analysis is available at http://www.ssd.noaa.gov/PS/SNOW/index.html (Ramsay, 1998). An additional NOAA product that can be found at this site provides higher resolution snow extent and percent coverage data using GOES data, specifically for North America (Romanov, *et al*, 2000) and South America (Romanov and Tarpley, 2003).

Beginning in 1999, moderate resolution imaging spectroradiometer (MODIS) snow products became available and include level 2 swath data at a 500 m resolution and level 3 gridded daily and eight-day composites in both a sinusoidal (500 m) grid and a "modelers grid" projection (0.05°) (Hall *et al.*, 2001) (http://nsidc.org/NASA/MODIS/). A comparison of three different methods to map snow cover using optical satellite data is contained in Bitner *et al.* (2002).

### 5.4.3 Snow water equivalent

Snow water equivalent (SWE) is determined at snow courses (North America) or along snow transects (Russia) at about 15–30- or 10-day intervals, respectively. The data are in the agency archives and many are not digitized. Snow water equivalent data are collected by other national, state, provincial, and private networks in many countries but are especially difficult to access (cost and other restrictions). Currently, no central archive exists and therefore many national and other databases are not readily accessible. A few examples of available SWE data sets are listed below.

The former Soviet Union hydrological snow surveys are based on observations at 1345 sites throughout the former Soviet Union between 1966 and 1990, and at 91 of those sites between 1991 and 1996. These observations include snow depths at World Meteorological Organization (WMO) stations and snow depth and snow water equivalent measured over nearby snow course transects at ten-day intervals. The transect snow depth data are the spatial average of 100–200 individual measuring points. The transect snow water equivalent is the spatial average of approximately twenty individual measuring points. These data are available from NSIDC (http://nsidc.org/data/g01170.html).

Weekly maps of SWE from passive microwave satellite remote sensing (SSM/I) over the Canadian prairies are available from http://www.socc.ca/snow/snow_current_e.cfm#prairies. Canadian snow course observations (~2000 sites, mostly in the period 1950–1995) are published on the Canadian Snow CD-ROM (MSC, 2000).

Weekly snow extent and water equivalent for the United States are available from the National Operational Hydrological Remote Sensing Center (NOHRSC) described above: http://www.nohrsc.nws.gov.

The United States Department of Agriculture (USDA) Natural Resources Conservation Service (NRCS) collects and distributes manual snow course and SNOTEL data for thirteen western states: http://www.wcc.nrcs.usda.gov.

## 5.5 Global–continental snow-cover climatology: measured and modeled

As outlined earlier, GCMs have generated a new need for continental and global-scale climatological information on the snow water equivalent and on the snow-cover extent to validate current climate simulations. Snow climatologies are also required for other purposes such as initializing atmospheric circulation models for sensitivity experiments.

Various attempts have been made to develop global snow depth climatologies from *in situ* observations (U.S. Army Corp of Engineers, 1954; Schutz and Bregman, 1975; Foster and Davey, 1988). The Foster and Davey (1988) snow depth climatology provides global monthly mean snow depth at a ~50 km resolution and has been used extensively in the validation of GCM snow simulations (e.g. Foster *et al.*, 1996). More recently, Brown *et al.* (2003) developed a higher resolution snow depth climatology for North America (~37 km grid) based on a modified version of the snow depth analysis scheme used at the Canadian Meteorological Centre (CMC) (Brasnett, 1999). The climatology, which was derived from *in situ* snow depth observations from the United States and Canada covering the snow seasons 1979/80–1996/97, was found to agree more closely with the observed snow-cover climatology generated from the NOAA weekly snow-cover product, and to have a more realistic pattern of snow accumulation over the western cordillera. Brown *et al.* (2003) were also able to develop an estimated SWE climatology for North America from the snow depth climatology by estimating snow density variation with a simplified snow ageing model. The SWE climatology agreed well with independent SWE observations over areas with a relatively dense network of snow depth observations (i.e. south of about 55°N), but was not as good at higher latitudes where the results were dominated by the simplified snow model used to generate the background field for the snow depth analysis. Grundstein (2003) followed a similar hybrid approach to develop a SWE climatology over the northern Great Plains using daily snow depth observations and snow density estimated from the detailed physical snowpack model SNTHERM (Jordan, 1991).

As described above in Section 5.3, the Satellite Analysis Branch (SAB) of NESDIS first began generating Northern Hemisphere weekly snow and ice cover analysis charts derived from the visible satellite imagery in November, 1966 (Dewey and Heim, 1982). The main issues involved in developing monthly snow extent information from the NOAA data set have been the correct partitioning of the weekly chart information into each month, and the correction of problems associated with the land/sea masks applied by NESDIS (Robinson *et al.*, 1993). The raw weekly data, data access software, and a land/sea mask can be downloaded from the NOAA Climate Prediction Center.

Synthetic snow cover climatologies can be generated from physical snow models forced with analyzed meteorological fields (Martin *et al.*, 1994) or coupled to a GCM (Brun *et al.*, 1997). A simulated five-year snow-cover climatology presented in Brun *et al.* (1997) from the CROCUS (Brun *et al.*, 1992) model coupled to the French ARPEGE GCM agreed closely with the NOAA-NESDIS snow-cover climatology. Foster *et al.* (1996) concluded that GCMs were able to provide, for the most part, reasonable simulations of seasonal and year-to-year variation in the continental-scale snow-covered area. Frei *et al.* (2005) evaluated 18 AGCMs over North America and concluded that most models simulated the seasonal timing and the relative spatial patterns of continental-scale SWE fairly well, although there were significant differences between models at regional scales. A major challenge for GCM snow-cover modelers is the inclusion of important physical processes, and topographic and landcover factors affecting the spatial and temporal distribution of snow cover and its properties. For example, Pomeroy *et al.* (1997) demonstrated that winter precipitation alone was insufficient to calculate snow accumulation in Arctic regions, and that blowing snow processes and landscape patterns were the key factors governing the spatial distribution of SWE in winter. Stochastic approaches can be used to account for subgrid-scale variations in vegetation (Woo and Steer, 1986) and topography (Walland and Simmonds, 1996; Stieglitz *et al.*, 1997). In the latter case, the parameterization of subgrid-scale topographic variability in a GCM snow submodel yielded more accurate simulations of seasonal variation in Northern Hemisphere snow-covered area, and solved the characteristic GCM problem of late spring snow-cover retreat noted by Frei and Robinson (1998) (Goodison *et al.*, 1998a). A fundamental problem for the realistic simulation of snow cover by GCMs is the realistic simulation of precipitation by atmospheric models. The evaluation of GCM simulations of winter regional precipitation (IPCC, 1996) documented errors ranging from −25% to +125% of observed precipitation over central North America. Major discrepancies in modeled precipitation have also been documented over high-elevation and high-latitude regions (e.g. Walsh *et al.*, 1998).

Global-scale information on average snow-cover properties (e.g. density, thermal conductivity, grain size) is important for many applications such as satellite algorithm development, estimation of SWE or snow depth from climate model output, snow trafficability, and ground frost studies. A number of seasonal snow-cover classifications have been developed for this purpose (see the review in Groisman and Davies, 2001). For example Sturm *et al.* (1995) classified snow cover into six distinct classes based on precipitation, air temperature, and wind speed. Each of the six classes was determined to have distinct snowpack properties (thickness, layers, thermal conductivity, density). The classification is available on a 0.5 degree latitude/longitude grid over the Northern Hemisphere from NSIDC.

## References

AES (1977). *MANOBS – Manual of Surface Weather Observations*, 7th edn. Downsview, Ontario: Atmospheric Environment Service.

Armstrong, R. (2001). *Historical Soviet Daily Snow Depth Version 2 (HSDSD)*. Boulder, CO: National Snow and Ice Data Center/World Data Center for Glaciology.

Armstrong, R. L. and Brodzik, M. J. (2001a). Recent Northern Hemisphere snow extent: a comparison of data derived from visible and microwave sensors. *Geophys. Res. Lett.*, **28**(19), 3673–3676.

Armstrong, R. L. and Brodzik, M. J. (2001b). Validation of passive microwave snow algorithms. In *Remote Sensing and Hydrology 2000*. IAHS Publication No. 267 (ed. Owe, M., Brubaker, K., Ritchie, J., and Rango, A.), pp. 87–92.

Armstrong, R. L. and Brodzik, M. J. (2002). Hemispheric-scale comparison and evaluation of passive microwave snow algorithms. *Ann. Glaciol.*, **34**, 38–44.

Armstrong, R. L. and Brodzik, M. J. (2005). *Northern Hemisphere EASE-Grid Weekly Snow Cover and Sea Ice Extent Version 3*. Boulder, CO: National Snow and Ice Data Center.

Armstrong, R. L., Brodzik, M. J., Knowles, K., and Savoie, M. (2005). *Global Monthly EASE-Grid Snow Water Equivalent Climatology*. Boulder, CO: National Snow and Ice Data Center.

Ball, T. F. (1992). Historical and instrumental evidence of climate: western Hudson Bay, Canada, 1714–1850. In *Climate since A. D. 1500* (ed. Bradley, R. S. and Jones, P. D.). London: Routledge, pp. 40–73.

Bamzai, A. S. and Shukla, J. (1999). Relation between Eurasian snow cover, snow depth, and the Indian summer monsoon: an observational study. *J. Climate*, **12**(10), 3117–3132.

Barnett, T. P., Dumenil, L., Schlese, U., Roeckner, E., and Latif, E. (1989). Effects of Eurasian snow cover on regional and global climate variations. *J. Atmos. Sci.*, **46**, 661–685.

Barry, R. G. (1995). Observing systems and data sets related to the cryosphere in Canada. A contribution to planning for the global climate observing system. *Atmos.–Ocean*, **33**, 771–807.

Barry, R. G. (1997). Cryospheric data for model validations: requirements and status. *Ann. Glaciol.*, **25**, 371–375.

Barry, R. G., Fallot, J.-M., and Armstrong, R. L. (1995). Twentieth-century variability in snow cover conditions and approaches to detecting and monitoring changes: status and prospects. *Prog. Phys. Geog.*, **19**, 520–532.

Bartelt, P. and Lehning, M. (2002). A physical SNOWPACK model for the Swiss avalanche warning; Part I: numerical model. *Cold Reg. Sci. Technol.*, **35**(3) 123–145.

Basist, A., Garrett, D., Ferraro, R., Grody, N., and Mitchell, K. (1996). A comparison between snow cover products derived from visible and microwave satellite observations. *J. Appl. Meteorol.*, **35**, 163–177.

Basist, A., Grody, N. C., Peterson, T. C., and Williams, C. N. (1998). Using the special sensor microwave imager to monitor land surface temperatures, wetness and snow cover. *J. Appl. Meteorol.*, **37**(9), 888–911.

Biswas, A. K. (1970). *History of Hydrology*. Amsterdam: Elsevier.

Bitner, D., Carroll, T., Cline D., and Romanov, P. (2002). An assessment of the differences between three satellite snow cover mapping techniques. *Hydrol. Process.*, **16**(18), 3723–3733.

Blöschl, G. (1999). Scaling issues in snow hydrology. *Hydrol. Process.*, **13**(14–15), 2149–2175.

Bradley, R. S., Diaz, H. F., Eischeid, J. K., *et al.* (1987). Precipitation fluctuations over Northern Hemisphere land areas since the mid-19th Century. *Science*, **237**, 171–275.

Brasnett, B. (1999). A global analysis of snow depth for numerical weather prediction. *J. Appl. Meteorol.*, **38**, 726–740.

Brodzik, M. J., Armstrong, R. L., Knowles, K., and Savoie, M. (2005). The effect of sensor differences in deriving long-term trends from satellite passive microwave snow extent. *EOS Trans. AGU*, **86**(52), Fall Meeting Suppl. Abstract U21A-0804.

Brodzik, M. J., Armstrong, R. L., Weatherhead, E. C., *et al.* (2006). Regional trend analysis of satellite-derived snow extent and global temperature anomalies. *EOS Trans. AGU*, **87**(52), Fall Meeting Suppl. Abstract U33A-0011.

Brown, R. D. (2000). Northern Hemisphere snow cover variability and change, 1915–1997. *J. Climate*, **13**, 2339–2355.

Brown, R. D. and Braaten, R. O. (1998). Spatial and temporal variability of Canadian monthly snow depths, 1946–1995. *Atmosphere-Ocean*, **36**, 37–45.

Brown, R. D., Brasnett, B., and Robinson, D. (2003). Gridded North American monthly snow depth and snow water equivalent for GCM evaluation. *Atmosphere-Ocean*, **41**, 1–14.

Brown, R. D. and Goodison, B. E. (1996). Interannual variability in reconstructed Canadian snow cover, 1915–1992. *J. Climate*, **9**, 1299–1318.

Brubaker, K. L., Jasinski, M., Chang, A. T. C., and Josberger, E. (2001). Interpolated sparse surface measurements for calibration and validation of satellite-derived snow water equivalent in Russian Siberia. In *Remote Sensing and Hydrology 2000*. IAHS Publication No. 267, pp. 93–104.

Brun, E., David, P., Sudul, M., and Brunot, G. (1992). A numerical model to simulate snow-cover stratigraphy for operational avalanche forecasting. *J. Glaciol.*, **38**, 13–22.

Brun, E., Martin, E., and Spiridonov, V. (1997). Coupling a multi-layered snow model with a GCM. *Ann. Glaciol.*, **25**, 66–72.

Carroll, S. S., Carroll, T. R., and Poston, R. W. (1999). Spatial modelling and prediction of snow–water equivalent using ground based, airborne, and satellite snow data. *J. Geophys. Res.*, **104**(D16), 19 623–19 629.

Carroll, T., Cline, D., Fall, G., *et al.* (2001). NOHRSC operations and the simulation of snow cover properties for the coterminous U.S. *Proc. Western Snow Conference*, Sun Valley, Idaho, April 16–19, 2001.

Chang, A. T. C., Foster, J. L., and Hall, D. K. (1987). Nimbus-7 SMMR derived global snow cover parameters. *Ann. Glaciol.*, **9**, 39–44.

Chang, A. T. C., Foster, J. L., and Hall, D. K. (1996). Effects of forest on the snow parameters derived from microwave measurements during the BOREAS winter field campaign. *Hydrol. Process.*, **10**, 1565–1574.

Cihlar, J., Barry, T. G., Ortega Gil, E., *et al.* (1997). *GCOS/GTOS Plan for Terrestrial Climate-related Observations*. Version 2.0, June 1997. GCOS-32, WMO/TD, No. 796.

Clark, M. P., Serreze, M. C., and Robinson, D. A. (1999). Atmospheric controls on Eurasian snow extent. *Int. J. Climatol.*, **19**, 27–40.

Cline, D., Elder, K., and Bales, R. (1998). Scale effects in a distributed snow water equivalence and snowmelt model for mountain basins. *Hydrol. Process.*, **12**, 1527–1536.

Colbeck, S., Akitaya, E., Armstrong, R., *et al.* (1990). *The International Classification for Seasonal Snow on the Ground*. International Commission on Snow and Ice (IAHS). Boulder, CO: World Data Center A for Glaciology, University of Colorado.

Derksen, C. and Walker, A. (2003). Identification of systematic bias in the cross-platform (SMMR and SSM/I) EASE-Grid brightness temperature time series. *IEEE Trans. Geosci. Remote Sens.* **41**(4), 910–915.

Derksen, C., Walker, A., LeDrew E., and Goodison, B. (2003). Combining SMMR and SSM/I data for time series analysis of central North American snow water equivalent. *J. Hydrometeorol*, **4**(2), 304–316.

Dewey, K. F. and Heim, R. Jr. (1982). A digital archive of Northern Hemisphere snow cover, November 1966 through December 1980. *Bull. Am. Met. Soc.*, **63**, 1132–1141.

Doesken, N. J. and Judson, A. (1997). *The SNOW Booklet: a Guide to the Science, Climatology and Measurement of Snow in the United States*. Ft. Collins, CO: Colorado State University.

Dozier, J. and Painter, T. H. (2004). Multispectral and hyperspectral remote sensing of alpine snow properties. *Ann. Rev. Earth Planet. Sci.*, **32**, 465–494.

Duguay, C. R. and Pietroniro, A. (eds.) (2005) *Remote Sensing in Northern Hydrology: Measuring Environmental Change*. Geophysical Monograph 163. Washington, DC: American Geophysical Union.

Easterling, D. R., Karl, T. R., Lawrimore, J. H., and Del Greco, S. A. (1999). *United States Historical Climatology Network Daily Temperature, Precipitation, and Snow Data for 1871–1997*. ORNL/CDIAC-118, NDP-070.

Erxleben, J., Elder, K., and Davis, R. (2002). Comparison of spatial interpolation methods for estimating snow distribution in the Colorado Rocky Mountains. *Hydrol. Process.*, **16**(18), 3627–3649.

Essery, R. (1997). Parameterization of fluxes over heterogeneous snow cover for GCMs. *Ann. Glaciol.*, **25**, 38–41.

Fitzharris, B. B., Owens, I., and Chinn, T. (1992). Snow and glacier hydrology. In *Waters of New Zealand*. Wellington: New Zealand Hydrological Society, pp. 76–94.

Föhn, P. M. B. (1985). Besonderheiten des Schneeniederschlages. In *Der Niederschlag in der Schweiz*. Beitr. Geol. Schweiz – Hydrol. Nr. 31. Ben: Kümmerly und Frey, pp. 87–96.

Föhn, P. M. B. (1990). Schnee und lawinen. In *Schnee, Eis und Wasser in der Alpen in einer wärmeren Aatmosphaere, Mitteil.* VAW/ETH Zürich, Nr. 10–8, S. 33–48.

Foster, D. J. Jr. and Davy, R. D. (1988). *Global Snow Depth Climatology*. USAF Environmental Technical Applications Center, USAFETAC/TN-88/006.

Foster, J. L. (1986). The snow cover record in Eurasia. In *Glaciological Data*. Boulder, CO: World Data Center A for Glaciology Snow and Ice, GD-18, pp. 79–88.

Foster, J., Liston, G., Koster, R., *et al.* (1996). Snow cover and snow mass intercomparisons of general circulation models and remotely sensed datasets. *J. Climate*, **9**, 409–426.

Foster, J., Owe, M., and Rango, A. (1983). Snow cover and temperature relationships in North America and Eurasia. *J. Climate Appl. Meteorol.*, **22**, 460–469.

Frei, A., Brown, R., Miller, J. A., and Robinson, D. A. (2005). Snow mass over North America: observations and results from the second phase of the Atmospheric Model Intercomparison Project. *J. Hydrometeorol.*, **6**(5), 681–695.

Frei, A., Hughes, M. G., and Robinson, D. A. (1999). North American snow extent: 1910–1994. *Int. J. Climatol.*, **19**, 1517–1534.

Frei, A. and Robinson, D. A. (1998). Evaluation of snow extent and its variability in the Atmospheric Model Intercomparison Project. *J. Geophys. Res. Atmos.*, **103**(D8), 8859–8871.

Frei, A. and Robinson, D. A. (1999). Northern Hemisphere snow extent: regional variability 1972–1994. *Int. J. Climatol.*, **19**, 1535–1560.

Gloersen, P., Cavalier, D. J., Chang, A. T., *et al.* (1984). A summary of results from the first Nimbus-7 SMMR observations. *J. Geophys. Res.*, **89**, 5335–5344.

Goodison, B. E. (1989). Determination of areal snow water equivalent on the Canadian prairies using passive microwave satellite data. In *IGARSS'89, 12th Canadian Symposium on Remote Sensing*, Vancouver, Canada, July 10–14, 1989, vol. 3, pp. 1243–1246.

Goodison, B. E., Brown, R. D., and Crane, R. G. (1998a). *Cryospheric Systems. EOS Science Plan*, ch. 6. (ed. King, M. D.). Greenbelt, MD: National Aeronautics and Space Agency, Goddard Space Flight Center. (CD-ROM NP-1999-01-006-GSFC).

Goodison, B. E., Ferguson, H. L., and McKay, G. A. (1981). Measurement and data analysis. In *Handbook of Snow* (ed. Gray, D. M. and Male, D. H.). Toronto: Pergamon Press, pp. 191–274.

Goodison, B. E., Louie, P. Y. T., and Yang, D. (1998b). *WMO Solid Precipitation Measurement Intercomparison*. WMO Instruments and Observing Methods Report No. 67, WMO/TD No. 872.

Goodison, B. E., Metcalfe, J. R., Wilson, R. A., and Jones, K. (1988). The Canadian automatic snow depth sensor: a performance update. *Proc. 56th Western Snow Conference*, pp. 178–181.

Goodison, B. E. and Walker, A. E. (1993). Use of snow cover derived from satellite passive microwave data as an indicator of climate change. *Ann. Glaciol.*, **17**, 137–142.

Grenfell, T. C. and Perovich, D. K. (1981). Radiation absorption coefficients of polycrystalline ice from 400–1400 nm. *J. Geophys. Res.*, **86**(C8), 7447–7450.

Grody, N. C. and Basist, A. (1996). Global identification of snow cover using SSM/I measurements. *IEEE Trans. Geosci. Remote Sens.*, **34**(1), 237–249.

Groisman P. Ya. and Davies, T. D. (2001). Snow cover and the climate system. In *Snow Ecology – an Interdisciplinary Examination of Snow-covered Ecosystems* (ed. Jones, H. J., Pomeroy, J., Walker, D. A., and Hoham, R.). Cambridge: Cambridge University Press, pp. 1–44.

Groisman, P. Ya. and Easterling, D. (1994). Variability and trends of total precipitation and snowfall over the United States and Canada. *J. Climate*, **7**, 184–205.

Groisman, P. Ya., Karl, T. R., Easterling, D. R., Sun, B., and Lawrimore, J. H. (2004). Contemporary changes of the hydrological cycle over the contiguous United States: trends derived from *in situ* observations. *J. Hydrometeorol.*, **5**, 64–85.

Groisman, P. Ya., Karl, T. R., and Knight, R. W. (1994). Changes of snow cover, temperature and radiative heat balance over the Northern Hemisphere. *J. Climate*, **7**, 1633–1656.

Groisman, P. Ya., Koknaeva, V. V., Belokrylova, T. A., and Karl, T. R. (1991). Overcoming biases of precipitation measurement: a history of the USSR experience. *Bull. Am. Meteorlog. Soc.*, **72**, 1725–1733.

Grundstein, A. (2003). Patterns of seasonal maximum snow–water equivalent over the Northern Great Plains of the United States analyzed using hybrid-modeled climatology. *Clim. Res.*, **24**, 119–128.

Gubler, H. (1981). An inexpensive snow depth gauge based on ultrasonic wave reflection from the snow surface. *J. Glaciol.*, **127**, 157–163.

Gutzler, D. S. and Preston, J. W. (1997). Evidence for a relationship between spring snow cover in North America and summer rainfall in New Mexico. *Geophys. Res. Lett.*, **24**, 2207–2210.

Gutzler, D. S. and Rosen, R. D. (1992). Interannual variability of wintertime snow cover across the Northern Hemisphere. *J. Climate*, **5**, 1441–1447.

Hall, D. K. (1988). Assessment of polar climate change using satellite technology. *Rev. Geophys.*, **26**, 26–39.

Hall, D. K., Kelly, R. E. J., Foster, J. L., and Chang, A. T. C. (2005). Hydrological application of remote sensing: surface states: snow. In *Encyclopedia of Hydrological Sciences* (ed. Anderson, M. G.). Chichester: John Wiley and Sons, pp. 811–830.

Hall D. K. and Martinec, J. (1985). *Remote Sensing of Ice and Snow*. New York: Chapman and Hall.

Hall, D. K., Riggs, G. A., Salomonson, V. V., and Scharfen, G. R. (2001). Earth Observing System (EOS) Moderate Resolution Imaging Spectroradiometer (MODIS) global snow cover products. In *Remote Sensing and Hydrology 2000*. IAHS Publication No. 267, pp. 55–60.

Hartman, R. K., Rost A. A., and Anderson, D. M. (1995). Operational processing of multi-source snow data. *Proc. Western Snow Conference*, 1995, pp. 147–151.

Heim, R. (1998). *Data Documentation for U.S. Snow Climatology, TD-9641*. Ashville, NC: National Climatic Data Center.
Hiltbrunner, D. (1996). Land surface temperature and microwave emissivity from SSM/I data. Ph.D. thesis, University of Bern. Institute of Applied Physics.
Hollinger, J. P., Pierce, J. L., and Poe, G. A. (1990). SSM/I instrument evaluation. *IEEE Trans. Geosci. Remote Sens.*, **28**(5), 781–790.
Hopkinson, C., Sitar, M., Chasmer, L., *et al.* (2001). Mapping the spatial distribution of snowpack depth beneath a variable forest canopy using airborne laser altimetry. *Proc. 58th Eastern Snow Conference*, Ottawa, Ontario, Canada. May 14–18, 2001, pp. 253–264.
IPCC (1996). *Climate Change 1995: the Science of Climate Change* (ed. Houghton, J. T., Meira Filho, L. G., Callander, B. A., *et al.*). Contribution of WGI to the Second Assessment Report of the Intergovernmental Panel on Climate Change. Cambridge: Cambridge University Press.
Jackson, M. C. (1978a). Sixty years of snow depth in Birmingham. *Weather*, **33**, 32–34.
Jackson, M. C. (1978b). Snow cover in Great Britain. *Weather*, **33**, 298–309.
Jones, H. J., Pomeroy, J., Walker, D. A., and Hoham, R., eds. (2001). *Snow Ecology – an Interdisciplinary Examination of Snow-covered Ecosystems*. Toronto: Cambridge University Press.
Jordan, R. (1991). *A One-dimensional Temperature Model for a Snow Cover*. Technical documentation for SNTHERM.89 Special Report 91–16. U.S. Army Corps of Engineers. Hanover, NH: Cold Regions Research and Engineering Laboratory.
Judson, A. and Doesken, N. (2000). Density of freshly fallen snow in the Central Rocky Mountains. *Bull. Am. Meteorlog. Soc.* **81**(7), 1577–1587.
Karl, T. R., Groisman, P. Y., Knight, R. W., and Heim, R. R., Jr. (1993). Recent variations of snow cover and snowfall in North America and their relation to precipitation and temperature variations. *J. Climate*, **6**, 1327–1344.
Kelly, R., Chang, A. T. C., Tsang, L., and Foster, J. (2003). A prototype AMSR-E global snow area and snow depth algorithm. *IEEE Trans. Geosci. Remote Sens.*, **41**(2), 230–242.
Krenke, A. (1998, updated 2004). *Former Soviet Union Hydrological Snow Surveys, 1966–1996*. Edited by NSIDC. Boulder, Colorado: National Snow and Ice Data Center/World Data Center for Glaciology. Digital media.
Kuusisto, E. (1984). *Snow Accumulation and Snowmelt in Finland*. Publications of the Water Research Institute, 55, Helsinki, Finland: National Board of Waters.
Lavoie, C. and Payette, S. (1992). Black spruce growth as a record of a changing winter environment at treeline, Québec, Canada. *Arctic Alpine Res.*, **25**, 40–49.
Leathers, D. J. and Robinson, D. A. (1993). The association between extremes in North American snow cover extent and United States temperatures. *J. Climate*, **6**, 1345–1355.
Lehning, M., Bartelt, P., Brown, B., and Fierz, C. (2002a). A physical SNOWPACK model for the Swiss avalanche warning; Part III: Meteorological forcing, thin layer formation and evaluation. *Cold Reg. Sci. Technol.*, **35**(3), 169–184.
Lehning, M., Bartelt, P., Brown, B., Fierz, C., and Satyawali, P. (2002b). A physical SNOWPACK model for the Swiss avalanche warning; Part II: Snow microstructure. *Cold Reg. Sci. Technol.*, **35**(3), 147–167.
Lobl, E. S., Spencer, R. W., Shibata, A., *et al.* (2002). AMSR and AMSR-E global climate monitoring with the Advanced Microwave Scanning Radiometer. In *Proc. SPIE*

*Third International Asia-Pacific Environmental Remote Sensing Symposium*, 233–27 October, 2002, Hangzhou, China.

Manley, G. (1969). Snowfall in Britain over the past 300 years. *Weather*, **24**, 428–437.

Martin, E., Brun, E., and Durand, Y. (1994). Sensitivity of the French Alps snow cover to the variation of climatic variables. *Ann. Geophysicae*, **12**, 469–477.

Massom, R. and Lubin, D. (2006). *Polar Remote Sensing: Volume II: Ice Sheets*. Berlin: Springer-Verlag.

Matson, M., Ropelewski, C. F., and Varnadore, M. S. (1986). *An Atlas of Satellite-derived Northern Hemisphere Snow Cover Frequency*. Washington, DC: NOAA.

Matson, M. and Wiesnet, D. R. (1981). New database for climate studies. *Nature* **268**, 451–456.

Mätzler, C. (1994). Passive microwave signatures of landscapes in winter. *Meteorol. Atmos. Phys.*, **54**, 241–260.

Mätzler, C. (2002). Relation between grain size and correlation length of snow. *J. Glaciol.*, **48**(152), 461–466.

Mätzler, C. and Schanda, E. (1984). Snow mapping with active microwave sensors. *Int. J. Remote Sens.*, **5**, 409–422.

McClung, D. and Schaerer, P. (2006). *The Avalanche Handbook*. Seattle, WA: The Mountaineers.

Mekis, E. and Hogg, W. (1999). Rehabilitation and analysis of Canadian daily precipitation time series. *Atmos. Ocean*, **37**, 53–85.

Metcalfe J. R. and Goodison, B. E. (1993). Correction of Canadian winter precipitation data. *Eighth Symposium on Meteorological Observations and Instrumentation*, 17–22 January 1993, Anaheim, California, pp. 338–343.

Metcalfe, J. R., Routledge, B., and Devine, K. (1997). Rainfall measurement in Canada: changing observational methods and archive adjustment procedures. *J. Climate*, **10**, 92–101.

Milewska, E. and Hogg, W. D. (2002). Continuity of climatological observations with automation – temperature and precipitation amounts from AWOS (Automated Weather Observing System). *Atoms.–Ocean*, **40**, 333–359.

Moltoch, N. P., Colee, M. T., Bales, R. C., and Dozier, J. (2004). Estimating the spatial distribution of snow water equivalent in an alpine basin using binary regression tree models: the impact of digital elevation data and independent variable selection. *Hydrol. Process.*, **19**(7), 1459–1479.

Mote, P. W., Hamlet, A. F., Clark, M. P., and Lettenmaier, D. P. (2005). Declining mountain snowpack in western North America. *Bull. Am. Meteor. Soc.*, **86**, 39–49.

MSC (2000). *Canadian Snow Data CD-ROM*. CRYSYS Project, Climate Processes and Earth Observation Division, Meteorological Service of Canada, Downsview, Ontario, January, 2000.

Nagler, T. (1991). Verfahren zur Analyse der Schneebedeckung aus Messungen des SSM/I. Diplomarbeit thesis, University of Innsbruck.

Nagler, T. and Rott, H. (2000). Retrieval of wet snow by means of multitemporal SAR data. *IEEE Trans. Geosci. Remote Sens.* **38**(2), 754–765.

Nghiem, S. V. and Tsai, W. Y. (2001). Global snow cover monitoring with spaceborne Ku-band scatterometer. *IEEE Trans. Geosci. Remote Sens.*, **39**(10), 2118–2134.

Nolin, A., Dozier, J., and Mertes, L. A. K. (1993). Mapping alpine snow using a spectral mixture modeling technique. *Ann. Glaciol.*, **17**, 121–124.

Painter, T. H., Dozier J., Roberts, D. A., Davis, R. E., and Green, R. O. (2003). Retrieval of subpixel snow-covered area and grain size from imaging spectrometer data. *Remote Sens. Environ.*, **85**, 64–77.

Piwowar, J. M., Walker, A. E., Chasmer, L. E., and Goodison, B. E. (1999). The derivation of snow-cover "normals" over the Canadian prairies from passive microwave satellite imagery. *Proc. 4th International Airborne Remote Sensing Conference and Exhibition/21st Canadian Symposium on Remote Sensing*, 21–24 June 1999, Ottawa, Ontario, vol. 1, pp. 596–603.

Pomeroy, J. W. and Gray, D. M. (1995). *Snowcover – Accumulation, Relocation and Management*. National Hydrology Research Institute Science Report No. 7, Saskatoon, Canada.

Pomeroy, J. W., Gray, D. M., Shook, K. R., *et al.* (1998). An evaluation of snow accumulation and ablation processes for land surface modelling. *Hydrol. Process.* **12**, 2339–2367.

Pomeroy, J. W., Marsh, P., and Gray, D. M. (1997). Application of a distributed blowing snow model to the Arctic. *Hydrol. Process.*, **11**, 1451–1464.

Pulliainen, J. and Hallikainen, M. (2001). Retrieval of regional snow water equivalent from space-borne passive microwave observations. *Remote Sens. Environ.*, **75**, 76–85.

Rallison, R. E. (1981). Automated system for collecting snow and related hydrological data in the mountains of the western United States. *Hydrol. Sci. Bull.*, **26**, 83–89.

Ramsay, B. H. (1998). The interactive multisensor snow and ice mapping system. *Hydrol. Process.* **12**, 1537–1546.

Rango, A. (ed). (1975). *Proc., Workshop on Operational Applications of Satellite Snowcover Observations*. NASA SP39. Washington, DC: National Aeronautics and Space Administration.

Rau, F., Braun, M., Saurer, H., *et al.* (2000). Monitoring multi-year snow cover dynamics on the Antarctic Peninsula using SAR imagery. *Polarforschung*, **67**(172), 27–40.

Robinson, D. A. (1989). Evaluation of the collection, archiving and publication of daily snow data in the United States. *Phys. Geog.*, **10**, 120–130.

Robinson, D. A. (1991). Merging operational satellite and historical station snow cover data to monitor climate change. *Palaeogeog., Paleoclim., Paleoecol.*, **90**, 235–240.

Robinson, D. A. (1999). Northern Hemisphere snow extent during the satellite era. *Preprints: Fifth Conference on Polar Meteorlogy and Oceanography*, Dallas, TX, American Meteorological Society, 255–260.

Robinson, D. and Dewey, K. F. (1990). Recent secular variations in the extent of Northern Hemisphere snow cover. *Geophys. Res. Lett.*, **17**, 1557–1560.

Robinson, D. A., Dewey, K. F., and Heim, R. R. (1993). Global snow cover monitoring: an update. *Bull. Am. Meteorol. Soc.*, **74**, 1689–1696.

Robinson, D. A. and Kukla, G. 1987: Comments on "Comparison of Northern Hemisphere snow cover datasets". *J. Climate*, **1**, 435–440.

Robinson, D. A., Tarpley, J. D., and Ramsay, B. H. (1999). Transition from NOAA weekly to daily hemispheric snow charts. *American Meteorlogical Society 10th Symposium on Global Change Studies*, Dallas, TX, January, 1999, pp. 487–489.

Romanov, P., Gutman, G., and Csiszar, I. (2000). Automated monitoring of snow cover over North America with multispectral satellite data. *J. Appl. Meteorol*, **39**, 1866–1880.

Romanov, P. and Tarpley, D. (2003). Automated monitoring of snow cover over South America using GOES Imager data, *Int. J. Remote Sens.*, **24**(5), 1119–1125.

Rosenthal, W. and Dozier, J. (1996). Automated mapping of montane snow cover at subpixel resolution from Landsat Thematic Mapper. *Water Resources Res.*, **32**(1), 115–130.

Rott, H. and Kunzi, K. F. (1985). Remote sensing of snow cover with passive and active microwave sensors. In *Hydrological Applications of Remote Sensing and Remote Data Transmission*. IAHS Publ. No. 145, pp. 361–369.

Rott, H. and Mätzler, C. (1987). Possibilities and limits of SAR for snow and glacier surveying. *Ann. Glaciol.*, **9**, 195–199.
Rott, H. and Nagler, T. (1994). Capabilities of ERS-1 SAR for snow and glacier monitoring in Alpine areas. In *Proc. Second ERS-1 Symp.* ESA SP No. 359, pp. 965–970.
Shapiro, L. H., Johnson, J. B., Sturm, M., and Blaisdel, G. L. (1997). *Snow Mechanics: Review of the State of Knowledge and Applications.* U.S. Army Corps of Engineers. Hanover, NH: Cold Regions Research and Engineering Laboratory. CRREL Report 97–3.
Schmidlin, T. W. (1990). A critique of the climatic record of "Water equivalent of snow on the Ground" in the United States. *J. Appl. Meteorol.*, **29**, 1136–1141.
Schneider, S. R., Wiesnet, D. R., and McMillan, M. C. (1976). *River Basin Snow Mapping at the National Environmental Satellite Service.* NOAA Technical Memorandom NESS 83. Washington, DC: U.S. Department of Commerce.
Schutz, C. and Bregman, L. D. (1975). *Global Snow Depth Data: a Monthly Summary.* Santa Monica, CA: The Rand Corporation. (Available from nsidc@kryos.colorado.edu. University of Colorado, Boulder, CO.)
Scialdone, J. and Robock, A. (1987). Comparison of Northern Hemisphere snow cover data sets. *J. Clim. Appl. Meteorol.*, **26**, 53–68.
Seligman, G. (1980). *Snow Structure and Ski Fields.* Cambridge: International Glaciological Society.
Serreze, M. C., Clark M. P., Armstrong, R. L., McGinnis, D. A., and Pulwarty, R. S. (1999). Characteristics of the western United States snowpack from snowpack telemetry (SNOTEL) data. *Water Resources Res.*, **7**, 2145–2160.
Sevruk, B. (1992). *Snow Cover Measurements and Areal Assessment of Precipitation and Soil Moisture.* Operational Hydrology Report No. 35, WMO-No. 749.
Shi, J., Dozier, J., and Rott, H. (1994). Snow mapping in alpine regions with synthetic aperture radar. *IEEE Trans. Geosci. Remote. Sens.* **32**(1), 152–158.
Shi, J., Hensley, S., and Dozier, J. (2001). Mapping snow with repeat pass synthetic aperture radar. In *Remote Sensing and Hydrology 2000.* IAHS Publication No. 267, pp. 339–342
Shiyan, T., Congbin, F., Zhaomei, Z., and Qingyun, Z. (1997). *Two Long-term Instrumental Climatic Data Bases of the People's Republic of China.* Carbon Dioxide Information Analysis Center Report NDP 039/R1. Oak Ridge, TN: Oak Ridge National Laboratory.
Solberg, R., Hiltbrunner, D., Koskinen, J., *et al.* (1997). *Snow Algorithms and Products, Review and Recommendations for Research and Development.* Report from SNOWTOOLS WP 410, Norwegian Computing Center, Oslo.
Staelin, D. H., Rosenkrnaz, P. W., Barath, F. T., Johnston, E. J., and Waters, J. W. (1977). Microwave spectroscopic imagery of the Earth. *Science*, **197**, 991–993.
Stieglitz, M., Rind, D., Famiglietti, J., and Rosenzweig, C. (1997). An efficient approach to modeling the topographic control of surface hydrology for regional and global climate modeling. *J. Climate*, **10**, 118–137.
Strozzi, T., Wegmueller, U., and Mätzler, C. (1999). Mapping wet snow covers with SAR interferometry. *Int. J. Remote Sens.* **20**(12), 2395–2403.
Sturm, M., Holmgren J., and Liston, G. E. (1995). A seasonal snow cover classification system for local to global applications. *J. Climate*, **8**, 1261–1283.
Tait, A. (1998). Estimation of snow water equivalent using passive microwave radiation data. *Remote Sens. Environ.*, **64**, 286–291.
Ulaby, F. T. and Stiles, W. H. (1980). The active and passive microwave response to snow parameters. *J. Geophys. Res.*, **85**, 1045–1049.

U.S. Army Corps of Engineers. (1954). *Depth of Snow Cover in the Northern Hemisphere*. Boston, MA: Arctic Construction and Frost Effects Laboratory, New England Division.

Vance, R. E., Mathewes, R. W., and Clague, J. J. (1992). 7000 year record of lake-level change on the northern Great Plains: a high resolution proxy of past climate. *Geology*, **20**, 879–882.

Verseghy, D. (1991). CLASS A Canadian land surface scheme for GCMS. I: Soil model. *Int. J. Climatol.*, **11**, 111–133.

Vinnikov, K. Ya., Groisman, P. Ya., and Lugina, K. M. (1990). Empirical data on contemporary global climate changes (temperature and precipitation). *J. Climate*, **3**, 662–677.

Walker, A. E. and Goodison, B. E. (1993). Discrimination of wet snow cover using passive microwave satellite data. *Ann. Glaciol.*, **17**, 307–311.

Walland, D. J. and Simmonds, I. (1996). Sub-grid-scale topography and the simulation of Northern Hemisphere snow cover. *Int. J. Climatol.*, **16**, 961–982.

Walsh, J. E., Kattsov, V., Portis, D., and Meleshko, V. (1998). Arctic precipitation and evaporation: model results and observational estimates. *J. Climate*, **11**, 72–87.

Wang, L., Sharp, M., Brown, R., Derksen, C., and Rivard, B. (2005). Evaluation of spring snow covered area depletion in the Canadian Arctic from NOAA snow charts. *Remote Sens. Environ.*, **95**, 453–463.

Warren, S. G. and Wiscombe, W. J. (1981). A model for the spectral albedo of snow. II: Snow containing atmosperic aerosols. *J. Atmos. Sci.* **37**, 2734–2745.

Watanabe, M. and Nitta, T. (1999). Decadal changes in the atmospheric circulation and associated surface climate variations in the Northern Hemisphere winter. *J. Climate*, **12**, 494–510.

Wiesnet, D. R., Ropelewski, C. F., Kukla, G. J., and Robinson, D. A. (1987). A discussion of the accuracy of NOAA satellite-derived global seasonal snow cover measurements. *Proc. Symposium on Large Scale Effects of Seasonal Snow Cover*, Vancouver, Canada. IAHS, pp. 291–304.

Wiscombe, W. J. and Warren, S. G. (1981). A model for the spectral albedo of snow. I. Pure snow. *J. Atmos. Sci.* **37**, 2712–2733.

Woo, M.-K. and Steer, P. (1986). Monte Carlo simulation of snow depth in a forest. *Water. Resources Res.*, **22**, 864–868.

World Meteorological Organization (1981). *Guide to Hydrological Practices*, 4th edn, vol. 1. WMO-No. 168. Geneva: World Meteorological Organization.

Yang, D., Goodison, B. E., Metcalfe, J. R., *et al.* (1998). Accuracy of NWS 8" standard nonrecrording precipitation gauge: results and application of WMO intercomparison. *J. Atmos. Oceanic Technol.*, **15**, 54–68.

Yang, D., Goodison, B. E., Metcalfe, J. R., *et al.* (1999). Quantification of precipitation measurement discontinuity induced by wind shields on national gauges. *Water Resources Res.*, **35**, 491–508.

Ye, H., Cho H.-R., and Gustafson, P. E. (1998). The changes in Russian winter snow accumulation during 1936–83 and its spatial patterns. *J. Climate*, **11**, 856–863.

# Appendix: Snow model questionnaire

1. Please write down the name (and abbreviation) of your snow model or land-surface model with snow component.
2. Name and address of model developer.
3. Name and address of model user.
4. Please indicate whether your model is developed for application:
   in understanding snow processes,
   in a runoff forecasting model,
   in a weather forecasting model,
   in a global climate model (GCM),
   or other (please specify)?
5. The first year when the model was used.
6. One paragraph description of your model (e.g. abstract from report or paper).
7. Please specify any known application range or restrictions.
8. What are the development data needs?
9. What are the operational data needs?
10. Please indicate with an "x" those meteorological variables used to DRIVE your snow model:
    precipitation;
    air temperature;
    wind speed;
    wind direction;
    humidity;
    downwelling shortwave radiation;
    downwelling longwave radiation;
    cloud cover;
    surface pressure.
11. List the state variables (e.g. snow temperature, snow water equivalent, etc) your snow model uses.
12. List the measurable/adjustable parameters (e.g. snow surface aerodynamic roughness, maximum albedo at visible wavelength, etc, excluding initial conditions) your snow model uses.
13. What are the output data?
14. What computer language does your model use?
15. How many subroutines (or functions) does your snow model have?
16. Number of lines for the snow code?

17. What is the recommended hardware?
18. How does your model determine the form of precipitation (i.e. snowfall and rainfall)?
    Please give the formulation.
19. Is your snow model one dimensional or multi-dimensional?
    Please specify.
20. If one dimensional, how many layers are there in your snow model?
    Please specify layering structure.
21. What is your snow model time step?
22. Does your model snow albedo allow its spectral differences (visible vs. near-IR)?
    Are there directional differences (direct vs. diffuse)?
23. Is your model snow albedo a function of:
    snow age;
    grain size;
    solar zenith angle;
    pollution;
    snow depth?
24. Does your snow model explicitly treat liquid water retention and percolation within the snowpack?
25. Does your snow model account for changes in the hydraulic and thermal properties of snow due to meltwater refreezing?
26. Is snow density in your snow model changing with time or fixed?
27. Is heat capacity and conductivity in your snow model changing with time or fixed?
28. Does your snow model simulate vapor transfer in the snowpack?
29. Does your snow model account for the heat transfer between the bottom of the snowpack and the underlying soil?
30. In snow energy balance, does your model consider heat convected by rain or falling snow?
31. Does your snow model include snow drifting and redistribution by wind (or avalanche)? If so, how?
32. How is areal snow distribution treated?
33. Does your snow model account for subgrid (or subwatershed) effects of topography? If so, how is temperature distributed?
    How is precipitation (spatial, elevation and corrections) distributed?
    How is solar radiation distributed?
    How is wind distributed?
    How are other meteorological variables distributed?
34. Does your snow model consider snow–vegetation interaction?
35. Does the snow–vegetation interaction account for:
    different vegetation types (grass vs. forest);
    different vegetation heights (short vs. tall);
    different vegetation densities (small vs. large LAI);
    different vegetation coverages (sparse vs. dense vegetation)?
36. Are snow interception, drip and melt on canopy surface allowed in your model?
37. How is the upper limit of the canopy interception determined?
38. In the presence of vegetation, how is snow surface albedo altered?
39. In the presence of vegetation, how is snow surface roughness altered?
40. In the presence of forest, does your snow model allow spatial variability of snow depth and water equivalent on forest floor?
41. (a) How does your model deliver snowmelt to the soil system (e.g. affecting soil moisture)?
    (b) Once snowmelt is generated, how does your model relate it to runoff?

42. How is frozen soil treated in your model?
43. Has your snow model been tested with the field data?
    Is so, what data?
    What are their temporal and spatial scales?
44. Has your snow model been used together with remote sensing data as input?
    If so, how?
45. If your snow model is coupled with a numerical weather forecasting model or climate model, has the model snow product been compared with satellite data?
    If so, what satellite data were used?
46. Please list any other previous applications.
47. Please specify verification criteria, if any.
48. What are the model fitting procedures, if any?
49. What are future plans for using/improving the model?
50. Please provide references relevant to the model description and use.

# Index